RADIO RESOURCE MANAGEMENT STRATEGIES IN UMTS

RADIO RESOURCE MANAGEMENT STRATEGIES IN UMTS

Jordi Pérez-Romero
Oriol Sallent
Ramon Agustí

All of Universitat Politècnica de Catalunya (UPC), Spain

Miguel Angel Díaz-Guerra

Telefónica Móviles España, S.A., Spain

John Wiley & Sons, Ltd

Other Wiley Editorial Offices

John Wiley & Sons Inc., 111 River Street, Hoboken, NJ 07030, USA

Jossey-Bass, 989 Market Street, San Francisco, CA 94103-1741, USA

Wiley-VCH Verlag GmbH, Boschstr. 12, D-69469 Weinheim, Germany

John Wiley & Sons Australia Ltd, 42 McDougall Street, Milton, Queensland 4064, Australia

John Wiley & Sons (Asia) Pte Ltd, 2 Clementi Loop #02-01, Jin Xing Distripark, Singapore 129809

John Wiley & Sons Canada Ltd, 22 Worcester Road, Etobicoke, Ontario, Canada M9W 1L1

Wiley also publishes its books in a variety of electronic formats. Some content that appears in print may not be
available in electronic books.

British Library Cataloguing in Publication Data

A catalogue record for this book is available from the British Library

ISBN-13 978-0-470-02277-1 (HB)
ISBN-10 0-470-02277-9 (HB)

Typeset in 9/11pt Times by Thomson Press (India) Limited, New Delhi.
Printed and bound in Great Britain by Antony Rowe Ltd, Chippenham, Wiltshire.
This book is printed on acid-free paper responsibly manufactured from sustainable forestry
in which at least two trees are planted for each one used for paper production.

Contents

Preface

It is more than a decade since GSM was first commercially available. After some unexpected delay, it seems that finally UMTS is here to stay as a 3G system standardised by 3GPP, at least for another ten years. UMTS will enable multi-service, multi-rate and flexible IP native-based mobile technologies to be used in wide area scenarios and also pave the way for a smooth transition from circuit switched voice networks to mobile packet services.

The scarcity of available spectrum, particularly as seen in the auctions and beauty contests that preceded the final licences allocation for UMTS operators, has revealed, to a larger extent than in the past, the importance of using the spectrum efficiently. Radio access systems such as UTRAN in UMTS certainly exploit higher system spectrum efficiencies than 1G and 2G by using advanced coding, multiple access, diversity schemes, etc.

On the other hand, the WCDMA technique adopted in UTRAN makes the accurate control of the inherent interference generated by this access a key issue in the good behaviour of the system. In addition, the inherent flexibility and high user bit rates provided by UMTS makes this interference control even more difficult. Therefore, manufacturers have to introduce, on a proprietary basis, much more involved Radio Resource Management (RRM) strategies than those used in the past, so that an efficient use of the available spectrum can be achieved. A complete picture of these RRM techniques has to include the retention of the QoS per service at the agreed values as an ultimate trade-off. Certainly, handling interference in UMTS will take the place of frequency planning in 1G and 2G systems to a much greater extent and will be one of the most important tasks if operators are to run the system efficiently.

This self-contained book, consisting of six chapters, intends to bring to the reader, in a comprehensive and systematic way, the material needed to understand the interiorities of the RRM strategies in the context of UMTS. This book is addressed to undergraduate students, engineers and researchers who would like to explore the UMTS world and learn how to run and improve its radio access part in an operative scenario. Although a short radio planning basis is provided, RRM concepts are actually exploited in different scenarios that go beyond the planning pre-operational stages so that eventually the radio resources can be efficiently exploited in a near real time operation.

The organisation of the book is represented schematically overleaf. In particular, Chapter 1 provides the introduction to the mobile communications sector and to UMTS, including the evolution towards the 4G systems. Also, it provides an overview of the QoS concept, which is key for the definition of Radio Resource Management strategies. After this introduction, the book is split into two different paths. The first path, which includes Chapters 2 and 4, is intended to provide the required theoretical fundamentals while the second, including Chapters 3, 5 and 6, presents to the reader how these theoretical aspects are translated into practical algorithms and systems. In that sense, Chapters 2 and 3 cover the characterisation of the radio access in UMTS. Specifically, Chapter 2 provides a brief description of the CDMA technique that constitutes the basis for the UMTS radio access network. In turn, Chapter 3 presents the

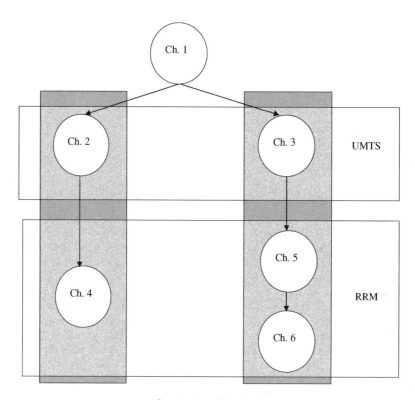

Organisation of the book

detailed description of the UMTS radio interface, focusing on the UTRAN FDD mode. After this characterisation, the following chapters focus on the Radio Resource Management concepts. In particular, Chapter 4 provides the theoretical background for the development of RRM strategies in WCDMA, which serves as a basis for the description of specific RRM algorithms in Chapter 5. Such algorithms are analysed in a variety of scenarios to identify the key parameters and factors that influence their performance. Finally, Chapter 6 provides the evolution of UMTS towards 'Beyond 3G' systems and explores the concept of Common RRM in heterogeneous networks, including some algorithm examples.

List of Acronyms

1G	First Generation
2G	Second Generation
2.5G	Generation between 2G and 3G
3G	Third Generation
3GPP	Third Generation Partnership Project
4G	Fourth Generation
AAA	Authentication, Authorisation and Accounting
ABC	Always Best Connected
AC	Admission Control
ACK	Acknowledgement
ACLR	Adjacent Channel Leakage power Ratio
AICH	Acquisition Indicator Channel
AM	Acknowledged Mode
AMD	Acknowledged Mode Data
AMR	Adaptive Multi Rate
AN	Access Network
AP	Access Preamble (in the context of Random Access) or Access Point (in the context of WLAN)
APC	Access Point Controller
ARFCN	Absolute Radio Frequency Channel Number
ARQ	Automatic Repeat Request
ARROWS	Advanced Radio Resource Management for Wireless Services
AS	Access Stratum (in the context of UMTS protocol stack) or Access Slot (in the context of PRACH channel)
ASC	Access Service Class
ASM	Advanced Spectrum Management
ATM	Asynchronous Transfer Mode
AuC	Authentication Centre
AWGN	Additive White Gaussian Noise
BCCH	Broadcast Control Channel
BCFE	Broadcast Control Function Entity
BCH	Broadcast Channel
BER	Bit Error Rate
BLER	Block Error Rate
BMC	Broadcast/Multicast Control
BPSK	Binary Phase Shift Keying
BRAN	Broadband Radio Access Network
BS	Base Station

BSC	Base Station Controller
BSIC	Base Station Identity Code
BSS	Base Station Subsystem (in the context of UTRAN and GSM/GPRS architecture) or Basic Service Set (in the context of WLAN)
BSSMAP	Base Station Subsystem Management Application Part
BTS	Base Transceiver Station
CA	Channel Assignment
CBR	Constant Bit Rate
CC	Call Control
CCC	CPCH Control Command
CCCH	Common Control Channel
CCK	Complementary Code Keying
CCTrCH	Coded Composite Transport Channel
CD	Collision Detection
CD/CA-ICH	Collision Detection/Channel Assignment Indicator Channel
CDF	Cumulative Distribution Function
CDMA	Code Division Multiple Access
CID	Context Identifier
CM	Connection Management
CN	Core Network
COST	Cooperation européenne dans le domaine de la recherche Scientifique et Technique
CPCH	Common Packet Channel
CPICH	Common Pilot Channel
CPU	Central Processor Unit
CQI	Channel Quality Indicator
CRC	Cyclic Redundancy Code
CRMS	Common Radio Resource Management Server
CRNC	Controlling Radio Network Controller
CRRM	Common Radio Resource Management
CS	Circuit Switched
CSD	Circuit Switched Data
CSICH	Channel Status Indicator Channel
CSMA/CA	Carrier Sense Multiple Access with Collision Avoidance
CTCH	Common Traffic Channel
DCCH	Dedicated Control Channel
DCF	Distributed Coordinated Function
DCFE	Dedicated Control Function Entity
DCH	Dedicated Channel
DCS	Digital Cellular System
DC-SAP	Dedicated Control Service Access Point
DL	Downlink
DNPM	Dynamic Network Planning and flexible network Management
DPCCH	Dedicated Physical Control Channel
DPCH	Dedicated Physical Channel
DPDCH	Dedicated Physical Data Channel
DRNC	Drift Radio Network Controller
DS	Distribution System
DS-CDMA	Direct Sequence Code Division Multiple Access
DSCH	Downlink Shared Channel
DSMA/CD	Digital Sense Multiple Access with Collision Detection

DSP	Digital Signal Processor
DS-SS	Direct Sequence Spread Spectrum
DTCH	Dedicated Traffic Channel
DTX	Discontinuous Transmission
Eb/No	Bit energy over noise power spectral density
Ec/No	Chip energy over noise power spectral density
ECSD	Enhanced Circuit Switched Data
EDGE	Enhanced Data Rates for GSM Evolution
EGPRS	Enhanced GPRS
EIR	Equipment Identity Register
EIRP	Equivalent Isotropic Radiated Power
ESS	Extended Service Set
ETSI	European Telecommunications Standards Institute
EVEREST	Evolutionary Strategies for Radio Resource Management in Cellular Heterogeneous Networks
FACH	Forward Access Channel
FBI	Feedback Information
FCC	Federal Communications Commission
FDD	Frequency Division Duplex
FDMA	Frequency Division Multiple Access
FFM	Fast Fading Margin
FH-SS	Frequency Hopping Spread Spectrum
FOMA	Freedom of Mobile Multimedia Access
FSD	Fuzzy Selected Decision
FTP	File Transfer Protocol
GBR	Guaranteed Bit Rate
GC-SAP	General Control Service Access Point
GERAN	GSM/EDGE Radio Access Network
GGSN	Gateway GPRS Support Node
GMM	GPRS Mobility Management
GMSC	Gateway Mobile Switching Centre
GMSK	Gaussian Minimum Shift Keying
GOP	Group of Pictures
GPRS	General Packet Radio Service
GSM	Global System for Mobile Communications
GTP	GPRS Tunnelling Protocol
HARQ	Hybrid Automatic Repeat Request
HCS	Hierarchical Cell Structure
HIPERLAN	High Performance Local Area Network
HLR	Home Location Register
HN	Home Network
HO	Handover
HPLMN	Home Public Land Mobile Network
HSDPA	High Speed Downlink Packet Access
HS-DPCCH	High Speed Dedicated Physical Control Channel
HS-DSCH	High Speed Downlink Shared Channel
HS-PDSCH	High Speed Physical Downlink Shared Channel
HSS	Home Subscriber Server
HS-SCCH	High Speed Shared Control Channel
HTML	Hyper Text Markup Language
IBSS	Independent Basic Service Set

IEEE	Institute of Electrical and Electronics Engineers
IETF	Internet Engineering Task Force
IMS	IP Multimedia Subsystem
IMSI	International Mobile Subscriber Identity
IMT-2000	International Mobile Telecommunications 2000
IP	Internet Protocol
IPTS	Institute for Prospective Technological Studies
IRNSAP	Inter Radio Network Subsystem Application Part
IS-95	Interim Standard 95
ISDN	Integrated Service Data Network
ISO	International Organisation for Standardisation
IST	Information Society Technologies
ITU	International Telecommunications Union
ITU-R	International Telecommunications Union – Radiocommunications sector
ITU-T	International Telecommunications Union – Telecommunications sector
L1	Layer 1
L2	Layer 2
L3	Layer 3
LAN	Local Area Network
LC	Load Control
LFSR	Linear Feedback Shift Register
LLC	Logical Link Control
LOS	Line of Sight
MAC	Medium Access Control
MAP	Mobile Application Part
MCL	Minimum Coupling Loss
MCS	Modulation and Coding Scheme
ME	Mobile Equipment
MGW	Media Gateway
MM	Mobility Management
MMS	Multimedia Messaging Service
MPEG	Moving Pictures Expert Group
MR	Maximum Rate
MRC	Maximum Ratio Combining
MSC	Mobile Switching Centre
MSDU	MAC Service Data Unit
MT	Mobile Termination
N/A	Not Applicable
NACK	Negative Acknowledgement
NAS	Non Access Stratum
NLOS	Non Line of Sight
NRT	Non Real Time
NS	Neighbour Set
Nt-SAP	Notification Service Access Point
NTT	Nipon Telephone and Telecommunications
OFDM	Orthogonal Frequency Division Multiplexing
OVSF	Orthogonal Variable Spreading Factor
PABAC	Power Averaged-Based Admission Control
PAN	Personal Area Network
PC	Personal Computer
PCCH	Paging Control Channel

P-CCPCH	Primary Common Control Physical Channel
PCF	Point Coordination Function
PCH	Paging Channel
PCPCH	Physical Common Packet Channel
PCU	Packet Control Unit
PDA	Personal Digital Assistant
PDC	Personal Digital Cellular
PDCH	Packet Data Channel
PDCP	Packet Data Convergence Protocol
pdf	probability density function
PDG	Packet Data Gateway
PDP	Packet Data Protocol
PDSCH	Physical Downlink Shared Channel
PDU	Protocol Data Unit
PER	Packet Error Rate
PHY	Physical layer
PI	Paging Indicator
PICH	Paging Indicator Channel
PL	Path Loss
PLEBAC	Path Loss Estimation-Based Admission Control
PLMN	Public Land Mobile Network
PN	Pseudo Noise
PNFE	Paging Notification Function Entity
PRACH	Physical Random Access Channel
PS	Packet Switched
PSK	Phase Shift Keying
PSTN	Public Switched Telephone Network
QAM	Quadrature Amplitude Modulation
QoS	Quality of Service
QPSK	Quadrature Phase Shift Keying
RAB	Radio Access Bearer
RACH	Random Access Channel
RAN	Radio Access Network
RANAP	Radio Access Network Application Part
RAT	Radio Access Technology
RB	Radio Bearer
RFC	Request for Comments
RFE	Routing Function Entity
RLA	Received Level Average
RLC	Radio Link Control
RM	Rate Matching
RNC	Radio Network Controller
RNS	Radio Network Subsystem
RNSAP	Radio Network Subsystem Application Part
RNTI	Radio Network Temporary Identity
ROHC	Robust Header Compression
RR	Radio Resource
RRC	Radio Resource Control
RREU	Radio Resource Equivalent Unit
RRM	Radio Resource Management
RRU	Radio Resource Unit

RSCP	Received Signal Code Power
RSSI	Received Signal Strength Indicator
RT	Real Time
RTP	Real Time Protocol
SACCH	Slow Associated Control Channel
SAP	Service Access Point
S-CCPCH	Secondary Common Control Physical Channel
SCH	Synchronisation Channel
SCr	Service Credit
SDCCH	Stand-alone Dedicated Control Channel
SDR	Software Defined Radio
SDU	Service Data Unit
SF	Spreading Factor
SFM	Slow Fading Margin
SGSN	Serving GPRS Support Node
SHO	Soft Handover
SIB	System Information Block
SIP	Session Initiation Protocol
SIR	Signal to Interference Ratio
SM	Session Management
SMS	Short Message Service
SN	Serving Network
SRB	Signalling Radio Bearer
SRNC	Serving Radio Network Controller
SRNS	Serving Radio Network Subsystem
SS7	Signalling System No. 7
SSDT	Site Selection Diversity Transmission
STTD	Space Time block coding based Transmit Diversity
TACS	Total Access Communications System
TB	Transport Block
TBF	Temporary Block Flow
TCH	Traffic Channel
TCP	Transport Control Protocol
TD/CDMA	Time Division Code Division Multiple Access
TDD	Time Division Duplex
TDMA	Time Division Multiple Access
TD-SCDMA	Time Division – Synchronous Code Division Multiple Access
TE	Terminal Equipment
TF	Transport Format
TFC	Transport Format Combination
TFCI	Transport Format Combination Indicator
TFCS	Transport Format Combination Set
TFS	Transport Format Set
TM	Transparent Mode
TMD	Transparent Mode Data
TME	Transfer Mode Entity
TMSI	Temporary Mobile Subscriber Identity
TN	Transit Network
TO	Time-Oriented
TPC	Transmit Power Control
TrCH	Transport Channel

TSTD	Time Switched Transmit Diversity
TTI	Transmission Time Interval
UARFCN	UTRA Absolute Radio Frequency Channel Number
UDP	User Datagram Protocol
UE	User Equipment
UL	Uplink
UM	Unacknowledged Mode
UMD	Unacknowledged Mode Data
UMTS	Universal Mobile Telecommunications System
URA	UTRAN Registration Area
URANO	UMTS Radio Access Network Optimisation
USIM	UMTS Subscriber Identity Module
UTRA	Universal Terrestrial Radio Access
UTRAN	Universal Terrestrial Radio Access Network
VBR	Variable Bit Rate
VLR	Visitor Location Register
VoIP	Voice over IP
WAG	WLAN Access Gateway
WAP	Wireless Application Protocol
WARC	World Administrative Radio Conference
WCDMA	Wideband Code Division Multiple Access
WLAN	Wireless Local Area Network
WPAN	Wireless Personal Area Network
WRC	World Radiocommunication Conference
WWW	World Wide Web

1

Introduction

After the successful global introduction during the past decade of the second generation (2G) digital mobile communications systems, it seems that the third generation (3G) Universal Mobile Communication System (UMTS) has finally taken off, at least in some regions. The plethora of new services that are expected to be offered by this system requires the development of new paradigms in the way scarce radio resources should be managed. The Quality of Service (QoS) concept, which introduces in a natural way the service differentiation and the possibility of adapting the resource consumption to the specific service requirements, will open the door for the provision of advanced wireless services to the mass market.

Within this context, this chapter introduces the basic framework for the development of the radio resource management strategies, which is the main object of this book. To this end, Section 1.1 analyses the evolution of the mobile communications sector and tries to identify the key socio-economical aspects that could enable a successful deployment of 3G systems. In turn, Section 1.2 provides a description of the basic features of UMTS from the architectural point of view, including the initial architectures of the first releases as well as the evolution towards all-IP networks. Finally, Section 1.3 presents the QoS model that is defined in UMTS, including the identified service classes and the main QoS attributes.

1.1 THE MOBILE COMMUNICATIONS SECTOR

The development of mobile communications has traditionally been viewed as a sequence of successive generations. The first generation of analogue mobile telephony was followed by the second, digital, generation. Then, the third generation was envisaged to enable full multimedia data transmission as well as voice communications. However, the high cost and technical difficulties faced in standardisation and development have led to delays in 3G deployment and, in the meantime, the model of a succession of generations began to break down, first with the intercalation of a 2.5G enabling basic Internet access from mobile terminals, and then with the emergence of public WLAN (Wireless Local Area Network) technologies as potential competitors of the 3G UMTS (Universal Mobile Telecommunications System). In this context, looking at the period 2010–2015, the concept of *beyond 3G* encompasses a scenario with a variety of interoperating systems, each filling a different niche in the mobile communications market.

Recommendation ITU-R M.1645 defines the framework and overall objectives of future development of IMT-2000 (International Mobile Telecommunications 2000) and systems beyond IMT-2000 for the radio access network. In this respect, the significant technology trends need to be considered. Depending on their development, evolution, expected capabilities and deployment cost, each of these technologies may or may not have an impact or be used in the future. Moreover, beyond 3G technology is still very immature and a range of alternative scenarios remain possible. As a result, all the forecasts are by definition open to criticism. How mobile communications will evolve over the forthcoming years will depend on the interaction of a number of factors. These include the progress made in developing the various technologies, the emergence of new applications, and the adoption of new services by users. Although the technology is an essential element, a viable business model is clearly a crucial factor.

Information and communication technologies play an important role in determining competitiveness, employment and economic growth. They create new opportunities that at the same time affect existing production, communication and distribution processes. No technological development is possible without an effect upon society. Clearly, no one will deny the evolving nexus between technological innovation and the human condition. Technical devices have never before played such an important role in our daily lives. The development of mobile technologies has been pivotal in this transformation and, consequently, some considerations are discussed in Section 1.1.1. Plausible key factors in future market developments are covered in Section 1.1.2. Furthermore, the complexity of the mobile communications sector is due to a mix of technologies, business models, socio-cultural influences, etc., and therefore we must take notice of market developments in early adopters, such as Japan, described in Section 1.1.3. From this standpoint, the situation and approaches in different regions are covered in Section 1.1.4. The role of technological advances is stressed in Section 1.1.5.

Much analysis covering technical, business and demand-related aspects of what the future mobile communications environment might be like have been produced in different fora. This section collects different perspectives and sources together in order to forecast and/or highlight the key issues in the wireless arena, with the aim of providing a self-contained framework and a broader perspective on the Radio Resource Management problem. In particular, technical reports of the Institute for Prospective Technological Studies (IPTS) of the European Commission [1][2] and ITU background papers [3] and draft reports [4][5] have been considered. The interested reader is directed to these references for more details on these topics.

1.1.1 THE MOBILE EXPERIENCE

The world has witnessed an explosion in the growth of mobile communications in recent years. Year 2002 marked a turning point in the history of telecommunications in that the number of mobile subscribers overtook the number of fixed-line subscribers on a global scale, and mobile became the dominant technology for voice communications.

As a technical device, the mobile phone has become an incredible important part of human life, and a powerful determinant of individual identity. Indeed, the mobile phone has moved beyond being a mere technical device to becoming a key social object present in every aspect of our daily lives. At the same time, the highly personalised nature of the mobile phone has meant that its form and use have become important aspects of the individuality of a phone user. The mobile phone has indeed become one of the most intimate aspects of a user's personal sphere of objects (e.g. keys, wallet, money, etc.). Both physical and emotional attachment to mobile handsets is increasing. The mobile phone has become somewhat of a status symbol. Mobiles are quickly becoming fashion accessories rather than simply communications devices. The introduction of the mobile phone has also facilitated the balancing of professional and domestic life. In this respect, the mobile phone has become metaphorically an extension of one's physical self, intrinsically linked to identity and accessibility.

1.1.2 THE BUSINESS CASE

With voice traffic over current GSM (Global System for Mobile communications) and other networks approaching saturation point in many European countries, there is a real opportunity for 3G networks to accommodate the capacity shortage that is likely to emerge in the medium-term. There is as yet a lack of 'killer applications' for the mobile Internet in Europe. While MMS (Multimedia Messaging Service) and adult entertainment have been attractive to consumers, operators may need to realise that simultaneous efforts must be made to obtain customer preferences from a wide range of demographic, social and economic backgrounds in order to define market segments of service offerings. A possible weakness, paradoxically, lies in the cultural and linguistic diversity of Europe, which could work against 3G take-up. This is because localisation of content could increase the cost of production and subscribers may have to absorb part of it.

Doubts about the market potential of mobile data and multimedia have lowered expectations for 3G, and the roll out of 3G services has run into difficulties. As the lack of demand for 3G has shown, it is extremely difficult to predict the likely market adoption of mobile wireless communications and the revenues that can be expected. Added to this uncertainty is the potential impact of public WLANs. However, although operators have been deploying public WLAN networks for some years, most have been unable to turn them into a profitable business. Some estimations suggest that standalone public WLAN services will probably not provide a sustainable business in the short-term, despite the free use of spectrum and the relatively small investments required compared to 3G. The intrinsic problem of achieving efficient usage of free un-coordinated bandwidth could become critical as more players enter the field. Nevertheless, WLANs may prove to be of high strategic value and an important source of competitive differentiation. Even if the direct revenue impact of public WLAN is low, they may be important for subscriber retention, or as the means by which a fixed line operator could enter the mobile market.

Viable business models for public WLAN will depend on the cost of access to the backbone network, security, and charging mechanisms. As a public mobile technology, it could potentially evolve as a separate competitor to cellular networks in the form of a network of hotspots or it could become more closely integrated within the cellular network. Although public WLANs cannot substitute entirely for 3G in terms of functionality, if they are able to offer most of the services users might want from 3G at lower cost, they may undermine 3G's business model. Nevertheless, WLANs might stimulate demand for mobile broadband and create a cohort of users willing to pay to upgrade to higher quality 3G when they tire of the limited coverage, high demands on battery power, patchwork of hotspot ownership and congestion of WLAN access points. What seems less likely today, however, in the light of the problems faced by 3G deployment and in the context of emerging technologies, is a smooth linear transition to a homogeneous and universal fourth generation (4G) at some point in the medium term.

Considering the length of time that 3G appears to be taking to rollout, it could be overtaken by alternative technologies such as WLAN, old technologies such as GPRS (General Packet Radio Service), and increasingly sophisticated pager technology. Licensing problems arising from the multiple patents held by various parties to the 3G technologies also pose a complex and expensive issue, recalling the GSM patent problem. Furthermore, since each generation of handheld gadgets contains more and more complex software, it could turn potential 3G users away because the general consumer is finding it harder to leverage his knowledge from one gadget to another.

It may also appear that competition between different technologies (in the case of 3G, CDMA2000 versus WCDMA) helps bring down prices. The obvious policy conclusion, therefore, would be to shape market conditions so as to encourage competition between standards. On the other hand, experiences from 1G and 2G point to the opposite conclusion. Too much competition between technologies/standards limits the possibilities of economies of scale, and so the right balance is needed. Similarly, the right balance is needed to harmonise operators' and vendors' diverging strategic visions. However, the fragile business case suggests efforts should concentrate on creating a dynamic and sophisticated market for advanced mobile data and voice services based on 3G technologies. If this can be achieved, at the same time as integrating new technologies to improve the user experience further, the evolutionary path towards 4G will become clearer and maintain its momentum.

The downturn in the telecommunications sector caused by excessive operator debt and disappointment over market growth, as well as the extreme cases of vendor financing, makes it highly likely that it will be more difficult to secure financial backing for new investments in a future generation of mobile communications systems. It has been suggested that several 3G operators may recoup their investments slowly, and this will reduce the likelihood of operators investing in 4G by 2011, the date tentatively set by several equipment vendors for its introduction. Instead, for most operators, this investment is likely to be postponed a long way into the future. However, before more accurate predictions of operator investments in 4G can be made, 3G adoption will have to take off. It does not seem likely that a very high-speed mobile data network will gain user acceptance unless successful mobile data applications have been developed and commercialised with 3G.

1.1.3 A LEARNING CASE STUDY: JAPAN

The Japanese market is far more advanced than other regions in terms of the extent of use of cellular mobile data services and terminals. Therefore, it provides one of the few learning experiences that can provide feedback into the design of future mobile communication systems.

In the 2G world, very few countries have been successful with the 'mobile Internet'. WAP (Wireless Application Protocol) in Europe suffered from low transmission speeds, paucity of content and disenchanted users. Japan, on the other hand, introduced a wide array of mobile Internet services, and witnessed phenomenal growth in usage and subscribers. In fact, Japan made mobile Internet services an integral part of mobile phone ownership, and even made charging for Internet content a reality. The country exhibits the highest total number of mobile Internet users in the world.

NTT DoCoMo launched its Internet connection service, 'i-mode', in February 1999. i-mode subscribers can connect to the Internet through special designated handsets. The main services are email, information services and applications such as Internet banking and ticket reservation. Other mobile operators in Japan also began competitive Internet connection services in 1999. In September 2003, there were 78.6 million cellular mobile subscribers in Japan, of which 84% were using some kind of Internet browsing service. In 2003, the average annual revenue per i-mode user was about 200 €, most of which stems from packet transmission charges. The primary use of mobile Internet in Japan is for email: over 83% of mobile subscribers use the mobile Internet for sending and receiving email. Downloading or listening to online music, such as ring tones or tunes, and purchasing online content are other examples of key usages.

Low PC penetration is one of the main factors contributing to the success of mobile networks for Internet access in Japan. Some analysts point to the large number of long-distance commuters using public transport as a stimulus for growth. Nevertheless, a large majority of japanese use their mobile phone at home to make calls and some surveys also show that the use of the mobile browser in Japan is highest at home (in fact the peak time period for browser usage is after working hours, between 19:00 and 23:00 on weekdays).The introduction of colour display handsets is claimed to be another major driver for the take-up of i-mode services.

Japan has carefully and successfully developed the 2.5G mobile Internet market, thus cultivating the whole innovation system (in terms of usage, operating networks, terminal supply, content development, etc.). This cultivation has not only prepared the Japanese market for 3G services, it has given them first-mover advantages that they can leverage on the international market. Thus, it is expected that market shares of Japanese handset manufacturers and other actors will increase when the transition to 3G (and mobile Internet) takes place elsewhere.

The policies on the introduction of higher-speed 3G services in Japan fixed the number of operators to three per region, due to the shortage of frequencies. The regulator had a total of 60 MHz available for 3G services (uplink and downlink). In order to allocate a minimum of 2×20 MHz blocks of spectrum, only 3 licences could be awarded. New as well as incumbent operators were eligible for the licences. Operators were required to cover 50% of the population in the first five years. Only the three incumbent operators, i.e. NTT DoCoMo Group, IDO and Cellular Group (KDDI), and J-Phone Group, applied, and obtained, the three available licences in each region.

NTT DoCoMo was the first operator to launch 3G services in Japan, under the brand name FOMA (Freedom of Mobile Multimedia Access), and based on WCDMA (Wideband CDMA). The full-scale commercial launch of FOMA was initially scheduled for 30 May 2001. Although DoCoMo postponed the launch until October 2001, it was one of the first operators to launch a 3G commercial service. However, due to the limited service coverage at the time of launch, the fact that the WCDMA system does not have backward compatibility with its 2G service based on the Personal Digital Cellular (PDC) system, relatively short battery life and lack of killer applications (the highly publicised video-phone capability was not a resounding success), it was only by the end of 2002 that 150 000 subscribers were reached. Then, the advent of a flat rate contributed to a very significant increase in the number of subscribers.

High-speed Internet access services based on WLAN were launched in 2002 in Japan. However, it seemed a challenging task to develop a sound business model, attracting a large number of paying users. There are also several WLAN access points offered free of charge by a number of providers. Nonetheless, other types of fixed wireless access services are being launched. A handful of companies are planning to offer a wireless IP (Internet Protocol) phone service for Personal Digital Assistants (PDAs) and WLAN service providers are hoping this will get them out of their current business plan conundrum, but it remains to be seen whether they will be successful or not.

The lack of profitability of WLAN services is likely to persist for some time to come, and for this reason, a number of providers are exploring options to combine or integrate WLAN services with other types of services, notably NTT Communications and NTT DoCoMo. A WLAN service is being offered in combination with its 3G or FOMA service, which typically provides speeds of 384 kb/s so far. Users can benefit from 3G data transmission rates when away from WLAN access points, through the 3G network.

One of the most distinguishing aspects of the japanese mobile industry is that it is operator-led. Equipment manufacturers and operators work very closely and supply the market with handsets and portable devices in a coordinated effort. The close relationship between manufacturers and operators in Japan accounts in part for the sophistication and availability of handset technology and the take-up of value-added services. Another peculiarity of the Japanese mobile market is the early agreement between content providers and operators. In principle, the mobile operator bills for content, retains a commission, and passes on the majority of the content fees to the content provider.

1.1.4 REGIONAL PERSPECTIVES IN MOBILE EVOLUTION TOWARDS 4G

The European roadmap encompasses a clear tendency towards the development of a future mobile system where heterogeneous technologies, complementing each other in terms of coverage, bit rate and other characteristics, work together in a seamless system to optimise usability for the end user. There is an emphasis on taking advantage of existing and emerging technologies to provide what is, from an end-user perspective, a seamlessly integrated communications environment, with software defined radio as an enabling technology.

Although a European consensus seems to exist on the future diversity of wireless technologies and on the development of services driven by user needs as opposed to technology push, these visions express uncertainty as to the industry structure that will deliver 4G services in the 2010–2015 timeframe, partially motivated by the emergence of new players and the possibility of a fragmented industry. In the short term, 3G in Europe will be driven by mobile operators and especially telecom equipment suppliers.

In Europe, limited experience of advanced mobile data communications is still observed and, for the time being, there are not yet signs of any increase in demand from users for these services (in contrast to Japan, which is the world's most advanced mobile market). There is clearly a need to abandon the technology push approach that has so far characterised European mobile communications in favour of a more user-focused perspective.

Europe runs the risks of being a late starter in the race to deploy 4G. In this situation, mobile telecommunications equipment will be built cheaply in Asia, causing Europe to fall behind in the production and deployment of mobile communications systems. The development and adoption of 4G in Europe will require the prior large-scale adoption of 3G. While European actors should certainly aim for a leading role in 4G in the future to avoid missing opportunities, efforts should also be made to consolidate 3G infrastructure as a means of supporting a multitude of coexisting applications and enable the continuous incorporation of emerging standards and technologies. The standardisation made possible by UMTS adoption is an opportunity, but does not mean that other emerging technologies and standards should be ignored. On the contrary, UMTS integration should be the priority in the coming years, encouraging other standards to be made compatible with UMTS, promoting its enhancement and ensuring the removal of any barriers to its adoption. It should include provisions for spectrum regulation harmonisation and interconnection issues, which would allow investments in 3G infrastructure to be recouped without missing the opportunities stemming from technological innovation in other areas.

The US appears to lack a shared industry-wide view of how mobile telecommunications are likely to develop; at the same time, there is no representative body that articulates US visions for 4G. The trend in the US is towards new proprietary technologies deployed over unlicensed spectrum, coexisting with new standards developed for use on both unlicensed and licensed spectrum. At the same time, more unlicensed spectrum is being made available and flexible spectrum management is supporting the interoperability of products and technologies offered by a more fragmented industry. Thus, the US is leading the way in the deployment of potentially disruptive technologies such as public WLAN. The push by some US actors to make further free spectrum available, and the increasing flexibility of the FCC (Federal Communications Commission) in the field of spectrum regulation, has important policy implications for the rest of the world. The future existence of more unlicensed frequency could speed up developments leading towards a more fragmented industry structure with a rapid entry of new service providers.

In Asia, several countries are showing a desire to take the lead in 4G through ambitious, long-range plans and by aiming to achieve the early introduction of public standards for 4G systems. Korea and Japan are taking a proactive approach to the introduction of 4G. China is pursuing a leading role in 4G. In order to achieve this, the country has started developing its own technological standards such as TD-SCDMA (Time Division – Synchronous Code Division Multiple Access). It has also launched a number of government-sponsored research projects on 4G. Furthermore, a crucial step for China is the establishment of many joint ventures between chinese and foreign companies, allowing chinese companies to get both knowledge and capital. China's large population, willingness to adopt new technologies and rapid economic growth means that 4G development there should be followed closely. If China succeeds in developing 4G systems, it can be anticipated that these will be offered at very competitive prices.

The main players in Asia are taking an entirely different approach by promoting a vision of a high data-rate public standard for the 4G system as a whole, building on strong demand for advanced data and entertainment services. Their 4G visions have many points in common with those of Europe, but on the whole, they tend to be more in line with the original linear vision of 4G developing as the next stage in the sequential evolution of mobile communications. They focus more on increasing mobile system data rates, and on developing new systems or system components, and less on the seamless operation of existing systems (though this latter strategy is increasingly included as the visions are further developed). These countries also envisage their governments taking an active role in driving domestic manufacturers to set early 4G standards.

1.1.5 TECHNOLOGY DEVELOPMENTS

The radio spectrum is a precious and scarce resource. Therefore, novel technologies for efficient spectrum utilisation to enhance the capacity of 3G and beyond systems are keenly anticipated. Factors that could have a significant impact on the deployment of mobile telecommunications technologies in this timeframe include radio access techniques enabling greater intelligence and flexibility to be built into transmitters and receivers. Some technology topics that appear relevant to some lesser or greater degree to the future development are: advanced radio resource management (RRM) algorithms; flexible frequency sharing methods; smart antennas; diversity techniques; coding techniques; space-time coding; efficient multiple access schemes or adaptive modulation.

Software Defined Radio (SDR) provides reconfigurable mobile communications systems that aim at providing a common platform to run software that addresses reconfigurable radio protocol stacks thereby increasing network and terminal capabilities and versatility through software modifications (downloads). Basically, SDR concerns all communication layers (from the physical layer to the application layer) of the radio interface and has an impact on both the user terminal and network side.

Future mobile user equipment may assume characteristics of general-purpose programmable platforms by containing high-power general-purpose processors and provide a flexible, programmable platform that can be applied to an ever-increasing variety of uses. The convergence of wireless connectivity and a general-purpose programmable platform might heighten some existing concerns and raise new ones; thus, environmental factors as well as traditional technology and market drivers influence the architecture of these devices. A well-designed embedded processor with a reconfigurable unit may enable user-defined instructions to be efficiently executed, since general-purpose processors such as CPUs or DSPs are not suitable for bit-level operation. This type of processor, which can handle many kinds of bit-level data processes, can be applied to various applications for mobile communication systems with efficient operation.

1.2 UMTS

3G mobile communications systems arose as a response to the challenge of developing systems that increased the capacity of the existing 2G systems. Simultaneously, they would provide a platform that allowed a seamless and ubiquitous access to the user of a wide range of new services, both circuit and packet switched, with higher requirements in terms of bit rate than those for which 2G systems were conceived. The development of 3G systems started in 1995, coordinated by the ITU-T (International Telecommunications Union – Telecommunications sector) under the generic terminology of IMT-2000 and so far different radio access technologies have been considered [6], leading to the development of several standards. Within this framework, the Universal Mobile Telecommunications System (UMTS) is the European proposal given by ETSI (European Telecommunications Standards Institute) to the 3G challenge. As a matter of fact, it is the dominant standard, resulting from the standardisation work done by the 3GPP (3rd Generation Partnership Project), an organisation formed by different regional standardisation bodies that include the presence of both manufacturers and operators from all around the world.

UMTS has been developed as the migration of the ETSI 2G/2.5G systems GSM/GPRS. The aim is to facilitate as much as possible the extension of the existing networks of these worldwide systems as well as the interoperability of the new UMTS system with the previous networks, thus allowing a progressive migration of the technology. As a result of this requirement, the most important changes introduced in the initial release of UMTS consist of a new radio access network based on a different radio access technology, while keeping the core network similar to that existing in GSM/GPRS systems. After this initial implementation, the subsequent releases of the UMTS system introduce important changes in the architecture of the core network, taking the Internet Protocol (IP) as the driving technology.

In the above context, this sub-section presents the main features of the UMTS network architecture, by defining the different elements that comprise it and that establish the basis over which Radio Resource Management (RRM) strategies, which are the main focus of this book, can be implemented.

1.2.1 UMTS ARCHITECTURE

The general UMTS network architecture from the physical point of view is presented in Figure 1.1 and it consists of an abstract model, applicable to any UMTS network, with independency of the specific release [7]. It is organised in domains, and each domain represents the highest level group of physical entities. Reference points are defined between the different domains. The basic split considers the User Equipment (UE) domain, used by the user to access the UMTS services, and the Infrastructure domain,

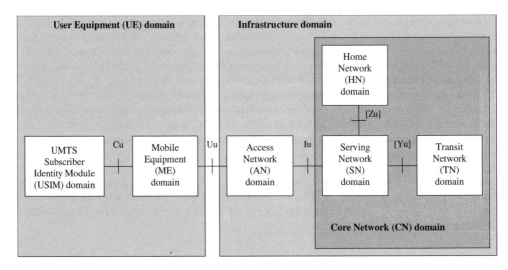

Figure 1.1   General UMTS architecture

composed of the physical nodes, belonging to the network operator, that support the service requirements and the interconnection with the entity at the other end (e.g. another UE from the same or another network) with whom the end-to-end service has to be established. Both domains are separated by means of the Uu reference point, which represents the radio interface, and their elements are explained in the following sub-sections.

1.2.1.1 User Equipment Domain

The User Equipment domain consists of the terminal that allows the user access to the mobile services through the radio interface. From an architectural point of view, it is split into two sub-domains, separated by the Cu reference point (see Figure 1.1):

- Mobile Equipment (ME) domain. This represents the physical entity (e.g. a handset) that in turn is sub-divided into the Mobile Termination (MT) entity, which performs the radio transmission and reception, and the Terminal Equipment (TE), which contains the applications. These two entities may be physically located at the same hardware device depending on the specific application. For example, in the case of a handset used for a speech application, both MT and TE are usually located in the handset, while if the same handset is being used for a web browsing application, the handset will contain the MT and the TE can reside in, for example, an external laptop that contains the web browser.
- UMTS Subscriber Identity Module (USIM) domain. Typically, the physical hardware device containing the USIM is a removable smart card. The USIM contains the identification of the profile of a given user, including his identity in the network as well as information about the services that this user is allowed to access depending on the contractual relationship with the mobile network operator. So, the USIM is specific for each user and allows him/her to access the contracted services in a secure way by means of authentication and encryption procedures regardless of the ME that is used.

1.2.1.2 Infrastructure Domain

The infrastructure domain in the UMTS architecture contains the physical nodes that terminate the radio interface allowing the provision of the end-to-end service to the UE. In order to separate the functionalities that are dependent on the radio access technology being used from those that are independent, the infrastructure domain is in turn split into two domains, namely the Access Network and the Core Network domains (see Figure 1.1), separated by the Iu reference point. This allows there to be a generic UMTS architecture that enables the combination of different approaches for the radio access technology as well as different approaches for the core network. As a matter of fact, notice that this architecture is the same used for GSM/GPRS networks so that the difference between a GSM/GPRS network and a UMTS network will mainly rely on the specific implementations of the access network and the core networks domains.

With respect to the core network, and in order to take into account different scenarios in which the user communicates with users in other types of networks (e.g. other mobile networks, fixed networks, Internet, etc.), three different sub-domains are defined (see Figure 1.1):

- Home Network (HN) domain. This corresponds to the network to which the user is subscribed, so it belongs to the operator that has the contractual relationship with the user. The user service profile as well as the user secure identification parameters are kept in the HN and should be coordinated with those included in the USIM at the UE.
- Serving Network (SN) domain. This represents the network containing the access network to which the user is connected in a given moment and it is responsible for transporting the user data from the source to the destination. Physically, it can be either the same HN or a different network in the case where the user is roaming with another network operator. The SN is then connected to the access network through the Iu reference point and to the HN through the [Zu] reference point. The interconnection with the HN is necessary in order to retrieve specific information about the user service abilities and for billing purposes.
- Transit Network (TN) domain. This is the core network part located on the communication path, between the SN and the remote party, and it is connected to the SN through the [Yu] reference point. Note that, where the remote party belongs to the same network to which the user is connected, the SN and the TN are physically the same network. Note also that, in general, the TN may not be a UMTS network, for example, in the case of a connection with a fixed network or when accessing the Internet.

According to this generic framework, several scenarios can be defined depending on the networks to which the UE and the remote party are connected, and it is even possible that the HN, the SN and the AN are physically the same network.

We will now describe the specific architectures of the AN for terrestrial UMTS networks, denoted as UTRAN (Universal Terrestrial Radio Access Network), and the UMTS generic Core Network (CN), which can be the HN, SN or TN.

Universal Terrestrial Radio Access Network (UTRAN)   The architecture of the UTRAN is shown in Figure 1.2 [8]. It is composed of Radio Network Subsystems (RNSs) that are connected to the Core Network through the Iu interface that coincides with the Iu reference point of the overall UMTS architecture. Each RNS is responsible for the transmission and reception over a set of UMTS cells. The connection between the RNS and the UE is done through the Uu or radio interface.

The RNSs contain a number of Nodes B or base stations and one Radio Network Controller (RNC), connected through Iub interfaces. RNCs belonging to different RNSs are interconnected by means of the Iur interface.

A node B is the termination point between the air interface and the network and it is composed of one or several cells or sectors. In the 3GPP terminology, a cell stands as the smallest radio network entity that has its own identification number, denoted as Cell ID. Conceptually, a cell is regarded as a UTRAN Access Point through which radio links with the UEs are established. From a functional point of view, the cell executes the physical transmission and reception procedures over the radio interface.

The RNC is the node responsible for controlling the use of the radio resources in the nodes B that are under its control, thus it is the main entity where UMTS Radio Resource Management (RRM) algorithms

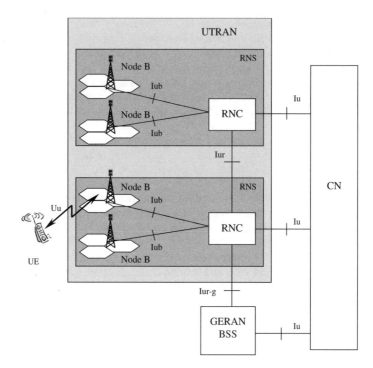

Figure 1.2   UTRAN architecture

are executed. The majority of functionalities related to the radio interface are executed in the RNC, with the exception of the physical transmission and reception processes and some specific Medium Access Control (MAC) functions that are executed in the Node B. On the network side, the RNC interoperates with the CN through the Iu interface and establishes, maintains and releases the connections with the CN elements that the UEs under its control require in order to receive the UMTS services.

Additionally, as is shown in Figure 1.2, it is also possible for a RNC to interoperate with the Base Station Subsystems (BSSs) that form the GERAN (GSM/EDGE Radio Access Network) by means of the Iur-g interface. This interoperation allows the execution of Common Radio Resource Management (CRRM) algorithms between UMTS and GSM/GPRS systems.

From a functional point of view, the RNC may take several logical roles:

- CRNC (Controlling RNC). This is the role with respect to the Node B, and refers to the control that the RNC has over a set of Nodes B.
- SRNC (Serving RNC). This role is taken with respect to the UE. The SRNC is the RNC that holds the connection of a given UE with the CN through the Iu interface, so it can be regarded as the RNC that controls the RNS to which the mobile is connected at a given moment. When the UE moves across the network and executes handover between the different cells, it may require a SRNS (Serving RNS, i.e. the RNS having the SRNC) relocation procedure when the new cell belongs to a different RNC. This procedure requires the communication between the SRNC and the new RNC through the Iur interface in order for the new RNC to establish a new connection with the CN over its Iu interface.
- DRNC (Drift RNC). This role is also taken with respect to the UE and is a consequence of a specific type of handover that exists with CDMA systems, denoted as soft handover. In this case, a UE can be simultaneously connected to several cells (i.e. it has radio links with several cells). Then, when a UE moves in the border between RNSs, it is possible that it establishes new radio links with cells

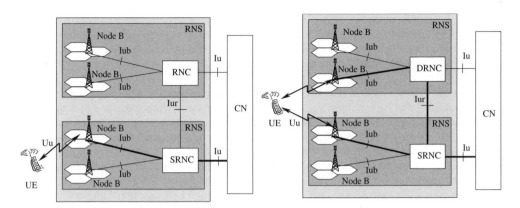

Figure 1.3   SRNC (left) and DRNC (right) roles of the RNC

belonging to a new RNC while at the same time keeping the radio link with some cells of the SRNC. In this case, the new RNC takes the role of DRNC, and the connectivity with the CN is not done through the Iu of the DRNC but still through the Iu of the SRNC, thus requiring it to establish resources for the UE in the Iur interface between SRNC and DRNC. Only when all the radio links of the old RNC are released and the UE is connected only to the new RNC, will the SRNS relocation procedure be executed. Figure 1.3 illustrates the difference between SRNC and DRNC roles.

Notice that all the RNCs are CRNC and that a given RNC may be SRNC for certain UEs and simultaneously DRNC for others.

Two different operation modes have been standardised for the UTRAN radio interface and can be supported with the architecture of Figure 1.2 simply by changing the radio access technology. These modes are:

- UTRAN FDD (Frequency Division Duplex) mode. In this case, the uplink and downlink transmit with different carrier frequencies, thus requiring the allocation of paired bands. The access technique being used is WCDMA (Wideband Code Division Multiple Access), which means that several transmissions in the same frequency and time are supported and can be distinguished by using different code sequences.
- UTRAN TDD (Time Division Duplex) mode. In this case, the uplink and downlink operate with the same carrier frequency but in different time instants, thus they are able to use unpaired bands. The access technique being used is a combination of TDMA and CDMA, denoted as TD/CDMA, which means that simultaneous transmissions are distinguished by different code sequences (CDMA component) and that a frame structure is defined to allocate different transmission instants (time slots) to the different users (TDMA component).

The initial frequency bands reserved for each of the two UTRAN modes are shown in Figure 1.4 for regions 1 and 3 (i.e. Europa and Asia). The two radio access technologies lead to the existence of cells

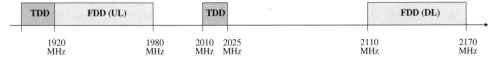

Figure 1.4   Frequency bands for the UTRAN FDD and TDD modes

supporting one or both of the two modes, as well as the ability to interoperate between them. Notice that, from the radio resource management point of view, the concept of radio resource is different for each mode. As a result, the RRM strategies in both cases lead to different types of algorithms. In the context of this book, only the RRM strategies for the UTRAN FDD mode are considered.

Core Network   The Core Network (CN) is the part of the mobile network infrastructure that covers all the functionalities that are not directly related with the radio access technology, thus it is possible to combine different core network architectures with different radio access networks. Examples of these functionalities are the connection and session management (i.e. establishment, maintenance and release of the connections and sessions for circuit switched and packet switched services) as well as mobility management (i.e. keep track of the area where each UE can be found in order to route calls to it).

While the access network in UMTS has suffered relatively few changes since the initial UMTS release (release 99), this is not the case with the Core Network. The reason is that the initial implementations of UMTS were seen simply as an extension of the GSM/GPRS networks because they maintained the existing core network for GSM/GPRS (with small modifications) in order to make it compatible with the new UMTS access network. This was due to the impression that the new radio interface technology posed the most critical challenges in the support of the expected UMTS services. The new releases of UMTS introduced the major changes in the architecture of the core network only after the radio access part was stabilised, therefore driving it towards the development of an all IP network.

Figure 1.5 shows the elements that compose the architecture of the UMTS core network as well as the interfaces between them [9]. The figure reflects the UMTS release 99, which is essentially the same as the GSM/GPRS system, and the evolution of this architecture in future releases will be explained in Section 1.2.2. As can be observed, the infrastructure of the core network is divided into two domains that differ in the way they support user traffic. They are the Circuit Switched (CS) and the Packet Switched (PS) domains. The CS domain supports the traffic composed by connections that require dedicated

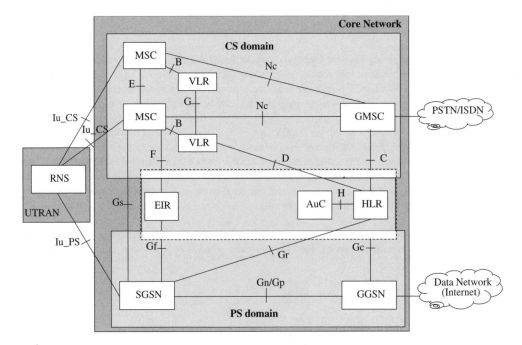

Figure 1.5   Architecture of the UMTS Core Network (Release 99)

network resources, and allows the interconnection with external CS networks like the PSTN (Public Switched Telephone Network) or the ISDN (Integrated Services Digital Network). In turn, the PS domain supports a traffic composed of packets, which are groups of bits that are autonomously transmitted and independently routed, so no dedicated resources are required throughout the connection time, since the resources are allocated on a packet basis only when needed. This allows a group of packet flows to share the network resources based on traffic multiplexing. The PS domain allows the interconnection of external PS networks, like the Internet. The division between CS and PS domains introduces the requirement to split the Iu reference point between core and access networks in two interfaces, denoted as Iu_CS and Iu_PS.

There are some entities in the CN that belong both to the CS and the PS domains. They are the HLR (Home Location Register), the AuC (Authentication Centre) and the EIR (Equipment Identity Register). The HLR is a database that stores information about the users that are subscribed in a given network, including the different user identifiers and the service profile. The AuC stores the identity keys of the subscribed users and is used by the HLR to perform security operations. In turn, the EIR is a database that stores the identifiers of the mobile terminals in order to detect those terminals whose access to the network must be denied due to different reasons (for example, because they have been stolen).

The CS domain is composed of three specific entities, namely the MSC (Mobile Switching Centre), the GMSC (Gateway Mobile Switching Centre) and the VLR (Visitor Location Register). The MSC interacts with the radio access network by means of the Iu_CS interface and executes the necessary operations to handle CS services. This includes routing the calls towards the corresponding transit network and establishing the corresponding circuits in the path. The MSC is the same as that which uses the GSM network with the difference that a specific interworking function is required between the MSC and the access network in UMTS. The reason is that in GSM the speech traffic delivered to the core network by the access network uses 64 kb/s circuits while in UMTS the speech uses adaptive multi-rate technique (AMR) with bit rates between 4.75 kb/s and 12.2 kb/s that are transported in the access network with ATM (Asynchronous Transfer Mode) technology [6]. This is why the term 3G MSC is sometimes used to differentiate between the MSC from GSM system and the MSC from UMTS networks.

The VLR is a database associated with a MSC that contains specific information (e.g. identifiers, location information, etc.) about the users that are currently in the area of this MSC, which allows the performing of certain operations without the need to interact with the HLR. The information contained in the VLR and the HLR must be coordinated.

The GMSC is a specific MSC that interfaces with the external CS networks and is responsible of routing calls to/from the external network. To this end, it interacts with the HLR to determine the MSC through which the call should be routed. In release 99, the communication between the entities of the CS domain is done by means of 64 kb/s circuits and uses SS7 (Signalling System No. 7) for signalling purposes.

The PS domain is composed of two specific entities, namely the SGSN (Serving GPRS Support Node) and GGSN (Gateway GPRS Support Node), which perform the necessary functions to handle packet transmission to and from the UEs. The SGSN is the node that serves the UE and establishes a mobility management context including security and mobility information. It interacts with the UTRAN by means of the Iu_PS interface. The GGSN, in turn, interfaces with the external data networks and contains routing information of the attached users. IP tunnels between the GGSN and the SGSN are used to transmit the data packets of the different users [10].

1.2.2 UMTS EVOLUTION

The development of the first UMTS specifications was done at a time when the Internet was becoming progressively more and more popular and the IP technology began to be used not only for the transport of data services but also for speech and video services, thus becoming a new paradigm for the deployment of multiservice networks. The response of the UMTS system to this expansion of IP technology is given

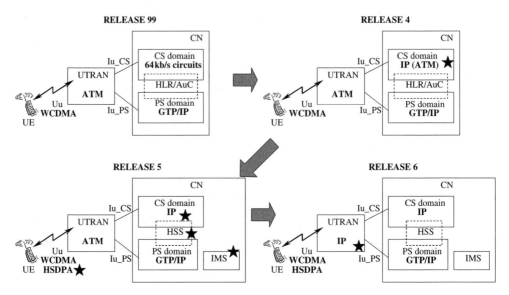

Figure 1.6   UMTS evolution

in the releases that followed the initial release 99, and whose main purpose was the progressive transformation of UMTS in an all IP network that was more efficient than the coexistence of two separated networks for the CS and PS core network domains.

Figure 1.6 tries to show in a schematic way the main changes, represented with black stars, which appear in the different UMTS releases. In release 99, the transport technology used between the elements of the UTRAN is ATM. In turn, in the CS domain of the core network, 64 kb/s circuits are used and in the PS domain transmissions are done by means of IP tunnels using GTP (GPRS Tunnelling Protocol). The first step in the evolution towards an all IP network is release 4, in which the CS domain of the CN is replaced by an IP or an IP/ATM backbone. The transmission of speech services over this IP backbone introduces important technological challenges that lead to the so-called voice over IP technology. This modification of the core network involves the evolution of MSC in two different components, namely the MSC server, which comprises the call control and mobility control parts of the MSC, thus handling only signalling, and the MGW (Media Gateway function), which handles the users' data flows.

Release 5 executes the final step to achieving a CN completely based on IP technology by removing the possibility of using ATM in the CS domain. In the new architecture, the provision of real time IP multimedia service is done by means of the inclusion of a new CN domain, namely the IP Multimedia Subsystem (IMS), which is connected to the GGSN and the MGW, and makes use of the Session Initiation Protocol (SIP) as a means of establishing multimedia sessions between users supporting user mobility and call redirection [11]. Another of the changes introduced by release 5 in the CN consists of the integration of the functionalities of the HLR and the AuC in the HSS (Home Subscriber Server), which contains the subscription related information for each user in order to support the call and session handling.

The release 5 does not limit the changes to the CN, and introduces an important modification at the radio interface as well. In particular, a new packet access mechanism over WCDMA, denoted as HSDPA (High Speed Downlink Packet Access) is defined. HSDPA supports much higher bit rates up to around 10 Mb/s by means of an additional modulation scheme and the implementation of fast packet scheduling and hybrid retransmission mechanisms, and coexists with the radio access mechanisms existing in previous releases [12].

The final objective of an all IP architecture like the one defined in release 5 is the inclusion of the IP technology in the radio access network as well. Due to the important modifications that such a change requires, this inclusion was postponed to the release 6. The existence of an all IP network including the radio access part facilitates the integration of different radio access technologies operating over a unique backbone technology and therefore enables the development of heterogeneous networks that integrate the UTRAN and GERAN technologies with others like WLAN.

1.3 QoS MODEL IN UMTS

In order to provide a service with specific QoS requirements, a bearer service with clearly defined characteristics and functionality needs to be set up from the source to the destination of the service. Since the end-to-end path extends across different system levels each having their own QoS properties, the QoS is handled and split in different parts taking into account the special characteristics of each component. In this framework, the UMTS QoS mechanisms shall provide a mapping between application require-ments and UMTS services.

The layered architecture of a UMTS bearer service is depicted in Figure 1.7. Each bearer service on a specific layer offers its individual services using the services provided by the layers below. The end-to-end service used by the TE (Terminal Equipment) will be realised using a TE/MT Local Bearer Service, a UMTS Bearer Service, and an External Bearer Service. The QoS mechanisms outside the UMTS network are not within the scope of 3GPP specifications and, consequently, the end-to-end bearer service is beyond the scope of specification TS 23.107 [13], which is mainly described in this section. Nevertheless, it is worth noting that the UMTS operator offers a wide variety of services by means of the UMTS Bearer Service. In turn, the UMTS Bearer Service consists of two parts, the Radio Access Bearer (RAB) Service and the Core Network Bearer Service. In this way, an optimised realisation of the UMTS Bearer Service over the respective segments is more feasible. The Radio Access Bearer Service is based

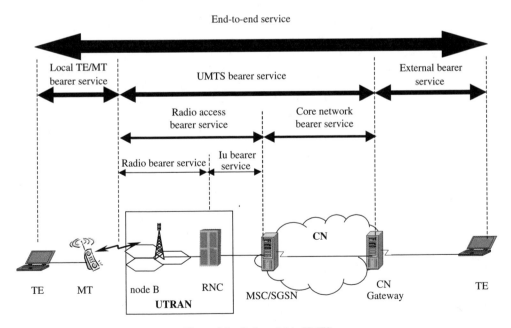

Figure 1.7   QoS model in UMTS

on the characteristics of the radio interface and is maintained for a moving MT. The role of the Core Network Bearer Service is to control and utilise efficiently the backbone network.

The Radio Access Bearer Service is realised by a Radio Bearer Service and an Iu-Bearer Service. The role of the Radio Bearer Service is to cover all the aspects of the radio interface transport. In the context of this book, it is considered that this bearer service uses the UTRAN FDD. The Iu-Bearer Service together with the Physical Bearer Service provides the transport between UTRAN and CN.

In UMTS, four different QoS classes have been identified:

- Conversational class. The Real time conversation scheme is characterised by a low transfer time because of the conversational nature of the scheme and fact that the time variation between information entities of the stream will be preserved in the same way as for real time streams. The maximum transfer delay is given by the human perception of video and audio conversation. The most well known use of this scheme is telephony speech. Nevertheless, with the Internet and multimedia, a number of new applications will require this scheme, for example voice over IP (VoIP) and video conferencing tools.
- Streaming class. This scheme is one of the newcomers in data communication, raising a number of new requirements in telecommunication systems. It is characterised by the fact that the time variation between information entities (i.e. samples, packets) within a flow will be preserved, although it does not have any requirements on low transfer delay. As the stream normally is time aligned at the receiving end (in the user equipment), the highest acceptable delay variation over the transmission media is given by the capability of the time alignment function of the application. Acceptable delay variation is thus much greater than the delay variation given by the limits of human perception.
- Interactive class. Interactive traffic is the other classical data communication scheme that on an overall level is characterised by the request response pattern of the end user. This scheme applies when the end user, which can be either a machine or a human, is online requesting data from remote equipment (e.g. a server). Examples of human interaction with the remote equipment are: web browsing, data base retrieval, server access. Examples of machines interaction with remote equipment are: polling for measurement records and automatic database enquiries (tele-machines). At the message destination, there is an entity expecting the response within a certain time. Round trip delay time is therefore one of the key attributes. Another characteristic is that the content of the packets are transparently transferred (i.e. with low bit error rate).
- Background class. When the end user, which typically is a computer, sends and receives data-files in the background, this scheme applies. Examples are background delivery of emails, SMS (Short Message Service), download of databases and reception of measurement records. Background traffic is one of the classical data communication schemes that on an overall level is characterised by the fact that the destination is not expecting the data within a certain time. The scheme is thus more or less delivery time insensitive. Another characteristic is that the content of the packets are transparently transferred (i.e. with low bit error rate)

The Radio Access Bearer Service attributes, which will be applied to both CS and PS domains, are:

- Traffic class ('conversational', 'streaming', 'interactive', 'background'). With this attribute, UTRAN can make assumptions about the traffic source and optimise the transport for that traffic type.
- Maximum bit rate. This is the maximum number of bits delivered by UTRAN or to UTRAN at a SAP (Service Access Point) within a period of time, divided by the duration of the period. The purpose of this attribute is mainly to limit the delivered bit rate to applications or external networks as well as to allow the maximum desired RAB bit rate to be defined for applications able to operate with different bit rates.
- Guaranteed bit rate. This is the guaranteed number of bits delivered at a SAP within a period of time (provided that there are data to deliver), divided by the duration of the period. This attribute may be used to facilitate admission control based on available resources, and for resource allocation within

UTRAN. Quality requirements expressed by, for example, delay and reliability attributes, only apply to incoming traffic up to the guaranteed bit rate. It is worth noting that the guaranteed bit rate at the RAB level may be different from that on the UMTS bearer level, for example due to header compression.

- Delivery order. This indicates whether the UMTS bearer shall provide in-sequence SDU (Service Data Unit) delivery or not and specifies if out-of-sequence SDUs are acceptable or not.
- Maximum SDU size used for admission control and policing. This corresponds to the maximum packet size that can be delivered at the top of the radio interface.
- SDU format information. This is the list of possible exact sizes of SDUs.
- SDU error ratio. This indicates the fraction of SDUs lost or detected as erroneous. This attribute is used to configure the protocols, algorithms and error detection schemes, primarily within UTRAN.
- Residual bit error ratio. This indicates the undetected bit error ratio in the delivered SDUs. It is used to configure radio interface protocols, algorithms and error detection coding.
- Delivery of erroneous SDUs. This indicates whether SDUs detected as erroneous will be delivered or discarded.
- Transfer delay. This indicates the maximum delay for the 95th percentile of the distribution of delay for all delivered SDUs during the lifetime of a bearer service, where delay of an SDU is defined as the time from a request to transfer an SDU at one SAP to its delivery at the other SAP. The attribute is used to specify the delay tolerated by the application and allows UTRAN to set transport formats and ARQ (Automatic Repeat Request) parameters.
- Traffic handling priority, specifying the relative importance of handling all SDUs belonging to the UMTS bearer compared to the SDUs of other bearers. In particular, there is a need to differentiate between bearer qualities within the interactive class. This is handled with this attribute, to allow UMTS to schedule traffic accordingly. By definition, priority is an alternative to absolute guarantees, and thus these two attributes cannot be used together for a single bearer.
- Allocation/Retention Priority. This specifies the relative importance, compared to other UMTS bearers, of allocation and retention of the UMTS bearer. In situations where resources are scarce, this attribute may be used to prioritise bearers when performing admission control.
- Source statistics descriptor. This specifies characteristics of the source of submitted SDUs and it may take the values 'speech' or 'unknown'. Since conversational speech has a well-known statistical behaviour, UTRAN may calculate a statistical multiplex gain for use in admission control on the radio and Iu interfaces.

It is worth remarking that, when establishing a UMTS bearer and the underlying Radio Access Bearer for support of a service request, some attributes typically have different values on both levels. For example, the requested transfer delay of the UMTS bearer will typically be larger than the requested transfer delay of the Radio Access Bearer, as the transport through the core network will use part of the acceptable delay. Similarly, SDU error ratio for Radio Access Bearer service will be reduced with the errors introduced in the core network, by the Core Network Bearer service. Furthermore, some attributes/ settings only exist on the Radio Access Bearer level, such as Source statistics descriptor.

REFERENCES

[1] J.C. Burgelman, G. Carat (editors), 'Prospects for Third Generation Mobile Systems', ESTO Project Report, IPTS Technical Report, June 2003
[2] C. Rodríguez-Casal, J.C. Burgelman, G. Carat (editors), 'The Future of Mobile Communications in the EU: Assessing the potential of 4G', ESTO Project Report, February 2004
[3] ITU-R, Working Party 8F, 'Technology trends (draft new report)', October 2003
[4] ITU, 'Social and human considerations for a more mobile world (background paper)', ITU/MIC workshop on shaping the future mobile information society, February 2004

[5] ITU, 'The case of Japan', ITU/MIC workshop on shaping the future mobile information society, February 2004

[6] J. Bannister, P. Mather, S. Coope, *Convergence Technologies for 3G Networks*, John Wiley & Sons, Ltd. 2004

[7] 3GPP TS 23.101 'General UMTS architecture'

[8] 3GPP TS 25.401 'UTRAN overall description'

[9] 3GPP TS 23.002 'Network architecture'

[10] 3GPP TS 23.060 'GPRS; Service Description; Stage 2'

[11] RFC 3261, 'SIP: Session Initiation Protocol', J. Rosenberg *et al.*, June 2002

[12] 3GPP TS 25.308 'High Speed Downlink Packet Access (HSDPA); Overall Description'

[13] 3GPP TS 23.107 'Quality of Service (QoS) concept and architecture (Release 5)'

2

CDMA Concepts

The multiple access technique in mobile communications systems determines the way that signals from different transmitters share the same radio transmission medium. For this purpose, it defines a set of available radio resource units so that each signal occupies a certain amount of these radio resource units. Consequently, radio resource management strategies for a given system must be devised to take into account the characteristics of the multiple access technique being used. Within this framework, the purpose of this chapter is to introduce the fundamentals of the multiple access technique used in UMTS, which is CDMA (Code Division Multiple Access). These fundamentals constitute the basis for the definition of the radio resource management algorithms in Chapters 4 and 5.

The chapter starts with a brief overview of the most commonly used multiple access techniques in mobile communications systems, trying to point out the relevant differences among them. It continues with an explanation of the procedures involved in the CDMA signal generation and reception and concludes with a description of the implications of using CDMA in a cellular system.

2.1 MULTIPLE ACCESS TECHNIQUES

The definition of a multiple access technique is based on the fact that the receiver must be able to separate the desired signal from the rest of the signals present at the antenna. This capability can be ensured provided that the different signals are orthogonal, so the multiple access techniques mainly differ in the way they achieve orthogonality. Taking this into account, the most commonly used multiple access techniques are:

- Frequency Division Multiple Access (FDMA). This technique ensures the orthogonality by using signals that do not overlap in the frequency domain. To this end, it divides the total available bandwidth B_T (Hz) into smaller pieces of bandwidth B_c (Hz), and assigns each one of them to one user, as depicted in Figure 2.1. At the receiver, signals can be separated simply by using a filter that selects the bandwidth allocated to the corresponding user. In practice, and in order to facilitate the implementation of the filters at the receiver, a certain guard band B_g (Hz) is usually left between the bandwidths of the different users. Taking this into account, the maximum number of users that can be allocated in the system is given by $K = B_T/(B_c + B_g)$. Naturally, it is interesting to keep the guard band to a minimum in order to increase the efficiency, since high values of this band lead to unused bandwidth. Due to its simplicity, this technique was the first one to be used in analogue mobile communications systems and current 2G and 3G systems still use it in combination with other techniques.

Radio Resource Management Strategies in UMTS   J. Pérez-Romero, O. Sallent, R. Agustí and M. A. Díaz-Guerra
© 2005 John Wiley & Sons, Ltd

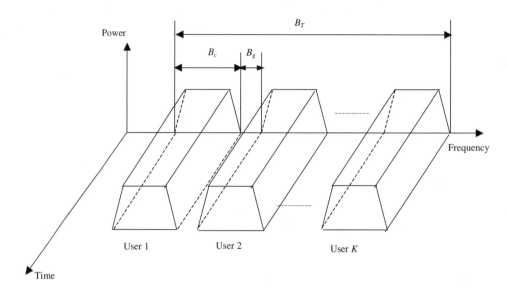

Figure 2.1 Frequency division multiple access technique

- Time Division Multiple Access (TDMA). In this case, the orthogonality among signals is achieved by avoiding that signals overlap in the time domain. To this end, the time axis is organised into frames of T_F (s) that are repeated periodically and are subdivided into K time slots of duration T_S (s). Each slot is allocated to a different user, so that it transmits once per frame occupying the whole bandwidth B_T (Hz) during one slot. Then, the resulting equivalent bandwidth allocated per user is B_T/K. Notice that simply by allocating several slots of the frame to the same user, it is also possible to support users with different bandwidth requirements.

  In practice, the transmitted signals consist of bursts that do not occupy the full slot duration T_S but do leave a certain guard time T_g in order not to overlap with the adjacent slots due to differences in the propagation delays (see Figure 2.2).

  Notice also that TDMA can only be supported by digital signals, due to the discontinuity in the transmission, which requires the information generated by the source to be buffered, waiting for the instant when it can be transmitted. Similarly, the buffering process at the receiver allows delivery of the information in a continuous way to the final user, even if the transmission at the radio interface has been discontinuous. In any case, there is a minimum delay that must be supported by the application given by the frame time.

  It is rare that systems make use of a pure TDMA; more usually, it is used in combination with FDMA to constitute hybrid TDMA/FDMA systems, such as the case with GSM/GPRS. In this way, the total system bandwidth is divided into several carrier frequencies and each one of them is organised into frames, so that the resource to be allocated to a given user is constituted by the pair frequency and time slot.
- Code Division Multiple Access (CDMA). The CDMA technique is related to the so-called spread spectrum techniques, in which the bandwidth of the signal is spread to a higher value than the original signal, resulting in the transmission of wideband signals with reduced power densities that can be even lower than the receiver noise spectral density [1][2]. Consequently, such signals are difficult to detect by receivers different than the desired one, which leads to a higher natural confidentiality in the transmission. As a matter of fact, this feature was the key aspect that introduced the first spread spectrum communication systems, mainly focused on military applications. Later on, the applicability of such systems was extended to the field of multiple access techniques [3][4].

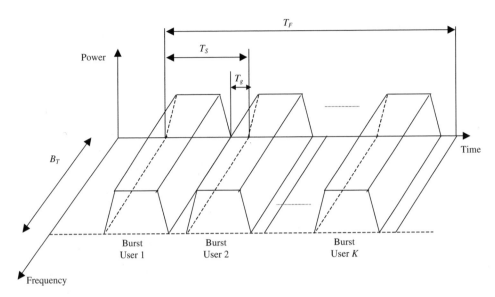

Figure 2.2   Time division multiple access technique

Essentially, there are two types of spread spectrum techniques, denoted as Frequency Hopping Spread Spectrum (FH-SS), in which the signals are transmitted by varying the carrier frequency in very short periods of time according to a predefined sequence, and Direct Sequence Spread Spectrum (DS-SS), in which the signals are multiplied by pseudo-random code sequences. The latter is the strategy used in mobile communication systems such as IS-95 and UMTS, so the multiple access technique is denoted as DS-CDMA (Direct Sequence CDMA), although it is usually referred to simply as CDMA.

In DS-CDMA, the signal orthogonality is achieved by multiplying the signal of each user by a different orthogonal code sequence. In the frequency domain, this process spreads the bandwidth of the original signal. Therefore, the receiver can recover the desired signal simply by making use of the same code sequence to despread the bandwidth. In CDMA, as depicted in Figure 2.3, all users transmit simultaneously and make use of the whole bandwidth. The maximum number of simultaneous users in this case is not fixed, as in FDMA and TDMA systems, but depends on different factors. In particular, if the code sequences are perfectly orthogonal the maximum number of users is equal to the number of available sequences. However, in practice, some interference remains after the despreading process and, as a result, the maximum number of users depends on the maximum interference that can be tolerated by the receiver. This lack of constant capacity in CDMA systems is known as 'soft capacity'.

IS-95 was the first 2G mobile communication system to make use of CDMA, occupying a total bandwidth of 1.25 MHz. In the development of 3G systems, UMTS also adopted CDMA but increased the bandwidth to 5 MHz, which was the reason for denoting the technique as Wideband Code Division Multiple Access (WCDMA) [5].

2.2 CDMA SIGNAL GENERATION

A simplified block diagram of a CDMA generator, for the case of BPSK (Binary Phase Shift Keying) modulation, is shown in Figure 2.4. Let us assume that a sequence of information bits $b(t)$ with bit rate

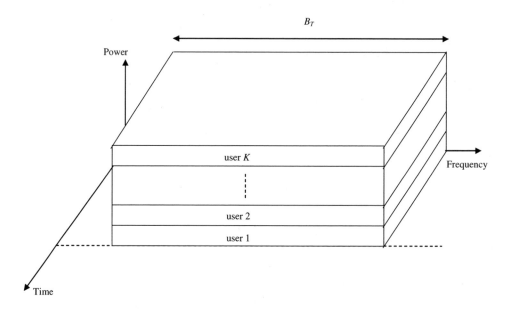

Figure 2.3 Code division multiple access technique

R_b (b/s), or equivalently bit period $T_b(\text{s}) = 1/R_b$, has to be transmitted. Let's also assume that these information bits are protected by means of some type of channel encoding and interleaving procedures that enable the correction of errors caused by the radio channel at the receiver. As a result of these procedures, which are not specific to CDMA systems, a sequence $d(t)$ of encoded bits is obtained with bit rate $R_s(\text{b/s}) = R_b/r$ or equivalently bit period $T_s(\text{s}) = 1/R_s$. r is the channel code rate, defined as the ratio between the number of information bits with respect to the number of encoded bits. Therefore, the sequence $d(t)$ represents the bits that must be transmitted through the channel, so they are referred as 'channel bits'.

Mathematically, the sequence $d(t)$ can be expressed as:

$$d(t) = \sum_{k=-\infty}^{\infty} d_k p_{Ts}(t - kT_s) \tag{2.1}$$

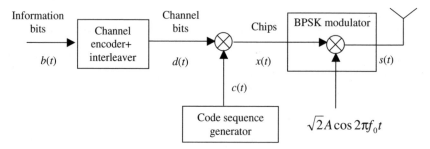

Figure 2.4 Scheme of the CDMA transmitter with BPSK modulation

where $d_k \in \{1, -1\}$ and $p_{Ts}(t)$ is a rectangular pulse of length T_s. Assuming that the bits are random with equal probability of the values $+1$ and -1, the spectral density of this signal can be expressed as:

$$S_d(f) = T_s \frac{\sin^2(\pi T_s f)}{(\pi T_s f)^2}$$

(2.2)

which presents a bandwidth:

$$BW_d(\text{Hz}) \approx 1/T_s = R_s$$

(2.3)

The CDMA signal is obtained by multiplying each bit of the sequence $d(t)$ by a code sequence $c(t)$, different for each transmitter, composed of N symbols $c_l \in \{1, -1\}$, known as 'chips', of duration T_c corresponding to a chip rate of W chips/s. Without lack of generality, it will be assumed that the sequence of N chips is repeated periodically. Then, it can be mathematically expressed as:

$$c(t) = \sum_{k=-\infty}^{\infty} \sum_{l=0}^{N-1} c_l p_{Tc}(t - lT_c - kT_s)$$

(2.4)

where $p_{Tc}(t)$ is a rectangular pulse of length T_c.

The relationship between the channel bit period and the chip period is $T_s = NT_c$, so that each bit d_k is multiplied exactly by N chips. The sequence $x(t)$ obtained after the multiplication will then be composed of T_c duration chips that will depend on the value of the chip c_l of the code sequence and on the channel bit d_k. It can be mathematically expressed as:

$$x(t) = d(t) \cdot c(t) = \sum_{k=-\infty}^{\infty} d_k \sum_{l=0}^{N-1} c_l p_{Tc}(t - lT_c - kT_s)$$

(2.5)

Each chip c_l takes the values $+1$ and -1 with equal probability. Consequently, the spectral density of the sequence $x(t)$ is given by:

$$S_x(f) = T_c \frac{\sin^2(\pi T_c f)}{(\pi T_c f)^2}$$

(2.6)

whose bandwidth is approximately:

$$BW_x(\text{Hz}) \approx 1/T_c = W$$

(2.7)

Actually, by making use of other pulse shapes different from the rectangular one, a flat spectral density in the bandwidth W could be achieved.

By comparing $S_x(f)$ with $S_d(f)$ in Equations (2.2) and (2.6), it can be observed that both have the same shape but the bandwidth of the CDMA signal BW_x is higher than the bandwidth BW_d of the original signal $d(t)$, so the multiplication of the code sequence has spread the bandwidth of the signal. The relationship between bandwidths is denoted as the 'spreading factor' or 'processing gain' and given by:

$$SF = \frac{BW_x}{BW_d} = \frac{T_s}{T_c} = \frac{W}{R_s} = N$$

(2.8)

Furthermore, and since the multiplication by a sequence $\{+1, -1\}$ does not change the total power of the signal $d(t)$, the spreading process also results in a reduction of the power spectral density of signal $x(t)$ with respect to $d(t)$ and the reduction factor is equal to the spreading factor SF:

$$\frac{S_d(0)}{S_x(0)} = \frac{T_s}{T_c} = N = SF \tag{2.9}$$

The spreading factor is one of the key parameters of any CDMA system since it relates the transmission bandwidth with the bit rate. One consequence of this relationship is the possibility of having transmissions with variable bit rate simply by changing the spreading factor but keeping a constant transmission bandwidth W. Particularly, from Equation (2.8), note that

$$W = SF \cdot R_s \tag{2.10}$$

can be obtained. Therefore, if the bandwidth W must remain constant, when the channel bit rate R_s is increased, the spreading factor SF must be decreased in the same proportion or, equivalently, if the channel bit rate is decreased the spreading factor must be increased in the same proportion. As a result, CDMA provides an inherent flexibility to accommodate services with variable bit rate, like those expected in UMTS.

An example CDMA signal generation in the time domain is shown in Figure 2.5, where the considered code sequence is $\{1, -1, 1, -1, -1, 1, -1, 1\}$ and thus has a spreading factor equal to $SF = N = 8$. Notice that the appearance of the spread signal $x(t)$ is similar to that of the code sequence $c(t)$. The same example in the spectral domain is shown in Figure 2.6, which presents the value of the spectral densities

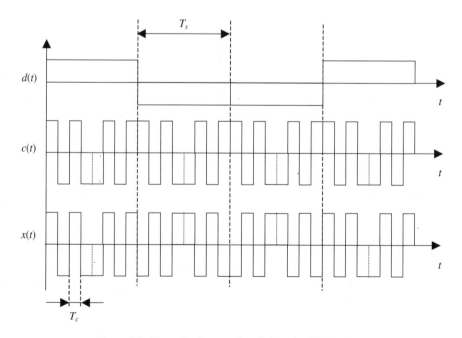

Figure 2.5   Example of temporal evolution of a CDMA signal

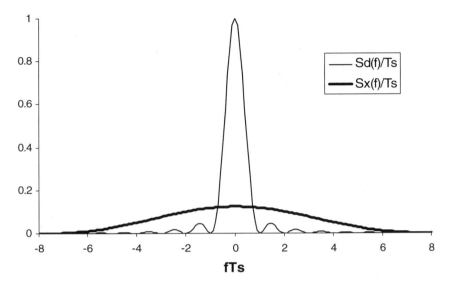

Figure 2.6   Spectral density of the signal before spreading $S_d(f)$ and after spreading $S_x(f)$

$S_x(f)$ and $S_d(f)$ normalised to the channel bit period T_s. It can be observed that the bandwidth of signal $x(t)$ is eight times higher than that of $d(t)$ while its power spectral density is eight times lower.

The above described procedure is common to any CDMA transmitter. In the next steps, the signal will be modulated into a specific carrier frequency f_0, amplified and sent to the antenna, and these procedures will depend on the specific modulation used. Furthermore, and in order to reduce the interference introduced in the adjacent channels, the pulse shape of the transmitted chips is usually changed by means of a pulse shaping filter from the rectangular pulse to other more efficient pulses, such as the root-raised cosine. In any case, it is worth mentioning that the main conclusions and relationships that can be derived with the analysis assuming rectangular pulses are also valid for other pulses.

In the example of the transmitter shown in Figure 2.4, where BPSK modulation and rectangular pulses are assumed, the transmitted signal is given by:

$$s(t) = \sqrt{2}A \sum_{k=-\infty}^{\infty} d_k \sum_{l=0}^{N-1} c_l p_{Tc}(t - lT_c - kT_s) \cos(2\pi f_0 t) \tag{2.11}$$

The power spectral density of the transmitted signal is given by:

$$S_s(f) = A^2 S_x(f - f_0) + A^2 S_x(f + f_0) \tag{2.12}$$

2.3 CDMA SIGNAL RECEPTION

This section describes the procedure to recover the original bit sequence from the received CDMA signal. The description starts with the situation in which there is only one CDMA signal at the antenna, corresponding to the desired user signal. The analysis is then extended to cases when interfering signals are present at the antenna, either narrowband interferences or interferences from other CDMA users.

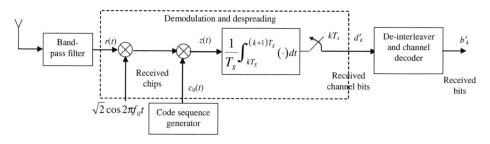

Figure 2.7 Scheme of a CDMA receiver with BPSK demodulation

2.3.1 SINGLE USER CASE

Figure 2.7 shows the simplified scheme of a CDMA receiver assuming BPSK modulation, rectangular pulses and coherent reception. Let us assume that this receiver corresponds to a user, denoted as 0th user, whose signal is spread with code sequence $c_0(t)$. Considering that no other users are present in the system, the received signal at the antenna is given by:

$$r(t) = \sqrt{2P_0}d_0(t)c_0(t)\cos(2\pi f_0 t) + n(t) \tag{2.13}$$

$$r(t) = \sqrt{2P_0} \sum_{k=-\infty}^{\infty} d_{0,k} \sum_{l=0}^{N-1} c_{0,l} p_{Tc}(t - lT_c - kT_s)\cos(2\pi f_0 t) + n(t) \tag{2.14}$$

where P_0 is the received power from the 0th user and $d_0(t)$ is its received channel bit sequence. It is assumed that the signal $r(t)$ is obtained at the output of a filter centred at the carrier frequency f_0 with bandwidth W used to cancel the rest of carriers in the system. As a result, the noise $n(t)$ will be a band-pass AWGN (Additive White Gaussian Noise) with two-sided power spectral density $N_0'/2$, so the total noise power can be approximated by $P_N = N_0'W$. Similarly, an ideal channel is assumed, so that no distortion is considered in the received signal.

As shown in Figure 2.7, after multiplication with the local carrier, the received signal is multiplied by the same code sequence $c_0(t)$ that was used by the transmitter. This sequence is generated locally at the receiver and it must be perfectly time-aligned with the received code sequence, which poses the requirement of having synchronisation mechanisms in the receiver. So, if perfect time alignment between sequences is assumed:

$$z(t) = \sqrt{P_0}d_0(t)c_0(t)c_0(t)(1 + \cos(2\pi 2f_0 t)) + \sqrt{2}n(t)c_0(t)\cos(2\pi f_0 t) \tag{2.15}$$

and, since $c_0(t) \cdot c_0(t) = 1$, then:

$$z(t) = \sqrt{P_0}d_0(t)(1 + \cos(2\pi 2f_0 t)) + n'(t)c_0(t) \tag{2.16}$$

where it has been defined that:

$$n'(t) = \sqrt{2}n(t)\cos(2\pi f_0 t) \tag{2.17}$$

Note from Equating (2.16) that the effect of multiplying by the code sequence at the receiver is to despread the received chips, thus obtaining the received bit sequence $d_0(t)$. This is why this multiplication process is called 'despreading'. In turn, the integration of the bit sequence during a channel bit period T_s will cancel the term at $2f_0$ thus providing the following output:

$$d'_k = \sqrt{P_0}d_k + n_k \qquad (2.18)$$

where n_k is given by the integration of the noise:

$$n_k = \frac{1}{T_s} \int_{kT_s}^{(k+1)T_s} n'(t) \sum_{l=0}^{N-1} c_{0,l} p_{Tc}(t - kT_s - lT_c)dt \qquad (2.19)$$

Note that the noise power spectral density N'_0 is not modified when multiplying by the code sequence. Consequently, and taking into account that the integrator is a low-pass filter with bandwidth $1/T_s$, the power of the noise n_k is given by N'_0/T_s.

The bit energy over noise spectral density for the channel bits d'_k is then given by:

$$\frac{E_s}{N_0} = \frac{P_0 T_s}{N'_0} = \frac{P_0 T_s T_c}{N'_0 T_c} = \frac{P_0}{P_N} \frac{T_s}{T_c} = SF \left(\frac{S}{N}\right) \qquad (2.20)$$

where (S/N) is the signal to interference ratio at the receiver input.

The resulting bits d'_k are de-interleaved and decoded in order to obtain the transmitted data bits b'_k, whose bit energy over noise spectral density can be related to the signal to interference ratio at the receiver input as follows:

$$\frac{E_b}{N_0} = \frac{P_0 T_b}{N'_0} = \frac{P_0 T_b T_c}{N'_0 T_c} = \frac{P_0}{P_N} \frac{1/T_c}{1/T_b} = \frac{W}{R_b} \left(\frac{S}{N}\right) \qquad (2.21)$$

On the other hand, and taking into account that the channel coding rate is $r = T_s/T_b$, the following relationship is obtained:

$$\frac{E_b}{N_0} = \frac{W}{R_b} \left(\frac{S}{N}\right) = \frac{SF}{r} \left(\frac{S}{N}\right) \qquad (2.22)$$

In order to achieve a given performance in terms of bit error rate or block error rate for a certain service, a minimum E_b/N_0 will be required depending on the channel coding scheme. In CDMA systems, Equation (2.21) allows the translation of this requirement in terms of a (S/N) or power requirement at the receiver input depending on the spreading factor or, equivalently, on the ratio W/R_b. Note the important fact that the higher the spreading factor, the lower the power requirement will be to ensure the same bit error rate. This relationship can be regarded also in terms of bit rate, in the sense that the lower the bit rate R_b, the lower the power requirement.

It should be pointed out that in the case of complex modulations (e.g. QPSK), it is possible to spread the symbols with complex code sequences. In this case, the despreading procedure in the receiver corresponding to the sequence $c_0(t)$ would multiply the received chip sequence by its complex conjugate $c_0^*(t)$.

2.3.2 PRESENCE OF NARROWBAND INTERFERENCE

In order to illustrate the capability of spread spectrum communication techniques like CDMA to coexist with other narrowband communications, let us consider that the received signal at the antenna of the receiver in Figure 2.7 includes the contributions of the desired user, the background noise and a

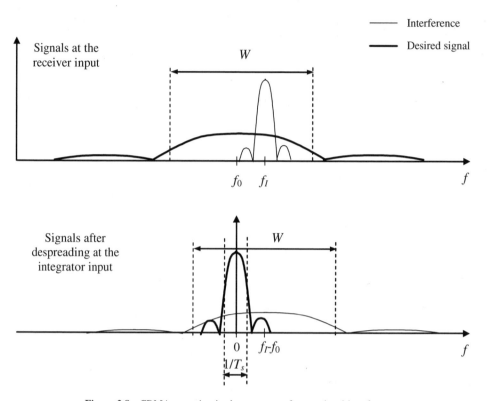

Figure 2.8 CDMA reception in the presence of narrowband interference

narrowband interference signal $i(t)$ centred at a carrier frequency f_I that falls inside the receiver bandwidth W (see the example in Figure 2.8). Then,

$$r(t) = \sqrt{2P_0}d_0(t)c_0(t)\cos(2\pi f_0 t) + n(t) + i(t) \tag{2.23}$$

$$i(t) = \sqrt{2P_I}\cos(2\pi f_I t + \varphi_I) \tag{2.24}$$

where P_I is the power of the interference signal at the receiver input and φ_I is a random phase uniformly distributed between 0 and 2π. Therefore, after the despreading process consisting of the multiplication with the code sequence, the contribution of the interference at the input of the integrator will be:

$$i'(t) = 2\sqrt{P_I}c_0(t)\cos(2\pi f_I t + \varphi_I)\cos(2\pi f_0 t) \tag{2.25}$$

Note that the multiplication with $c_0(t)$ despreads the desired signal and at the same time it spreads the narrowband interference $i(t)$ to occupy the total bandwidth W (see Figure 2.8). Therefore, the spectral power density of the interference $i'(t)$ can be approximated by $I_0 \approx P_I/W$, and the power of the interference at the output of the integrator will be:

$$P_{I,out} = I_0\frac{1}{T_s} = P_I\frac{T_c}{T_s} = \frac{P_I}{SF} \tag{2.26}$$

So, the CDMA receiver reduces the power of the narrowband interference at the receiver input in a factor equal to the spreading factor SF, so that the higher the spreading factor the higher will be the protection of the CDMA signal against interferences.

On the other hand, the bit energy over noise and interference spectral density at the receiver output will be:

$$\frac{E_b}{N_0} = \frac{P_0 T_b}{N_0' + I_0} = \frac{P_0 T_b}{P_N/W + P_I/W} = \frac{W}{R_b} \frac{P_0}{P_N + P_I} = \frac{W}{R_b} \left(\frac{S}{I}\right) \tag{2.27}$$

where (S/I) represents the signal to noise and interference ratio at the receiver input. As in the case with only background noise given by Equation (2.21), the spreading factor or equivalently the ratio W/R_b reflects the gain introduced in the despreading process.

2.3.3 MULTIPLE USER CASE

This section extends the previous analysis of the CDMA signal reception to the most usual situation in which the received signal at the antenna input contains the contribution of the desired user and of the rest of users that are simultaneously transmitting in the same carrier frequency. Therefore, let the signal at the input of the 0th user receiver in Figure 2.7 be:

$$r(t) = \sqrt{2P_0} d_0(t) c_0(t) \cos(2\pi f_0 t) + \sum_{i=1}^{K-1} \sqrt{2P_i} d_i(t - \tau_i) c_i(t - \tau_i) \cos(2\pi f_0 t + \varphi_i) + n(t) \tag{2.28}$$

where a total of $K - 1$ interfering users are considered, all of them connected to the same cell. We are considering a single cell here, and so no interference from other cells exists. The power received from the ith user at the input of the 0th user receiver is P_i. In turn, $d_i(t)$ and $c_i(t)$ represent the bit sequence and the code sequence of the ith user, respectively. Furthermore, τ_i represents the fact that the bit sequences of the different users do not necessarily have to be synchronised neither at bit nor at chip level, while φ_i represents the phase difference in the received signals.

Note that the formulation of Equation (2.28) represents a general situation that would be valid both for the uplink and for the downlink receiver. In any case, note that in the downlink case, since all the signals come from transmitters located at the same position, it is easy to have perfect synchronisation between all of them, so that $\tau_i = 0$ and $\varphi_i = 0$ because the propagation delay is the same for all the signals. Furthermore, if the transmitted power is the same for all the users in the downlink, it will also be the same at the receiver input. In the uplink direction, however, if the transmitted power was the same for all the users, the powers at the receiver of the base station would be different due to the different propagation losses. Similarly, and due to the different propagation delays, τ_i and φ_i will be different from zero in the uplink direction unless specific synchronisation strategies were used.

The despreading process at the receiver of the 0th user shown in Figure 2.7 makes use of the code sequence $c_0(t)$ generated locally and perfectly time-aligned with the received code sequence of the 0th user. Consequently, the multiplication with $c_0(t)$ will result only in the despreading of the 0th user signal, but not of the signals of the rest of users, since $c_0(t) \cdot c_i(t - \tau_i) \neq 1$ which in general provides a random chip sequence with bandwidth W. Therefore, the output of the integration over a channel bit period T_s will contain three contributions, the one of the desired 0th user, the contribution of the rest of interfering users and the background noise contribution:

$$d_k' = \sqrt{P_0} d_k + \sum_{i=1}^{K-1} \frac{\sqrt{P_i} \cos \varphi_i}{T_s} \int_{kT_s}^{(k+1)T_s} d_i(t - \tau_i) c_i(t - \tau_i) c_0(t) dt + n_k \tag{2.29}$$

It can be observed from Equation (2.29) that the interference from the other users at the receiver output depends on the following factors:

(a) The cross-correlation between the code sequences of the different users, reflected in the integral in Equation (2.29). This depends on the specific family of code sequences being used.
(b) The received power from each interfering user, P_i.
(c) The phase difference between the carriers of the received signals, φ_i.
(d) The number of interfering users, $K - 1$.

The above factors will now be explored in order to obtain the final E_b/N_0 at the receiver output.

2.3.3.1 Code Sequences

The selection of a family of code sequences to be used in CDMA systems is achieved based on the following premises:

1. It should be easy for the receiver to synchronise the local code sequence generator in order to ensure a perfect alignment between the local and the received code sequences. The synchronisation procedure is based on computing the correlation between the local and the received sequences and finding the maximum of this correlation. Consequently, for an easy synchronisation, it is required that when both signals are perfectly aligned (i.e. if the delay between them is $\tau = 0$) the correlation presents a maximum, while when the alignment is not perfect (i.e. the delay τ is different from zero) the correlation presents very low values. So, from the synchronism point of view, the ideal autocorrelation function of any code sequence $c_i(t)$ belonging to the family and repeated periodically every T_s would be:

$$R_{c_i}(\tau) = \frac{1}{T_s} \int_{T_s} c_i(t + \tau)c_i^*(t)dt = \delta(\tau) \tag{2.30}$$

2. In order to reduce the interference between simultaneous transmissions using different code sequences, the cross-correlation function between any two code sequences of the same family must take low values for any delay τ between them. The ideal condition for any pair of code sequences $c_i(t)$ and $c_j(t)$ with period T_s would be:

$$R_{c_i c_j}(\tau) = \frac{1}{T_s} \int_{T_s} c_i(t + \tau)c_j^*(t)dt = 0 \tag{2.31}$$

3. Any family of code sequences must consist of a large number of sequences, so that a high number of simultaneous transmissions is possible without having code limitations.

From a practical point of view, however, it is difficult to fulfil simultaneously the above three conditions, so the selection of the family of code sequences is based on a trade-off between these requirements.

Orthogonal Code Sequences A family of code sequences is orthogonal whenever the following mathematical condition is satisfied for any two sequences $c_i(t)$ and $c_j(t)$ of the family:

$$\frac{1}{T_s} \int_{T_s} c_i(t)c_j^*(t)dt = 0 \tag{2.32}$$

Returning to Equation (2.29), if the considered real code sequences $c_0(t)$ and $c_i(t)$ are orthogonal, and perfect time alignment at the bit level between the signals from the different users is assumed (i.e. $\tau_i = 0$), the interference from the rest of users is totally cancelled, since:

$$\sum_{i=1}^{K-1} \frac{\sqrt{P_i}\cos\varphi_i}{T_s} \int_{kT_s}^{(k+1)T_s} d_i(t)c_i(t)c_0(t)dt = \sum_{i=1}^{K-1} \frac{\sqrt{P_i}\cos\varphi_i}{T_s} d_i \int_{kT_s}^{(k+1)T_s} c_i(t)c_0(t)dt = 0 \qquad (2.33)$$

In this case, the maximum number of simultaneous transmissions would be limited by the number of available orthogonal code sequences. An example of a family of orthogonal code sequences is the OVSF (Orthogonal Variable Spreading Factor) code family [6][7], which is used in UMTS as will be explained in Chapter 3. These codes have the property that, for a given *SF*, there are exactly *SF* orthogonal codes, which may be a quite reduced number of codes in the case of low spreading factors.

The main problem of using orthogonal sequences is that the condition of orthogonality in Equation (2.32) does not assure a low cross-correlation for time shifts between sequences $\tau \neq 0$, so that, if perfect synchronisation does not exist between the signals, the resulting interference from the other users could be very high. An example of this situation is shown in Figure 2.9, which presents the time

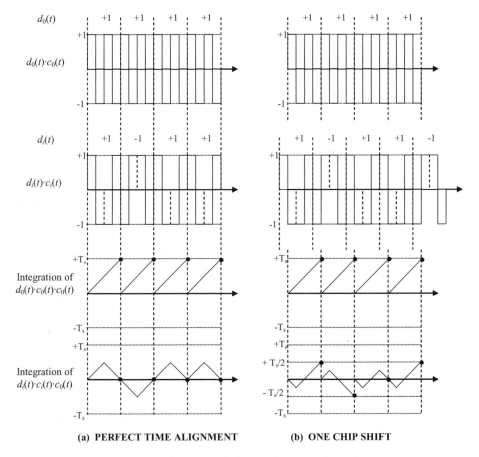

(a) **PERFECT TIME ALIGNMENT**   (b) **ONE CHIP SHIFT**

Figure 2.9 Impact of the time misalignment between orthogonal sequences

evolution of the signals of the 0th user and the ith user in the 0th user receiver when they are time-aligned and when a delay of one chip exists between them. The 0th user transmits with the OVSF code sequence $\{1, -1, 1, -1\}$ and the ith user with OVSF sequence $\{1, -1, -1, 1\}$. When the bits of the two users are aligned (Figure 2.9a), the output of the integration at the decision instants (represented with black points) is $+1$ for the desired signal of the 0th user and it takes the value of zero for the interference from the ith user in all the cases, revealing the orthogonality of the two code sequences. However, when one chip delay exists between the two signals (Figure 2.9b), the integration of the interference at the decision instants may take high values such as $+0.5$ or -0.5 depending on the specific bit sequence.

The above drawback means that the use of OVSF sequences is limited to cases when perfect time alignment between interfering signals can be assured. This occurs when all the interferer transmitters are located at the same position. Therefore, OVSF codes can be used in the uplink direction to separate different flows of the same user while in the downlink direction they can separate different users belonging to the same base station.

Nevertheless, and even when the code sequences are orthogonal and perfectly synchronised, in practice the interference from the rest of users is not completely cancelled due to the multipath propagation, which causes several shifted replicas of the same signal to be received at the antenna input (see Section 2.3.4). Consequently, the interference from the rest of the users will consist of some perfectly aligned replicas and others with some shift delay, which will cause some residual interference at the receiver output. This effect is modelled by considering that the power spectral density of the interference at the receiver output I_0 is a fraction ρ of the interference spectral density at the input I_I:

$$I_0 = \rho I_I = \rho \frac{P_{\text{intra},0}}{W} \tag{2.34}$$

where $P_{\text{intra},0}$ is the total power from the interferer users of the same cell (i.e. intracell users) at the receiver input of the 0th user, given by:

$$P_{\text{intra},0} = \sum_{i=1}^{K-1} P_i \tag{2.35}$$

The fraction ρ is called the *orthogonality factor* and takes values between 0 and 1, where 0 means perfect orthogonality and 1 means non-orthogonal signals. The specific value depends on the considered environment in terms of multipath. Typical values for the case of macrocells are around 0.4–0.6 while for the case of microcells lower values around 0.06 are usual [8].

With these considerations, and when orthogonal codes are used, the bit energy over noise and interference spectral density at the receiver output is given by:

$$\frac{E_b}{N_0} = \frac{P_0 T_b}{N_0' + I_0} = \frac{P_0 T_b}{P_N/W + \rho P_{\text{intra},0}/W} = \frac{W}{R_b} \frac{P_0}{P_N + \rho P_{\text{intra},0}} \tag{2.36}$$

Non Orthogonal Code Sequences   Taking into account the reduced number of available orthogonal code sequences and the requirement to have perfectly synchronised signals at the receiver, families of non-orthogonal code sequences have been extensively studied in the past during the evolution of DS-SS communications [7],[9–13].

An important family of such codes is constituted by the PN (Pseudo Noise)-sequences, also called m-sequences, which are generated by means of Linear Feedback Shift Registers (LFSR) [9][10]. These sequences have good auto-correlation properties, with very low values for time shifts different from zero, which makes them suitable from the point of view of synchronisation. In turn, from the point of view of cross-correlation, it is also possible to find specific sets of sequences with low cross-correlation values. Unfortunately, these sets of appropriate sequences are very small, which would limit the system capacity due to code availability. As a result, they are not usually used in mobile communication systems. However, they constitute the basis for the family of Gold codes, which are generated as the sum of two PN-sequences and are widely used in current systems like UMTS [11][12].

Gold sequences have the advantage of providing large sets of sequences with low cross-correlation properties but at the expense of having to tolerate autocorrelation functions somewhat worse than those of PN-sequences. Other examples of sequences that are suitable for CDMA communications, and that are also generated from PN-sequences, are the Gold-like and Dual-BCH sequences [13] as well as the large and the small sets of Kasami sequences [7][13].

When making use of non-orthogonal sequences, the interference at the receiver output corresponding to the integral in Equation (2.29) depends on the specific code family and its cross-correlation values as well as on the existence or not of bit level synchronisation between interfering signals and the phase differences between the different carrier frequencies. In this context, several studies exist in the literature coping with this problem [14–19]. In general, the interfering signals remain spread after the despreading process and it is widely accepted that in this case the total interference can be seen as an additional noise contribution, so the power spectral density at the receiver output is given by:

$$I_0 = \frac{P_{\text{intra},0}}{W} \tag{2.37}$$

where $P_{\text{intra},0}$ is the total power from the interferer users at the receiver input, defined in Equation (2.35). As a matter of fact, the same model used for orthogonal sequences in Equation (2.34) would be valid with non-orthogonal sequences simply by considering an orthogonality factor $\rho = 1$. Then, the bit energy over noise and interference spectral density at the receiver output will be given by:

$$\frac{E_b}{N_0} = \frac{P_0 T_b}{N_0' + I_0} = \frac{P_0 T_b}{P_N/W + P_{\text{intra},0}/W} = \frac{W}{R_b} \frac{P_0}{P_N + P_{\text{intra},0}} = \frac{W}{R_b} \left(\frac{S}{I} \right) \tag{2.38}$$

Note that the interference power at the output of the receiver is reduced by a factor equal to the ratio $W/R_b = SF/r$, which allows operating with reduced signal to interference ratios at the input (S/I) while maintaining a good performance.

2.3.3.2 Power Control

The previous section has shown that, since the usual code sequences are not orthogonal, each user represents an increase in the interference seen by the rest of the users that transmit in the same carrier, as reflected in Equation (2.35). Consequently, it is very important to keep this interference under control by adjusting the transmitted power to the minimum required value that ensures the E_b/N_0 requirement at the receiver output of each user. This is the function of the power control procedure, which constitutes one of the key aspects in any CDMA communication. Therefore, we now explain the basis for the power control mechanism, distinguishing between the uplink and the downlink.

Uplink Direction   For simplicity, let us assume a CDMA system with only two users, namely user 1 and user 2, located at two different distances from the base station, as depicted in Figure 2.10. Let's assume initially that both users transmit with the same bit rate R_b, and denote as L_1 and L_2 the total path loss from each user to the base station, including the shadowing and the antenna gains. Furthermore, let's assume that both users require the same minimum bit energy over noise and interference spectral density

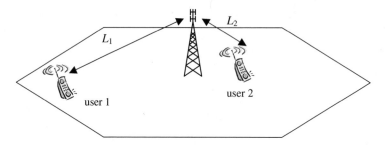

Figure 2.10   Illustration of the near-far effect

at the output, denoted as $(E_b/N_0)_{min}$. Then, by means of Equation (2.38), it is possible to compute the E_b/N_0 at the output of the receiver of each user as a function of the transmitted powers P_{T1} and P_{T2}:

$$\left(\frac{E_b}{N_0}\right)_1 = \frac{W}{R_b}\left(\frac{S}{I}\right)_1 = \frac{W}{R_b}\frac{\frac{P_{T,1}}{L_1}}{P_N + \frac{P_{T,2}}{L_2}} \tag{2.39}$$

$$\left(\frac{E_b}{N_0}\right)_2 = \frac{W}{R_b}\left(\frac{S}{I}\right)_2 = \frac{W}{R_b}\frac{\frac{P_{T,2}}{L_2}}{P_N + \frac{P_{T,1}}{L_1}} \tag{2.40}$$

It is clear from Equations (2.39) and (2.40) that, for the situation shown in Figure 2.10, where $L_1 \gg L_2$, if $P_{T1} = P_{T2} = P_T$ then user 2, located closer to the base station, would generate excessive interference for user 1. As a result, the performance observed by user 2 would be superior to the performance observed by user 1, as seen in Equation (2.41). This phenomenon is known as the 'near-far effect', revealing the fact that users near to the base station would degrade the reception of users far from the base station.

$$\left(\frac{E_b}{N_0}\right)_1 = \frac{W}{R_b}\frac{\frac{1}{L_1}}{\frac{P_N}{P_T}+\frac{1}{L_2}} \ll \frac{W}{R_b}\frac{\frac{1}{L_2}}{\frac{P_N}{P_T}+\frac{1}{L_1}} = \left(\frac{E_b}{N_0}\right)_2 \tag{2.41}$$

Consequently, the transmitted power must be optimised in order that the mutual interference between users is kept to a minimum, which is assured by setting the transmitted power in order to obtain exactly the required $(E_b/N_0)_{min}$ at the output of both receivers. Then, by imposing the conditions $(E_b/N_0)_1 = (E_b/N_0)_{min}$ and $(E_b/N_0)_2 = (E_b/N_0)_{min}$ in Equations (2.39) and (2.40), it is possible to find the optimum transmitted powers, given by:

$$P_{T,1} = \frac{L_1 P_N}{\frac{W}{\left(\frac{E_b}{N_0}\right)_{min} R_b} - 1} = \frac{L_1 P_N}{\frac{1}{\left(\frac{S}{I}\right)_{min}} - 1} \tag{2.42}$$

$$P_{T,2} = \frac{L_2 P_N}{\frac{W}{\left(\frac{E_b}{N_0}\right)_{min} R_b} - 1} = \frac{L_2 P_N}{\frac{1}{\left(\frac{S}{I}\right)_{min}} - 1} \tag{2.43}$$

where $(S/I)_{min}$ is the signal to noise and interference ratio required at the input of the receiver to ensure the $(E_b/N_0)_{min}$. The last two expressions indicate that the path loss is the only differential parameter between the powers transmitted by both users, because they have the same requirements in terms of bit rate and E_b/N_0. Consequently, in this case, the power received at the base station would be the same for the two users:

$$P_1 = \frac{P_{T1}}{L_1} = P_2 = \frac{P_{T2}}{L_2} = \frac{P_N}{\dfrac{W}{\left(\dfrac{E_b}{N_0}\right)_{min} R_b} - 1} \tag{2.44}$$

However, in a multiservice environment, it will be usual to have users with different bit rates and/or different bit error rate requirements, corresponding to different $(E_b/N_0)_{min}$ values. In this case, denoting $R_{b,1}$, $R_{b,2}$ as the bit rates of the two users and $(E_b/N_0)_{min,1}$, $(E_b/N_0)_{min,2}$ as their requirements, the optimum power allocation would be obtained from Equations (2.39) and (2.40) yielding:

$$P_{T,1} = \frac{L_1 P_N}{\dfrac{W^2}{\left(\dfrac{E_b}{N_0}\right)_{min,1}\left(\dfrac{E_b}{N_0}\right)_{min,2} R_{b,1} R_{b,2}} - 1} \left(\frac{W}{\left(\dfrac{E_b}{N_0}\right)_{min,2} R_{b,2}} + 1\right) \tag{2.45}$$

$$P_{T,2} = \frac{L_2 P_N}{\dfrac{W^2}{\left(\dfrac{E_b}{N_0}\right)_{min,1}\left(\dfrac{E_b}{N_0}\right)_{min,2} R_{b,1} R_{b,2}} - 1} \left(\frac{W}{\left(\dfrac{E_b}{N_0}\right)_{min,1} R_{b,1}} + 1\right) \tag{2.46}$$

Note that in this case the received power from the two users at the base station is not the same, and the ratio between powers reflects the differences in service requirements, that is:

$$\frac{P_1}{P_2} = \frac{P_{T,1}/L_1}{P_{T,2}/L_2} = \frac{\dfrac{W}{\left(\dfrac{E_b}{N_0}\right)_{min,2} R_{b,2}} + 1}{\dfrac{W}{\left(\dfrac{E_b}{N_0}\right)_{min,1} R_{b,1}} + 1} \approx \frac{R_{b,1}}{R_{b,2}} \frac{\left(\dfrac{E_b}{N_0}\right)_{min,1}}{\left(\dfrac{E_b}{N_0}\right)_{min,2}} \tag{2.47}$$

As well as the fact that the higher the $(E_b/N_0)_{min}$ requirements, the higher the required power, Equation (2.47) also reveals the important relationship in CDMA systems between the received power requirement and the service bit rate, in the sense that the higher the bit rate, the higher the required power. This is because high bit rates are achieved in CDMA without modifying the total bandwidth W but at the expense of a reduction in the spreading factor. Consequently, since the spreading factor is a measurement of the protection of the desired signal against the interferences in the despreading process, low spreading factor values require higher power levels in order to compensate for the lower protection capability.

Downlink Direction To analyse the impact of power control in the downlink direction, let us continue with the same CDMA system with two users as in the previous section and shown in Figure 2.10. The bit rates of the two users are $R_{b,1}$, $R_{b,2}$ and their requirements are $(E_b/N_0)_{min,1}$, $(E_b/N_0)_{min,2}$. Furthermore, it can be assumed that in the downlink orthogonal codes are used, as explained on page 32, so that the model in Equation (2.36) based on the orthogonality factor ρ is considered. Note that if codes were not orthogonal, the same results would apply simply by setting $\rho = 1$. On the other hand, and for simplicity, the same background noise power P_N is assumed in both mobile receivers. Then, the E_b/N_0 at the output of the receiver of each user as a function of the transmitted powers P_{T1} and P_{T2} is given by:

$$\left(\frac{E_b}{N_0}\right)_1 = \frac{W}{R_{b,1}} \frac{\dfrac{P_{T,1}}{L_1}}{P_N + \rho \dfrac{P_{T,2}}{L_1}} \tag{2.48}$$

$$\left(\frac{E_b}{N_0}\right)_2 = \frac{W}{R_{b,2}} \frac{\dfrac{P_{T,2}}{L_2}}{P_N + \rho \dfrac{P_{T,1}}{L_2}} \tag{2.49}$$

When comparing these expressions with those of the uplink in Equations (2.39) and (2.40), an important difference is observed: the interference from the other users is affected by the same path loss as the desired signal, since both transmitters are located at the base station and therefore both signals have travelled the same radio electrical distance. Consequently the 'near-far' effect existing in the uplink is no longer present in the downlink. However, it must be considered that in the downlink the users share the available power at the base station, so if one user consumes more power than is required, less power remains for the rest of the users, which may limit seriously the cell capacity. As a result, the power control in the downlink is not required to compensate for the near-far effect but to keep the total power consumption to an appropriate level to ensure the requirements of each user.

Then, by imposing the conditions $(E_b/N_0)_1 = (E_b/N_0)_{\min,1}$ and $(E_b/N_0)_2 = (E_b/N_0)_{\min,2}$ in Equations (2.48) and (2.49), it is possible to find the optimum power allocation, given by:

$$P_{T,1} = \frac{\dfrac{R_{b,1}P_N}{W}\left(\dfrac{E_b}{N_0}\right)_{\min,1}\left(L_1 + \rho L_2 \dfrac{R_{b,2}}{W}\left(\dfrac{E_b}{N_0}\right)_{\min,2}\right)}{1 - \rho^2 \dfrac{R_{b,1}R_{b,2}}{W^2}\left(\dfrac{E_b}{N_0}\right)_{\min,1}\left(\dfrac{E_b}{N_0}\right)_{\min,2}} \tag{2.50}$$

$$P_{T,2} = \frac{\dfrac{R_{b,2}P_N}{W}\left(\dfrac{E_b}{N_0}\right)_{\min,2}\left(L_2 + \rho L_1 \dfrac{R_{b,1}}{W}\left(\dfrac{E_b}{N_0}\right)_{\min,1}\right)}{1 - \rho^2 \dfrac{R_{b,1}R_{b,2}}{W^2}\left(\dfrac{E_b}{N_0}\right)_{\min,1}\left(\dfrac{E_b}{N_0}\right)_{\min,2}} \tag{2.51}$$

As in the uplink direction, the last two expressions show that the power requirement depends on the bit rate, so that the highest bit rate users are those that demand the most power, due to the reduction in the spreading factor and the consequent lowest protection against interference.

Power Control Strategies   At this point, it is clear that the power control must operate both in the uplink and the downlink to ensure the required $(E_b/N_0)_{\min}$ at the receiver output for each user transmitting at a given bit rate R_b. Furthermore, since the mobile environment will be dynamic, due to variations in the mobile positions and the service characteristics, in practice the power control cannot be based on solving numerically a set of equations, as is done in the previous sections from a theoretical perspective. Rather, it must include dynamic mechanisms that converge to the optimal solution in each situation depending on some measurements.

Ideally, these measurements would require monitoring the path loss as well the measured E_b/N_0, which is hard to measure in practice. This is why the usual power control strategies are based on measurements of the signal to noise and interference (S/I) ratio at the receiver input, which is an easier measurement. Then the power control algorithms try to ensure a certain $(S/I)_{\min}$ target, which is related with the $(E_b/N_0)_{\min}$ and bit rate requirements as follows:

$$\left(\frac{S}{I}\right)_{\min} = \frac{R_b}{W}\left(\frac{E_b}{N_0}\right)_{\min} \tag{2.52}$$

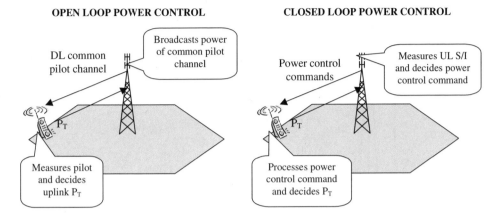

Figure 2.11   Open loop and closed loop power control strategies

Power control algorithms are divided into the following two types, which are illustrated in Figure 2.11:

(a) Closed loop power control. This strategy consists of adjusting the transmitted power based on the measured (S/I). Therefore, if the receiver measures that (S/I) is below the $(S/I)_{\min}$ requirement, it sends a power control command to the transmitter indicating that it has to increase the transmitted power by a certain amount. If (S/I) is higher than $(S/I)_{\min}$, however, the power control command indicates that the transmitter must decrease the transmitted power. If the periodicity of these power control commands is faster than the channel variations, this strategy allows the compensation of the path loss including fast fading, so it is able to achieve a good power adjustment and it is the preferred power control strategy in CDMA systems for both uplink and downlink. However, it requires a dedicated signalling channel in the opposite direction to transmit power control commands.

(b) Open loop power control. This strategy consists of adjusting the transmitted power of one link based on measurements from the opposite link. Then the transmitted power of the uplink is set based on the path loss of the downlink direction, which can be obtained by the receiver simply by measuring the received power $P_{R,pilot}$ of a downlink pilot channel (e.g. the CPICH channel in UTRAN FDD, see Section 3.3.4.2) whose transmitted power $P_{T,pilot}$ is known. Then, given the measured path loss L, the measured noise plus interference I_m, which is broadcast by the network, and the $(S/I)_{\min}$ requirement, the transmitted power will be:

$$P_T = L \cdot I_m \cdot \left(\frac{S}{I}\right)_{\min} = \frac{P_{T,pilot}}{P_{R,pilot}} \cdot I_m \cdot \left(\frac{S}{I}\right)_{\min} \qquad (2.53)$$

In contrast to the closed loop power control, the open loop does not need to send power control commands, so no dedicated control channel in the opposite direction is necessary. Nevertheless, in FDD systems where the uplink and downlink operate at different frequencies, both links are not reciprocal in terms of fast fading variations, which means that the instantaneous path loss of the uplink may be different than that of the downlink. Consequently, the open loop power control is only valid to follow long-term variations, and it cannot be used to compensate fast fading. Therefore, its use is limited to those cases in which a dedicated channel in the opposite direction is not available, as would be the case with the initial access of the mobile to the network.

2.3.4 EFFECT OF THE MOBILE RADIO CHANNEL

The channel propagation in mobile radio communications is characterised by different phenomena such as multiple reflections and diffraction due to obstacles (e.g. buildings or mountains) in the surrounding

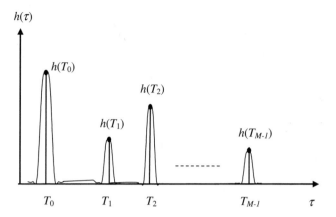

Figure 2.12 Example of a mobile radio channel impulse response

environment where the mobile transmitter and the receiver are located. As a result of this, the signal at the receiver consists of different replicas of the transmitted signal, each one corresponding to a different propagation path.

This multipath propagation can be modelled by means of a channel impulse response similar to the one shown in Figure 2.12, in which there are M propagation paths, each one characterised by a different propagation delay T_i and a different amplitude $h(T_i)$. Each of these paths consists of several contributions of multiple instances of the same signal that arrive with very small time differences, so that they are approximately simultaneous for the receiver but may differ substantially in terms of phase. Consequently, depending on the phase differences between these quasi-simultaneous signals, their combination may cause signal cancellations leading to high variations of the received signal when the receiver moves even across short distances. This phenomenon is called fast fading and the amplitudes $h(T_i)$ of the different propagation paths are characterised by means of statistical distributions (e.g. Rayleigh or Rice distributions) that depend on the specific environment in terms of mobile speed or the existence of line of sight between transmitter and receiver [20][21]. Note also that, since the phase differences between the combined signals depend on the specific frequency, the mobile radio channel is also frequency selective, which means that the channel behaviour is not the same for all the frequencies, so while some frequencies may suffer a deep fading, this may not be the case for other frequencies.

Multipath propagation can be especially critical in narrowband systems, in which the channel behaviour is approximately the same for all the frequency components of the transmitted signal, because in the case of a deep fading all the components will suffer. However, in the case of CDMA systems, deep fading is likely to affect only some frequency components of the transmitted signal because of the high bandwidth resulting from the spreading process. As a result, CDMA offers an inherent frequency diversity that makes the system more robust in front of channel variations. Additionally, the properties of the transmitted CDMA signal can also be used to take some benefit from the different multipath components by means of a constructive combination of all of them at the receiver. This can be done because the auto-correlation properties of the transmitted code sequences allow the distinguishing of the different propagation paths if a proper receiver structure is used. Such a structure is called a rake receiver and was proposed initially in 1958 by Price and Green [22]. A schematic representation is shown in Figure 2.13, by extending the initial basic receiver of Figure 2.7, which now is denoted as a rake finger in the overall structure.

Assuming the channel impulse response given in Figure 2.12, the received base band signal will consist of M shifted replicas of the signal of the desired user, as follows:

$$v(t) = \sum_{j=0}^{M-1} \sqrt{P_0} h(T_j) d_0(t - T_j) c_0(t - T_j) + n_I(t) \qquad (2.54)$$

where the AWGN noise and the interference from the rest of users are included in the term $n_I(t)$.

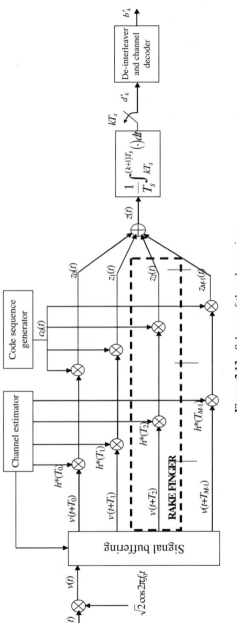

Figure 2.13 Scheme of the rake receiver

As shown in Figure 2.13, the rake receiver consists of M rake fingers, and each one is adapted to one channel propagation path, characterised by the delay T_i and the amplitude $h(T_i)$. So the received signal $v(t)$ is initially buffered and M shifted replicas are generated. Then the input to the mth finger is the signal:

$$v(t + T_m) = \sum_{j=0}^{M-1} \sqrt{P_0} h(T_j) d_0(t + T_m - T_j) c_0(t + T_m - T_j) + n_I(t + T_m) \quad (2.55)$$

Each rake finger multiplies the signal by the complex conjugate of the path amplitude and by the code sequence of the user, so the output of the mth finger will be:

$$z_m(t) = \sqrt{P_0} d_0(t) |h(T_m)|^2 + \sum_{\substack{j=0 \\ j \neq m}}^{M-1} \sqrt{P_0} h(T_j) h^*(T_m) d_0(t + T_m - T_j) c_0(t + T_m - T_j) c_0(t) + n_{I,m}(t) \quad (2.56)$$

where:

$$n_{I,m}(t) = n_I(t + T_m) h^*(T_m) c_0(t) \quad (2.57)$$

As can be seen in Equation (2.56), the mth finger despreads the desired signal $d_0(t)$ that has been received through the mth propagation path, while the rest of paths remain spread. Then, provided that the autocorrelation of the code sequence $c_0(t)$ is close to zero for $(T_m - T_j) \neq 0$, the integration of the signal of the mth finger will cancel the contribution of the rest of paths. The detected channel bits at the output of the receiver will then be:

$$d_k' \approx \sqrt{P_0} d_k \sum_{j=0}^{M-1} |h(T_j)|^2 + \sum_{j=0}^{M-1} \frac{1}{T_s} h^*(T_j) \int_{kT_s}^{(k+1)T_s} n_I(t + T_j) c_0(t) dt \quad (2.58)$$

The total noise power spectral density is a function of the power spectral densities of the background noise N_0' and the interference from other users I_0, given by:

$$N_0 = (N_0' + I_0) \sum_{j=0}^{M-1} |h(T_j)|^2 \quad (2.59)$$

Finally, the bit energy over interference and noise spectral density at the receiver output will be:

$$\frac{E_b}{N_0} \approx \frac{P_0 \left(\sum_{j=0}^{M-1} |h(T_j)|^2 \right)^2 T_b}{(N_0' + I_0) \sum_{j=0}^{M-1} |h(T_j)|^2} = \sum_{j=0}^{M-1} |h(T_j)|^2 \frac{P_0 T_b}{N_0' + I_0} = \sum_{j=0}^{M-1} \left(\frac{E_b}{N_0} \right)_j = \frac{W}{R_b} \sum_{j=0}^{M-1} \left(\frac{S}{I} \right)_j \quad (2.60)$$

Therefore, the rake receiver is able to constructively combine the different propagation paths by achieving a signal to noise and interference ratio that is the sum of the signal to noise and interference ratios from all the paths, which corresponds to the same performance that is obtained by Maximum Ratio Combiner (MRC) diversity schemes in narrowband systems [21].

From a practical point of view, the rake receiver requires a strict control of the available M fingers in order to adapt them to the specific channel conditions. This adaptation is achieved by means of the channel impulse response estimation and the identification of the time shifts and amplitudes of the most relevant paths based on known bit pilot sequences transmitted jointly with the data bits. Since the number of paths may, in general, be different from M, the finger management operation needs to react in front of changes in the channel impulse response to ensure that the available M fingers are always allocated to the best propagation paths (i.e. those having the highest amplitudes) thus achieving the highest gains in the combination. More details about the operations of management of rake fingers can be found in Reference 23.

2.3.4.1 Outer Loop Power Control

At this point, the E_b/N_0 at the output of a CDMA receiver has been obtained as a function of all the parameters that affect the CDMA transmission, namely the noise, the interference and the multipath channel. In any case, it is clear that this E_b/N_0 is indeed a random variable whose distribution depends on the statistical distribution of the channel impulse response. The distribution of the E_b/N_0 impacts over the final performance that is observed by the user in terms of bit error rate (BER) or block error rate (BLER), taking into account the dependency with the interleaving and channel coding scheme being used. This means that, in order to ensure a certain BER or BLER performance for a given service, the required $(E_b/N_0)_{min}$ will depend on the specific channel existing in the link between the mobile terminal and the base station, which varies with time as the user moves. Therefore, in order to cope with the long-term fluctuations of the channel statistics, the $(E_b/N_0)_{min}$ must indeed be varied dynamically depending on the measured BER or BLER performance. This is the role of the so-called outer loop power control, whose function is to set the $(E_b/N_0)_{min}$ that afterwards will be ensured by the open or closed loop power control mechanisms explained on page 37, corresponding to the so-called inner loop power control.

2.4 CDMA IN CELLULAR SYSTEMS

One of the main advantages of cellular CDMA mobile communication systems with respect to cellular FDMA/TDMA is that all the cells can operate with the same carrier frequency, provided that transmissions are carried out with different code sequences. Therefore, a complete frequency reuse pattern is possible in CDMA systems, which is more efficient than the frequency reuse based on clusters of cells used in FDMA/TDMA systems [24]. This means that the cellular planning concepts in CDMA are quite different from those of FDMA/TDMA, and are based on controlling the interference for a certain desired coverage area and on code planning strategies devoted to deciding the proper code sequences to be allocated to each cell. It is worth mentioning that, although code planning is similar to the frequency planning of FDMA/TDMA, it should be taken into account that the availability of non-orthogonal code sequences is much higher than the availability of carriers, so the CDMA code planning is not as critical as the FDMA/TDMA frequency planning.

Users in a given CDMA cell will perceive the transmissions in an adjacent cell operating at the same frequency simply as a wideband interference, denoted as intercell interference P_{inter}, to be added to the intracell interference from the rest of the users of the same cell. The E_b/N_0 at the CDMA receiver output can be expressed by extending Equation (2.38) as follows:

$$\frac{E_b}{N_0} = \frac{W}{R_b} \frac{P_0}{P_N + P_{intra,0} + P_{inter}} \tag{2.61}$$

2.4.1 INTERCELL INTERFERENCE

Although intracell and intercell interferences in CDMA are both wideband interference and therefore can be modelled as an additional noise, there is an important difference between them. The reason is that the power control algorithm in a given cell is able to regulate the received power of the users in this cell (i.e. the intracell interference), but not the received power at the adjacent cells (i.e. the intercell interference).

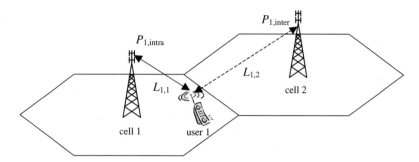

Figure 2.14   Intercell interference in cellular CDMA systems

As an example, consider the situation shown in Figure 2.14, in which there is one mobile and two base stations, the path loss between user 1 and cell 1 is $L_{1,1}$ and between user 1 and cell 2 it is $L_{1,2}$. Let N_1 be the total noise and interference measured in the uplink receiver of the user in cell 1, and $(S/I)_{\min}$ the requirement for the user. The uplink power control will adjust the transmitted power of the user to:

$$P_{T,1} = L_{1,1} N_1 \left(\frac{S}{I}\right)_1 \tag{2.62}$$

so the intracell interference generated by this user to the rest of the users in cell 1 will be:

$$P_{1,\text{intra}} = N_1 \left(\frac{S}{I}\right)_1 \tag{2.63}$$

and the intercell interference generated by this user to the rest of the users in cell 2 will be:

$$P_{1,\text{inter}} = \frac{L_{1,1}}{L_{1,2}} N_1 \left(\frac{S}{I}\right)_1 \tag{2.64}$$

From Equations (2.63) and (2.64), it can be seen that, provided that the noise and interference N_1 do not change, the power control in cell 1 assures that the intracell interference caused by user 1 remains constant no matter what the position of this user in the cell. However, the intercell interference depends on the fraction $L_{1,1}/L_{1,2}$, or equivalently on the relative position between cell 2 and user 1.

In order to cope with the above dependency between the intercell interference and the specific user positions, it is usual to model, at least for CDMA planning purposes, the intercell interference as a certain fraction f of the total intracell power in a given cell [5][25], that is:

$$f = \frac{P_{\text{inter}}}{P_{\text{intra}}} \tag{2.65}$$

where P_{intra} includes the total received power from the users connected to the cell. This fraction depends on the user distribution and on the propagation conditions, and typical values are around 0.6 [25], although in practice this fraction exhibits high variations along time [26].

It is worth mentioning that there exists an important difference between uplink and downlink intercell interference. In particular, the uplink intercell interference coming from neighbouring cells is the same for all the users' receivers in a given reference cell, since all these receivers are located in the same place.

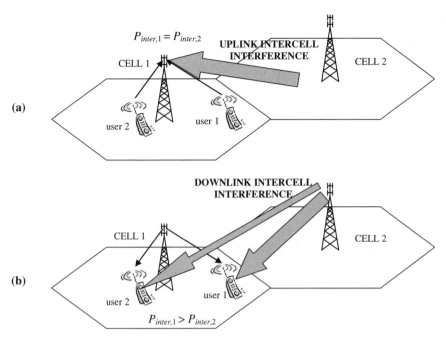

Figure 2.15   Uplink and downlink intercell interference in cellular CDMA systems

In the downlink direction, however, a user located far from its base station will perceive a higher intercell interference than a user located closer to the base station. This situation is illustrated in Figure 2.15. Specifically, the uplink case is shown in Figure 2.15 (a), where the intercell interference in cell 1 comes from the mobiles in cell 2. Therefore, the intercell interference $P_{inter,1}$ measured at the receiver of user 1 is the same as the interference $P_{inter,2}$ measured at the received of user 2. In the downlink case, however, shown in Figure 2.15 (b), the intercell interference is generated by cell 2, so the user 1 that is located closer to this cell measures a higher intercell interference $P_{inter,1}$ than user 2.

2.4.2 SOFT HANDOVER

The handover procedure is critical in any cellular mobile communication system since it allows keeping service continuity as the user moves across the network coverage area by changing the cell to which the mobile is connected. In CDMA systems, taking into account that all the cells operate at the same frequency, this procedure is revealed to be even more critical than in FDMA/TDMA systems, because a user that is not connected to the cell that requires the lowest power will generate more intercell interference over the neighbouring cells.

At the same time, the fact that CDMA cells operate with the same frequency has an interesting implication for the handover procedure, allowing a mobile to be simultaneously connected to more than one cell. This special type of handover is called 'soft handover' and has a strong relationship with the rake receiver presented in Section 2.3.4. It is implemented simply by programming the available rake fingers (see Figure 2.13) to propagation paths corresponding to different cells. In this way, when a user moves from cell 1 to cell 2 (see Figure 2.16), initially all the rake fingers will be connected to propagation paths from cell 1 and as the border area between both cells is entered it will progressively remove paths from cell 1 and add others from cell 2, until it is finally connected only to cell 2. Note that, in this way, the handover is done in a soft or progressive way [27].

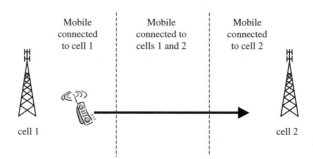

Figure 2.16 Soft handover procedure

In UMTS terminology, the set of cells that the mobile is simultaneously connected to during soft handover is denoted as the Active Set. Smart handover mechanisms are required to detect when one cell should be added or removed from this active set.

Significant differences arise when applying soft handover in the uplink and in the downlink direction. Specifically, in the downlink, since the rake receiver is located at the mobile terminal, the only task it has to do when adding a new cell to the active set is to programme some rake fingers to the code sequence of the new cell and to its corresponding propagation paths and perform the combination. Note that in this case the mobile will be consuming one code sequence for each of the cells of the active set and the total downlink power transmitted to this user will be shared between these cells.

In the uplink direction, however, the receivers are located in different cells, so if soft handover has to be implemented, this requires that the signals be transported from the cells to some entity deeper in the network (e.g. the RNC in UMTS) that performs the rake combination. This solution would consume an excessive amount of network resources and so it is not suitable in practice. Therefore, the solution adopted for soft handover in the uplink is to perform macrodiversity simply by selecting the signal coming from the cell of the active set that receives the highest power from the user, without performing rake combination. In this case, the power control algorithm would adjust the mobile transmitted power to ensure the S/I requirements are met in at least one cell of the active set.

A special case of soft handover is the so-called 'softer handover', which occurs when the mobile is simultaneously connected to two different sectors of the same cell site [28]. In this case, rake combination may be supported even in the uplink because the receivers are located at the same site and therefore there is no need to route the physical signals to other entities.

REFERENCES

[1] M. Simon, J. Omura, *Spread Spectrum Communications*, Computer Science Press, Maryland, USA, 1985
[2] J.K. Holmes, *Coherent Spread Spectrum Systems*, John Wiley & Sons Inc., 1982
[3] A.J. Viterbi, *CDMA: Principles of Spread Spectrum Communications*, Addison-Wesley, 1995
[4] K.S. Zigangirov, *Theory of Code Division Multiple Access Communication*, John Wiley & Sons Ltd., 2004
[5] H. Holma, A. Toskala, *WCDMA for UMTS*, John Wiley & Sons Ltd, 2nd edition, 2002
[6] 3GPP TS 25.213 '*Spreading and modulation (FDD)*'
[7] E.H. Dinan, B. Jabbari, 'Spreading Codes for Direct Sequence CDMA and Wideband CDMA Cellular Networks', *IEEE Communications Magazine*, pp. 48–54, September, 1998
[8] 3GPP TR 25.942, 'Radio Frequency (RF) System Scenarios'
[9] D.V. Sarwate, 'Bounds on Crosscorrelation and Autocorrelation of Sequences', *IEEE Transactions on Information Theory*, **IT-25**(6), pp. 720–724, November, 1979
[10] M. Sidelnikov, 'On Mutual Correlation of Sequences', *Soviet Mathematics Doklady*, **12**, pp. 197–201, 1971

[11] R. Gold, 'Optimal Binary Sequences for Spread Spectrum Multiplexing', *IEEE Transactions on Information Theory*, **13**(4), pp. 619–621, October, 1967

[12] R. Gold, 'Maximal Recursive Sequences with 3-Valued Recursive Cross-Correlation Functions', *IEEE Transactions on Information Theory*, **14**(1), pp. 154–156, January, 1968

[13] D.V. Sarwate, M. B. Pursley, 'Crosscorrelation Properties of Pseudorandom and Related Sequences', *Proceedings of the IEEE*, **68**(5), pp. 593–619, May, 1980

[14] M.B. Pursley, 'Performance Evaluation for Phase-Coded Spread Spectrum Multiple-Access Communication – Part I: System Analysis', *IEEE Transactions on Communications*, **COM-25**(8), pp. 795–799, August, 1977

[15] M.B. Pursley, D.V. Sarwate, 'Performance Evaluation for Phase-Coded Spread Spectrum Multiple-Access Communication – Part II: Code Sequence Analysis', *IEEE Transactions on Communications*, **COM-25**(8), pp. 800–803, August, 1977

[16] D.V. Sarwate, M.B. Pursley, T.Ü. Basar, 'Partial Correlation Effects in Direct-Sequence Spread-Spectrum Multiple-Access Communication Systems', *IEEE Transactions on Communications*, **COM-32**(5), pp. 567–573, May, 1984

[17] R.K. Morrow, Jr., 'Accurate CDMA BER Calculations with Low Computational Complexity', *IEEE Transactions on Communications*, **46**(11), pp. 1413–1417, November, 1998

[18] R.K. Morrow, Jr., and J.S. Lehnert, 'Bit-to-bit Error Dependence in Slotted DS/SSMA Packet Systems with Random Signature Sequences', *IEEE Transactions on Communications*, **COM-37**, pp. 1052–1061, October, 1989

[19] R.K. Morrow, J.S. Lehnert, 'Packet Throughput in Slotted ALOHA DS/SSMA Radio Systems with Random Signature Sequences', *IEEE Transactions on Communications*, **40**(7), pp. 1223–1230, July, 1992

[20] W.C.Y. Lee, *Mobile Communications Engineering*, McGraw-Hill, 1997

[21] J.G. Proakis, *Digital Communications*, McGraw-Hill, 4th edition, 2000

[22] R. Price, P.E. Green, 'A communication Technique for Multipath Channels', *Proceedings of the IRE*, **46**, pp. 555–570, 1958

[23] R. Tanner, J. Woodard, *WCDMA Requirements and Practical Design*, John Wiley & Sons Ltd, 2004

[24] K.S. Gilhousen, I.M. Jacobs, R. Padovani, A.J. Viterbi, L.A. Weaber, C.E.Wheatley III, 'On the Capacity of a Cellular CDMA System', *IEEE Transactions on Vehicular Technology*, **40**(2), pp. 303–312, May, 1991

[25] A.J. Viterbi, A.M. Viterbi, E. Zehavi, 'Other-Cell Interference in Cellular Power-Controlled CDMA', *IEEE Transactions on Communications*, **42**(2/3/4), pp. 1501–1504, February, March, April, 1994

[26] F. Adelantado, O. Sallent, J. Pérez-Romero, R. Agustí, 'Time Correlation of the Intercell to Intracell Interference Ratio in a W-CDMA Network', *IEE Electronics Letters*, **38**(25), pp. 1735–1737, December, 2002.

[27] A.J. Viterbi, A.M. Viterbi, K.S. Gilhousen, E. Zehavi, 'Soft Handoff Extends CDMA Cell Coverage and Increases Reverse Link Capacity', *IEEE Journal on Selected Areas in Communications*, **12**(8), pp. 1281–1287, October 1994

[28] C.C. Lee, R. Steele, 'Effect of Soft and Softer Handoffs on CDMA System Capacity', *IEEE Transactions on Vehicular Technology*, **47**(3), pp. 830–841, August 1998

3

UMTS Radio Interface Description

This chapter presents the characterisation of the protocols that define the procedures for transferring user information and signalling through the UMTS radio interface according to what has been standardised in 3GPP specifications. In the context of this book, this characterisation is required because it establishes the framework for the definition and operation of Radio Resource Management strategies and imposes specific constraints in terms of the available information and signalling to be used by these strategies.

The chapter starts with an introduction to set the UMTS radio interface protocols in the framework of the overall UMTS architecture. This allows the identification of the interrelationships between the radio interface and the rest of protocols. After this introduction, the radio interface layered protocol architecture is presented, including the definition of the different types of channels that allow the interoperation among the protocol layers. The detailed characterisation of each protocol layer is then presented in the subsequent sections, starting from the lowest layer and focusing on the UTRAN FDD mode. The last section includes some specific examples of the configuration of the protocol layers for different signalling and data services.

3.1 THE UMTS PROTOCOLS

The communication among the different entities of the UMTS architecture described in Chapter 1 involves several protocol stacks that are defined for each interface and are depicted in Figure 3.1. A protocol stack defines a set of layers that specify the communication procedures between two network entities. Each layer in a network entity (e.g. the UE) communicates with the same layer of the other network entity (e.g. the node B) by means of a specific protocol that includes a set of procedures involving a number of messages transferred between both entities. From a vertical perspective, a given layer provides the means for the transfer of the messages originated at the above layers. In turn, from a horizontal perspective, the concatenation of several protocol stacks allows the communication between non-adjacent entities (e.g. between the UE and the Core Network) [1].

As shown in Figure 3.1, the UMTS protocols are divided into two different strata, denoted as Non Access Stratum (NAS) and Access Stratum (AS) [1][2]. The provision of a UMTS service requires the exchange of information between the corresponding UE and a node of the CN (i.e. the MSC for circuit switched services or the SGSN for packet switched services) that will provide the means for the establishment of the end-to-end communication; for example, by routing the communication towards another UE in the same UMTS network or in an external network. The NAS protocols enable the transfer

Radio Resource Management Strategies in UMTS   J. Pérez-Romero, O. Sallent, R. Agustí and M. A. Díaz-Guerra
© 2005 John Wiley & Sons, Ltd

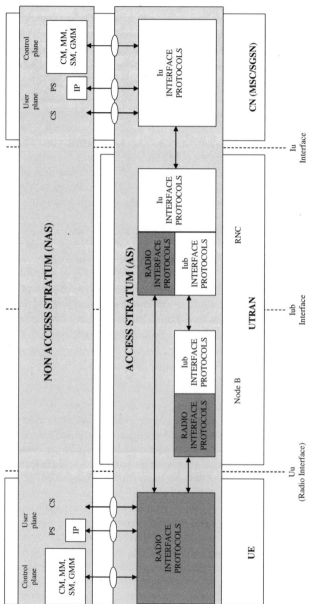

Figure 3.1 UMTS protocols

of this information between the UE and CN. The contents can be either user or control information carrying all the signalling required to set-up and release the service as well as to perform other functionalities specific to a mobile network (e.g. mobility management). In any case, this type of information is to some extent independent of the underlying layers of the protocol architecture and of the elements of the access network that are traversed in the path between the UE and the Core Network. As a matter of fact, this information would essentially be the same when having an access network different from UMTS as it would in the case of a GSM/GPRS access network.

Some examples of NAS protocols in the control plane are the Connection Management (CM) and Session Management (SM) protocols – responsible for the establishment and release of connections or sessions for an UE, respectively – or the Mobility Management (MM) and GPRS Mobility Management (GMM) protocols, responsible for dealing with mobility functions at the network layer (e.g. location area updating, routing area updating, paging, etc.). In turn, in the user plane, the main NAS protocol at the network layer for packet switched services is the IP protocol, while for circuit services information comes directly from the source without the need for a network protocol.

NAS protocols rely on the AS protocols to exchange information between the UE and the CN, as depicted in Figure 3.1. The AS consists of a group of protocols that are specific to the access network being used [2][3][4]. This means that even if the NAS protocols are the same for a UMTS or a GSM/ GPRS access network, the AS protocols that allow the transfer of these messages through the different nodes will be different. In the UMTS architecture, the AS includes three different protocol stacks, namely the radio interface protocols, the Iub interface protocols and the Iu interface protocols. In particular, the radio interface protocol stack allows communication between the UE and the UMTS access network (UTRAN). Note that the protocols at the upper layers terminate in the UE and RNC, while the lower layers terminate in the UE and Node B. With respect to the Iub interface protocols, they involve the communication of the lower layers of the RNC and the Node B. Finally, the Iu interface protocols allow communication between the RNC and the CN, distinguishing between the Iu-CS for communication between RNC and MSC and the Iu-PS for communication between RNC and SGSN.

The AS provides the NAS with a service of information transfer between the UE and the CN in what is named a Radio Access Bearer (RAB) [3]. A RAB consists of two parts, the Radio Bearer, corresponding to the Radio Access Network between the UE and the RNC, and the Iu Bearer, defined between the RNC and the MSC or SGSN (see Figure 3.2) As a result, in order for the communication to be established, resources should be provided for the RAB in the Uu and Iub interfaces of the Radio Access Network as well as in the Iu interface. In this context, this book is focused on the management of the resources at the radio interface, whose scarcity constitutes in most cases the bottleneck for a proper communication

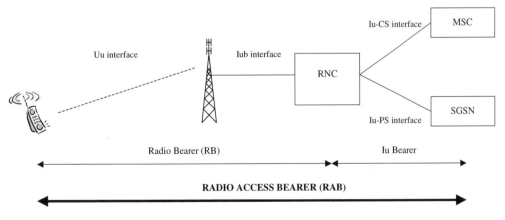

Figure 3.2 Radio Access Bearer concept

to be carried out. Consequently, the focus of this chapter is the characterisation of the radio interface protocols leading to the specification of the parameters defining a RAB.

The concept of Radio Bearer (RB) for transmission between the UE and the RNC refers exclusively to the transmission of user data either in CS or PS mode. In the control plane, the signalling information is transferred through the so-called Signalling Radio Bearers (SRB). Signalling information can be generated either at the NAS protocols or internally at the upper layers of the radio interface protocols, as will be described in the next section. Therefore, a RAB may include a combination of a RB and a SRB and also the combination of different data flows.

Although Iub protocols are not detailed in this book, it should be noted that they are responsible for the transfer of the information units that are delivered by layer 2 of the Radio Interface protocols to the physical layer (the so-called transport blocks, which will be explained in more detail in Section 3.2.2) between the RNC and the Node B. ATM is the transport technology being used in 3GPP releases up to 5. For more details, the reader is referred to 3GPP specifications [5]. ATM is also the technology used by Iu protocols. In this case, the reader is referred to Reference 6.

3.2 RADIO INTERFACE PROTOCOL STRUCTURE

This section details the reference protocol structure that defines the organisation of the radio interface in UTRAN. This structure follows the OSI layered reference model, as shown in Figure 3.3, for both the UE and the UTRAN parts, the latter including the Node B and the RNC. The protocol architecture is subdivided into the control plane, responsible for the transmission of signalling information, and the user plane, responsible for the transmission of user data. In particular, three protocol layers are considered: the physical layer (L1), the Data Link Layer (L2) and the Network Layer (L3). In turn, Layer 2 is split into two sub-layers, namely the Radio Link Control (RLC) and the Medium Access Control (MAC). With respect to layer 3, only the lowest sub-layer, denoted as Radio Resource Control

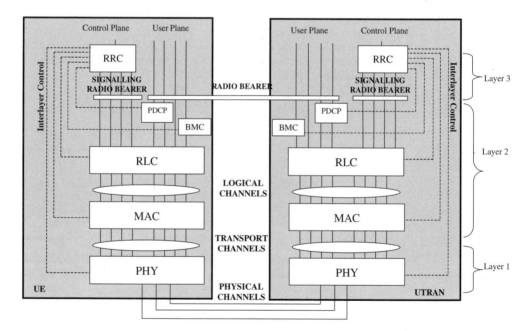

Figure 3.3 Radio interface protocol reference architecture

(RRC), terminates in the UTRAN control plane. Two other sub-layers are considered in the user plane: the Packet Data Convergence Protocol (PDCP), responsible for header compression of data packets for PS services, and the Broadcast/Multicast Control Protocol (BMC) that adapts the transmission for broadcast services. Note also that connections exist between the RRC and the lower layers. Such connections provide inter-layer control services and allow the configuration of the lower layers [2].

Each layer communicates with the same layer at the peer entity (e.g. the RRC layer at the UE communicates with the RRC at the RNC), and this communication is defined by the specific protocol of each layer. At the UTRAN side, the RLC and above radio protocols are located in the RNC. In turn, with respect to the MAC protocol, some of its functionalities are located in the Node B and others in the RNC. The layered structure is constructed upon the assumption that each layer provides message transfer services to the upper layers. At one extreme network entity (e.g. UE or RNC), a given layer receives Service Data Units (SDUs) containing the messages from its upper layer, processes them adding the required headers and control elements and eventually delivers them in the form of Protocol Data Units (PDUs) to its lower layer. Note that the PDU delivered by a given layer corresponds to the SDU seen by its lower layer. At the lowest layer, the information is finally transferred through the channels existing in the physical layer (i.e. the specific code sequences, time slots and frequency bands). At the other extreme network entity (e.g. RNC or UE), the information is received at the physical layer and delivered to the upper layers until reaching the destination layer. Note that this transfer of information requires the definition of adequate interfaces between adjacent layers specifying the path that information follows depending on its nature.

Layer 2 offers to the upper layers the service of information transmission between the UE and the UTRAN by means of the Radio Bearers (RBs) and Signalling Radio Bearers (SRBs). The former provide the transmission of user data while the latter are intended to transfer control information that can be originated either in the Radio Resource Control protocol or in upper layers. Whenever a service is provided to a given UE (e.g. a voice service, a videoconference service, an interactive web browsing service, etc.) it should be associated to a specific Radio Bearer that specifies the configuration and the parameters of the sub-layers in layer 2 and the physical layer depending on the characteristics of the service being provided. The information flow associated to a RB or a SRB is mapped into different types of channels depending on the position in the layered protocol architecture. They are the logical, transport and physical channels [2], respectively, as shown in Figure 3.3. This channel differentiation enables a flexible architecture that allows the provision of services by making use of different configurations of the radio interface, thus it becomes possible to accommodate different degrees of quality of service.

Logical channels allow communication between the RLC and MAC layers, and they are characterised by the type of information that is being transferred across these layers. As a result, there are logical channels for the transfer of user traffic, and also logical channels for the transfer of control information, which can be either dedicated to specific users or common to a set or to all of them. A detailed description of the available logical channels will be given in Section 3.2.1.

Logical channels are mapped onto transport channels in the MAC layer. Transport channels are defined between MAC and PHY layers and they specify how the information from logical channels should be adapted to get access to the radio transmission medium. As a result, they define the format used for the transmission in terms of, for example, channel coding, interleaving or bit rate. Different transport channels are defined, mainly distinguishing between transport channels operating in dedicated mode (i.e. allocated to a specific user) and in common mode (i.e. users should contend for the access to such channels whenever they have some information to be transmitted). It is worth mentioning that a logical channel can be mapped to different transport channels depending on how the system is configured. A detailed description of the available transport channels will be given in Section 3.2.2.

Finally, physical channels are defined in the physical layer and they specify the nature of the signals that are transmitted either in the uplink or in the downlink direction and that are code, time and frequency multiplexed with the signals coming from other users and nodes B. Physical channels include also physical signals, which serve as a support for the transmission on the physical channels (e.g. supporting the random access procedures) but do not contain information from upper layers.

3.2.1 LOGICAL CHANNELS

Different types of logical channels are specified in the interface between the RLC and MAC layers depending on the type of information that is being transferred (i.e. on the type of data transfer service offered by the MAC). The usual classification of logical channels is based on whether they carry information corresponding to the control (i.e. Control Channels) or to the user plane (i.e. Traffic Channels) [2].

3.2.1.1 Control Logical Channels

The control channels defined for the FDD mode are the following:

- BCCH (Broadcast Control Channel). This channel carries control information that is broadcast to all the users of a given cell in the form of System Information messages. Such information includes cell specific parameters (e.g. cell identifiers, code sequences, timers, etc.) that must be known by the UE before trying to camp on a given cell. This channel is only defined in the downlink direction.
- PCCH (Paging Control Channel). This logical channel is used to notify incoming calls or messages to the users in a given area. All the UEs in idle mode should listen to this channel periodically. This channel is only defined in the downlink direction.
- DCCH (Dedicated Control Channel). This logical channel transfers signalling information corresponding to a specific UE. It is a point-to-point bi-directional channel that exists for each UE that has a RRC connection with the RNC. Examples of messages that are transferred through this logical channel are connection establishment messages, radio resource control messages or measurement reports.
- CCCH (Common Control Channel). This channel would be equivalent to the DCCH channel but for UEs that do not yet have a RRC connection with the RNC and by users executing cell reselection procedures while transmitting in common channels. As a result, the UE should make use of shared physical and transport channels including the corresponding UE identity in the transmitted messages. An example of the utilisation of the CCCH would be the initial message that is transmitted by a UE during a connection establishment and the corresponding channel allocation response from the network.

3.2.1.2 Traffic Logical Channels

Two types of traffic channels are defined:

- DTCH (Dedicated Traffic Channel). This logical channel is defined in the user plane and transfers the information corresponding to a given service dedicated to a single user. It exists both in the uplink and downlink direction. Different DTCH channels may coexist for a given UE whenever several services are provided simultaneously (e.g. data and voice connections).
- CTCH (Common Traffic Channel). This logical channel carries dedicated user information to a group of UEs in a given cell (e.g. for the transfer of SMS Cell Broadcast Messages providing information depending on the geographical area). It is a point-to-multipoint unidirectional channel.

3.2.2 TRANSPORT CHANNELS

Transport channels are defined by how and with what characteristics the data from the MAC layer are transferred through the radio interface. It is worth mentioning that transport channels do not take into consideration the information that is being transferred, which corresponds to the concept of a logical channel, as described in Section 3.2.1. As a result, a given transport channel may carry information from different logical channels, if their transmission through the radio interface should have the same characteristics [2][7].

The introduction of the transport channel concept in UMTS, as a difference from the specifications of 2G systems like GSM, where only logical and physical channels were defined, responds to the desired flexibility of the radio interface to accommodate different types of services with transmission requirements of a very different nature. For example, data services do not have the same requirements in terms of delay or error tolerance as voice or video streaming services. Therefore, logical DTCH channels carrying such different services should be mapped to transport channels with different characteristics: while data traffic could be transmitted in some cases in common channels based on contention schemes, voice traffic would require a dedicated channel. The adequate mapping between each logical channel to the appropriate transport channel should respond to the trade-off that exists between resource consumption and service requirements. It is a key point in ensuring an efficient use of the scarce radio resources while at the same time keeping the QoS constraints.

As stated previously, the classification of the logical channels takes into account the nature of the information (i.e. control or user planes). However, the transport channels are classified into two sets depending on whether transmission is done in dedicated or in common mode. Dedicated transport channels are characterised by the allocation of physical radio resources (i.e. frequency and code sequence) to a particular UE, which retains them during its connection lifetime in a given cell, so that no other UE can transmit by making use of the same resources. In turn, common transport channels are mapped onto a pool of physical radio resources that are either shared by a set of UEs according to certain predefined rules or not addressed to any particular UE. As a result, when a common transport channel transmits information to/from a specific UE, explicit UE identification is required in the transmission, in contrast to dedicated channels, where the UE identification is implicit.

3.2.2.1 Dedicated Transport Channels

Only one dedicated transport channel for the FDD mode is defined in the UTRAN specifications. This is the DCH (Dedicated Channel), which carries information coming from upper layers of a given UE. This includes both user and control information. The DCH exists both in the uplink and downlink directions, and can be transmitted over the entire cell or only over a part of it by making use of beam-forming antennas. A single UE may have several DCH channels simultaneously allocated, which allows for the provision of multiple simultaneous services. However, even in the case of a single service, a usual situation is that two DCHs are allocated, one for the transfer of a traffic logical channel (DTCH) and another for the transfer of signalling (i.e. a DCCH).

The DCH channel is flexible enough to accommodate variable bit rate services thanks to the ability to change the transmission bit rate at given periods, called Transmission Time Intervals (TTI), which may range from 10 ms to 80 ms. The value of this parameter, together with other aspects such as the set of allowed bit rates, are specified in the allocation of the specific DCH channel, thus providing a high degree of flexibility.

3.2.2.2 Common Transport Channels

With respect to common transport channels, we now describe more possibilities that are defined in the UTRAN FDD specifications:

- BCH (Broadcast Channel). This transport channel only exists in the downlink direction and is not addressed to any UE in particular but to all the users in a given cell. It provides the transport for the BCCH logical channel, which includes system information. Therefore, it should be received in the whole coverage area of the cell and all the terminals must be able to decode its information before they are allowed to transmit. This requirement has a direct impact on the transmission characteristics of this channel, resulting in a fixed low bit rate and relatively high power level.
- PCH (Paging Channel). Like the BCH, this transport channel only exists in the downlink direction and should be decoded by all the users in the coverage area of a given cell. It provides the transport for the PCCH logical channel, which contains pagings and notifications from the network to specific

users. As a result, the messages included in the PCH are addressed to specific UEs, which requires the appropriate identification of the UE. In order to ensure low terminal power consumption, some physical layer signals, called Paging Indicators, exist so that UEs do not continuously listen to and decode the PCH for possible incoming messages.

- RACH (Random Access Channel). This transport channel exists only in the uplink direction and should be received from any location in the entire cell coverage area. Transmission in the RACH follows the rules of the S-ALOHA/CDMA protocol and therefore collision between different terminals may occur. Consequently, it is only used by services with very low quality requirements (e.g. transmission of short packets without delay constraints) or for the transfer of signalling information during the initial access to the system, before another type of channel can be allocated. Different transmission bit rates may be used.

- FACH (Forward Access Channel). This transport channel exists only in the downlink direction and should be received over the entire cell. Transmission may be carried out with different bit rates. Typically, it carries logical control channels addressed to specific UEs, which requires explicit UE identification. As an example, the response to a Random Access message transmitted in the RACH would be transmitted through the FACH. Additionally, it may also transport short data packets for services with low quality requirements.

- CPCH (Common Packet Channel). This channel is defined in the uplink direction and is devised as an extension of the RACH channel for the transmission of longer data packets. Transmission through CPCH follows the rules of the DSMA/CD (Digital Sense Multiple Access / Collision Detection) protocol. As a result, collisions may exist in the initial transmission on the channel. However, a collision resolution method allows a single user finally to acquire the channel and keep it for the transmission of the whole packet. Different transmission bit rates may be applied.

- DSCH (Downlink Shared Channel). This transport channel only exists in the downlink direction and is mapped over a set of physical resources (i.e. code sequences) that are allocated on a transmission time interval basis to different users according to some packet scheduling policy. All the users that transmit through the DSCH channel must also have an associated bi-directional low bit rate DCH channel allocated through which control information is sent. This information includes the indication of the instants when the DSCH is allocated to the specific UE as well as power control commands that allow the use of closed-loop power control. Depending on the packet scheduling algorithm, different users may transmit simultaneously in different code sequences of the DSCH channel and the transmission bit rate may be changed from TTI to TTI. Due to the characteristics of the channel, transmission through DSCH is normally reserved for non real time packet data services, since only statistical bounds for the transmission delay can be ensured depending on the packet scheduling algorithm.

- HS-DSCH (High Speed Downlink Shared Channel). This transport channel is an extension of the DSCH to provide the HSDPA (High Speed Downlink Packet Access), a new feature included in the specifications of Release 5 [8]. Transmission through the HS-DSCH is based on the application of link adaptation mechanisms including adaptive modulation schemes and type II hybrid ARQ according to packet scheduling policies. In this way, very high peak transmission bit rates in the order of 10 Mb/s may be achieved.

Different transport channels may coexist simultaneously for a single user. In such a case, the transport channels are multiplexed together over the same physical resources forming what is named a Coded Composite Transport Channel (CCTrCH).

3.2.2.3 Transport Channel Parameters

Apart from the transport channel type, there is another set of parameters that allow the appropriate configuration of the physical channel in each instant and complete the definition of the allocated transport channel by specifying all the possibilities of transmission through the radio interface [9]. These parameters are the following ones:

- Transmission Time Interval (TTI). This is the minimum amount of time in which the configured parameters for a physical channel remain the same. In other words, it specifies the instants in which the different transmissions of a transport channel can be carried out. Possible values for the TTI are 10, 20, 40 or 80 ms, and it is always a multiple of the frame time 10 ms.
- Transport Block (TB). This is the minimum amount of information that can be exchanged between the physical and MAC layers. It includes a data part and the MAC and RLC headers. The number of bits of a Transport Block is denoted as TBS (Transport Block Size).
- Transport Format (TF). This is the format that the physical layer offers to the MAC layer for the transmission of a set of transport blocks in a TTI. A transport format is then defined by the number of transport blocks that are transmitted in the corresponding TTI, which consequently defines the instantaneous bit rate or equivalently the spreading factor that should be used in the physical layer. Other aspects included in the definition of the Transport Format are the channel coding type being used (e.g. convolutional or turbo-code), as well as the code rate and the number of CRC bits.
- Transport Format Set (TFS). This defines the set of allowed Transport Formats for a given transport channel. As a result, it limits the maximum number of transport blocks that can be transmitted in a TTI, or equivalently, it limits the maximum instantaneous bit rate. This parameter can be dynamically adjusted to reduce the amount of interference during congestion situations.
- Transport Format Combination (TFC). This concept appears exclusively when several transport channels are multiplexed together onto a CCTrCH, and defines the number of transport blocks that are transmitted from each of the multiplexed transport channels in a given TTI.
- Transport Format Combination Set (TFCS). This concept is equivalent to the TFS when several transport channels are multiplexed into a CCTrCH. In this case, it specifies the allowed transport format combinations that can be used in a TTI.

During a given TTI, the selected TFC must remain the same, but it can be changed by the transmitter from one TTI to the next. Such a change involves alterations in the physical layer parameters such as the spreading factor or the time slot format being used. Consequently, the receiver must have mechanisms to detect the transport format combination that is being used at a given moment. This can be done either explicitly, by means of the TFCI (Transport Format Combination Indicator), a specific indicator included in the control physical channels, or by means of blind detection procedures, which do not require the transmission of the TFCI.

An example of these concepts is given in Table 3.1, where the characteristics of a transport channel for an interactive service in the uplink through a DCH channel are given. It corresponds to one of the Radio Access Bearers defined in Reference 10. As detailed in Table 3.1, the Transport Block size delivered by the MAC layer is 336 bits, which include both user information and RLC/MAC

Table 3.1 Example of the transport characteristics of an interactive service through DCH

TrCH type	DCH
TB size (bits)	336
TFS	
TF0 (bits)	0×336
TF1 (bits)	1×336
TF2 (bits)	2×336
TF3 (bits)	3×336
TF4 (bits)	4×336
TTI (ms)	20
Coding type	Turbo Code (1/3)
CRC (bits)	16

Table 3.2 Example of the transport characteristics of a Signalling
Radio Bearer through DCH

TrCH type	DCH
TB size (bits)	148
TFS	
TF0 (bits)	0×148
TF1 (bits)	1×148
TTI (ms)	40
Coding type	Convolutional Code (1/3)
CRC (bits)	16

headers. The transport format set contains five transport formats, corresponding to five different bit rates, which range from no transmission (i.e. TF0) to transmission at the maximum bit rate obtained with four transport blocks in one TTI of 20 ms (i.e. TF4). Once a transport format is selected, 16 bits of CRC are added to each transport block and the resulting bits are encoded according to a turbo code with code rate 1/3.

Continuing with the same example, the previous channel can be multiplexed together with a signalling logical channel DCCH that has two possible transport formats, TF0 corresponding to no transmission, and TF1, corresponding to the transmission of a transport block with 148 bits, as presented in Table 3.2, according to one of the Signalling Radio Bearers defined in Reference 10.

This leads to the definition of a CCTrCH with the following Transport Format Combination Set, which includes 10 TFC in which each combination is represented by a pair (TFA,TFB) where TFA is the transport format of the DTCH channel and TFB is the transport format of the DCCH channel:

$$(TF0, TF0), (TF1, TF0), (TF2, TF0), (TF3, TF0), (TF4, TF0)$$
$$(TF0, TF1), (TF1, TF1), (TF2, TF1), (TF3, TF1), (TF4, TF1)$$

3.2.3 PHYSICAL CHANNELS

The physical channels are constituted by the physical signals that are transmitted through the radio channel in both the uplink and downlink directions. Depending upon the characteristics and requirements of the information that is being transmitted, different types of physical channels are defined to specify how the transmission media is shared in an ordered way by each UE. The specification of a physical channel must include the characteristics in terms of frequency, time and code sequence that are being used.

UTRAN FDD is defined to operate in three paired bands [11][12], denoted as band I, II and III, as depicted in Figure 3.4, which allows for a certain flexibility to accommodate UMTS depending on the spectrum availability in each country. The total available bandwidth in band I, to be used in regions 1 and 3 (i.e. Europa and Asia), is defined between 1920 and 1980 MHz for the uplink direction and between 2110 and 2170 MHz in the downlink, so the frequency separation between transmission and reception is 190 MHz. For band II, to be used in region 2 (i.e. the Americas), the frequencies range from 1850 to 1910 MHz in the uplink and from 1930 to 1990 MHz in the downlink, with a frequency separation between uplink and downlink of 80 MHz. Bands I and II were the initial bands identified at the WARC-92 (World Administrative Radio Conference, 1992), and they are the ones that are used in most of these countries. More recently, following its identification at WRC-2000 (World Radiocommunication Conference, 2000), band III has been included for those countries not having bands I and II available. Band III ranges in the uplink from 1710 to 1785 MHz and in the downlink from 1805 to 1880 MHz, with a frequency separation between both transmission directions of 95 MHz. This band is already used by some 2G networks, so it is expected that in the long term it will allow the migration from 2G to 3G

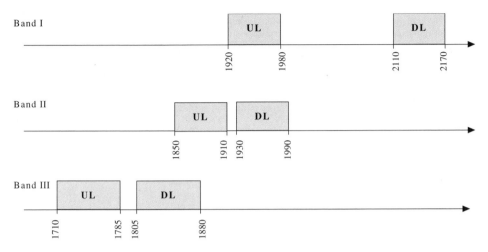

Figure 3.4 Frequency bands defined for UTRAN FDD

systems. In all the cases, the carrier separation is 5 MHz, thus leading to a total of 12 carriers in both bands I and II and 15 carriers in band III, which are numbered by making use of the UARFCN (UTRA Absolute Radio Frequency Channel Number).

The transmission in each frequency is based on the DS/CDMA multiple access scheme [13]. To this end, different code sequences are used to spread the signal spectra of the different transmissions up to a chip rate of 3.84 Mc/s. Therefore, the variable bit rate nature of the signals that are transmitted requires the allocation of code sequences with variable spreading factors. In UTRAN FDD, the spreading process is done in two steps, namely the channelisation and the scrambling processes, as will be detailed in Section 3.3.2. The spreading factor ranges from 4 to 256 in powers of 2 in the uplink and from 4 to 512 in the downlink. This corresponds to a raw bit rate ranging from 150 bits/frame to 9600 bits/frame in the uplink and from 150 bits/frame to 19200 bits/frame in the downlink.

On the other hand, the transmissions over a given physical channel are carried out in specific time intervals, defined by start and stop times, measured in integer numbers of chips. The main temporal reference in UTRAN FDD is the 10 ms radio frame, which contains 38400 chips and is subdivided into 15 slots of 0.666 ms each one with 2560 chips [7]. It should be pointed out that these time slots should not be understood as part of a TDMA access but rather they define the closed loop power control period for transmissions of a given user, thus leading to 15 periods over 10 ms (i.e. a power control rate of 1500 Hz).

Other transmission intervals for other channels are also defined based on the radio frame and the time slot reference. These are the Access Slots for RACH and CPCH, with 2 time slots duration, and the Sub-Frame, which is the basic time interval for the transmission with HS-DSCH, and that corresponds to a duration of three time slots.

The combinations of different frequency, code and time periods lead to the definition of a set of physical channels in the UTRAN FDD mode, in which the corresponding transport channels are mapped. As in the case of transport channels, they are classified as dedicated and common physical channels.

3.2.3.1 Dedicated Physical Channels

The existing dedicated physical channels in UTRAN FDD are:

- DPDCH (Dedicated Physical Data Channel). This channel carries the information of a DCH transport channel. It exists both in the uplink and downlink directions and makes use of closed loop power control.

- DPCCH (Dedicated Physical Control Channel). This channel is related to a DPDCH and transmits physical layer signalling information (e.g. power control commands and synchronisation sequences). In the uplink, it is code multiplexed with the DPDCH, while in the downlink DPDCH and DPCCH are time multiplexed and the resulting channel is usually called the DPCH (Dedicated Physical Channel).
- HS-DPCCH (High Speed Dedicated Physical Control Channel). This channel is defined in the uplink direction and it carries physical layer control information related to the HS-DSCH transport channel.

3.2.3.2 Physical control channels

The physical control channels in UTRAN FDD are:

- P-CCPCH (Primary Common Control Physical Channel). This physical channel only exists in the downlink direction and has a fixed channel bit rate of 30 kb/s, corresponding to a spreading factor of 256. It is used to carry the BCH transport channel.
- S-CCPCH (Secondary Common Control Physical Channel). This physical channel is used to carry the PCH and FACH transport channels and only exists in the downlink direction. Depending upon the amount of data that is being transmitted, different bit rates can be used in this channel.
- SCH (Synchronisation Channel). This channel is used for cell search and is the first channel that a terminal must detect before being able to measure a new cell. It allows the synchronisation at frame and time slot levels as well as the determination of the code sequence being used by the P-CCPCH channel that contains the BCH.
- CPICH (Common Pilot Channel). This channel transmits a pre-defined bit sequence in the downlink direction at 30 kb/s and is used by the terminals to make power measurements of the different cells. The measured level of this channel determinates whether or not the corresponding cell can be used and, therefore, it is possible to adjust the cell coverage area by adjusting the CPICH transmitted power.
- PRACH (Physical Random Access Channel). This channel is used in the uplink direction and carries the RACH transport channel. Transmission on the PRACH is subject to S-ALOHA / CDMA protocol with fast acquisition indication. To this end, the transmission consists of two steps: code acquisition, in which terminals select a code sequence, and message transmission, in which terminals that have succeeded in the acquisition step transmit their information. Open loop power control is being used, and different bit rates can be used in the message part.
- PCPCH (Physical Common Packet Channel). This channel exists in the uplink direction and trans-mits the CPCH transport channel. Transmissions are subject to DSMA/CD with fast acquisition indication. Unlike the PRACH, there are three steps in the PCPCH transmission: code acquisition, where terminals select an available code sequence, collision detection, where only those terminals that have succeeded in the acquisition transmit in order to avoid two or more terminals selecting the same code sequence, and finally message transmission, where longer messages and higher bit rates than in the PRACH are possible. An associated physical channel in the downlink direction allows the use of closed-loop power control in the transmissions of PCPCH, which ensures better quality in the transmissions than the PRACH channel.
- PDSCH (Physical Downlink Shared Channel). This physical channel carries the DSCH transport channel. It consists of a pool of code sequences with different bit rates that support transmissions based on a packet scheduling policy. It uses closed loop power control thanks to an associated dedi-cated channel. Different bit rates are available.
- HS-PDSCH (High Speed Physical Downlink Shared Channel). This channel carries the HS-DSCH transport channel in the downlink direction. It allows multicode transmission with a spreading factor of 16. It is organised into 2 ms sub-frames, each with three time slots. It may use either QPSK or 16-QAM modulations.

- HS-SCCH (HS-DSCH-related Shared Control Channel). This channel transmits at a fixed bit rate of 60 kb/s in the downlink direction and is used to carry downlink signalling related to the HS-DSCH channel (e.g. the allocation of the corresponding HS-PDSCH channel to specific UEs). It has the same frame structure as the HS-PDSCH channel.

Furthermore, there is a set of physical channels that do not have transport channels mapped onto them and that are associated with other physical channels to support their functionality. These channels are:

- AICH (Acquisition Indicator Channel). This is used in the acquisition phase of the PRACH and indicates, in the downlink direction, whether an uplink code acquisition has been successful or not.
- AP-AICH (CPCH Access Preamble Acquisition Indicator Channel). This signal is equivalent to the AICH but for the PCPCH channel in the acquisition phase.
- CD/CA-ICH (CPCH Collision Detection/Channel Assignment Indicator Channel). This downlink signal supports transmission in the PCPCH in the collision detection phase, by indicating whether a collision has occurred or not. In the case where the PCPCH operates with Channel Assignment mode, the corresponding indications are also included.
- CSICH (CPCH Status Indicator Channel). This downlink signal indicates the availability of the different code sequences used in the PCPCH at a given moment.
- PICH (Paging Indicator Channel). This is associated with a S-CCPCH to which a PCH channel is mapped. It carries paging indicators that inform the different UEs about the instant when they should decode the information of the PCH channel. This allows for an efficient battery consumption of terminals in idle mode.

In Section 3.3, a more thorough description of the characteristics of the previous channels will be presented, after providing an overview of the basic aspects of the physical layer.

3.2.4 MAPPING BETWEEN LOGICAL, TRANSPORT AND PHYSICAL CHANNELS

The organisation of the radio interface protocol stack in logical, transport and physical channels constitutes a flexible architecture so that the network operator can handle in a different way the different data flows that circulate through the different layers of the stack by establishing the appropriate mappings between the three types of channels. As a result, and depending on the nature of the transmitted information, its required quality and the volume of data, different transport and physical channels can be used for the same logical channel. This section addresses the possible mappings that exist among the three types of channels. Furthermore, some suitable mappings are highlighted depending on the considered services.

In the uplink direction, the mapping between logical, transport and physical channels is given in Figure 3.5. With respect to the CCCH, it is always mapped to the RACH transport channel, which is the only uplink common transport channel that carries signalling (e.g. the initial message that the mobile sends whenever it intends to communicate with the network, for example to establish a call). However, for the transmission of dedicated information, either DCCH or DTCH, three possibilities exist, namely the use of the RACH, the CPCH and the DCH channels [2].

With respect to the mapping among transport and physical channels, a one-to-one mapping exists between RACH-PRACH, CPCH-PCPCH and DCH-DPDCH. With respect to the dedicated physical channel, note that the DCH is mapped only in the DPDCH and not in the DPCCH, which carries only physical layer information and consequently does not belong to any transport channel.

Note that although CPCH and RACH channels are common transport channels, they can transport dedicated logical channels. This reflects the fact that transport channels do not take into consideration the type of information transmitted but rather the way the information should be transmitted in the physical layer. Thus, the term 'dedicated' for logical channels indicates that information belongs to a given user, while for transport/physical channels, the term 'dedicated' indicates that the user owns the channel

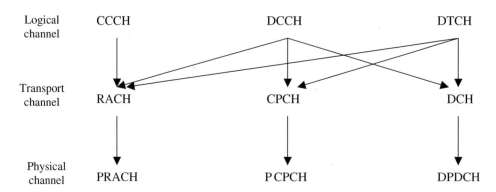

Figure 3.5 Mapping between logical, transport and physical channels in the uplink

and, consequently, no explicit user identification is required in the transmissions. However, the term 'common' for logical channels indicates that the information is not specific to a given user while for transport/physical channels it indicates that the channel is shared by different users and transmissions are subject to the rules of a MAC protocol, requiring explicit user identification in each transmission.

The selection between the three alternatives may be operator dependent and should respond to the characteristics of the service being provided, in combination with the features of each transport channel.

The DCH channel is characterised by a relatively slow initial establishment procedure. Nevertheless, after this procedure, strict timing references are guaranteed. Furthermore, it is mapped onto the DPDCH that makes use of closed loop power control, thus ensuring a higher efficiency in the power consumption and a better performance in terms of block error rate. Consequently, the DCH channel would be especially suited to services with strict timing constraints (e.g. real time services) that transmit relatively continuous flows of information. However, when the flows are discontinuous, the DCH may not be the most efficient solution because of the requirement to transmit the DPCCH even during inactivity periods, thus creating interference and increasing power consumption.

For discontinuous traffic, the use of RACH or CPCH alternatives may be a better solution since these channels do not require any establishment procedure and no transmission is carried out during inactivity periods. Nevertheless, it is difficult to ensure strict time constraints for transmissions in these channels, since the access is contention based and therefore only statistic delay bounds can be ensured. Consequently, normally only non real time services are provided through RACH or CPCH.

The selection between RACH and CPCH would normally be dependent on the length of the packets that should be transmitted and the reliability of the information. In particular, the RACH is suited to short packets that should tolerate high block error rates due to open loop power control. The CPCH, however, is suited to longer packets with higher reliability, provided by the closed loop power control.

In the downlink direction, the corresponding channel mapping [2] is shown in Figure 3.6. With respect to the common logical signalling channels, the BCCH is mapped to the BCH and also to the FACH, while the PCCH is mapped to the PCH transport channel, and the CTCH and CCCH are mapped to the FACH. More possibilities exist in the case of dedicated logical channels DCCH and DTCH, as occurs in the uplink direction. Specifically, the DCH, DSCH, HS-DSCH and FACH channels can be selected. The criteria to select one transport channel ahead of another would be similar to the uplink case, mainly based on traffic volume and delay constraints. As a result, real time services will be normally mapped to DCH channels, while discontinuous flows with softer delay restrictions (e.g. typically non-real time services) would be mapped either to FACH or to shared channels (DSCH or HS-DSCH).

With respect to the mapping between transport and physical channels, a direct relationship exists except in the case of the S-CCPCH, which can carry either the PCH or the FACH.

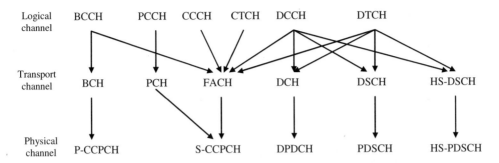

Figure 3.6 Mapping between logical, transport and physical channels in the downlink

The mapping between channels can be dynamic and vary during a connection lifetime. This allows a higher flexibility and a more efficient use of the scarce radio resources while at the same time keeping service constraints. As an example, for discontinuous flows without stringent delay constraints (e.g. non-real time services like Web browsing), it is possible to switch between the use of DCH during activity periods (e.g. when the user is downloading a Web page or a file) and RACH/FACH during inactivity periods (e.g. while the user is reading the downloaded page and no transmission is required). This is especially important from the point of view of code consumption, because it is not efficient to have a code sequence allocated to an inactive user, since this may block the assignment of new connections due to the code availability in the downlink direction. The procedure that allows such change of transport channel is the Transport Channel Type Switching, and it is normally triggered based on traffic volume monitoring. For instance, when the buffer of a UE exceeds a certain amount of data the DCH channel is allocated while when the buffer is below a given threshold for a certain period of time, the DCH channel is de-allocated and the UE must use the RACH/FACH channels to transmit new information.

Finally, it should be mentioned that a number of physical channels exist that do not have a correspondence with transport and logical channels because they only transmit information from the physical layer. This is the case with the pilot and synchronisation channels as well as the physical channels that are used as a support in the different access procedures in uplink common channels.

3.3 PHYSICAL LAYER

This section presents the characterisation of the physical layer, whose mission is to transform the flow of information coming from the different transport channels into physical radio signals transmitted by the antenna [14]. As shown in Figure 3.7, the physical layer at the transmitter side receives Transport Blocks (TBs) from the MAC layer. These transport blocks may belong either to one or to several transport channels that are simultaneously multiplexed. Then, the physical layer executes a set of procedures over the received transport blocks to generate the radio signal that is sent to the antenna. At the receiver side, the reverse procedures are carried out to recover the transport blocks from the received physical signal at the antenna and to deliver them to the MAC. The first set of procedures processes the TBs to multiplex the different transport channels and to introduce the required level of redundancy to overcome the effects of the radio channel, thus obtaining a flow of bits. These procedures will be presented in Section 3.3.1. The bits are then spread and modulated, as will be explained in Section 3.3.2. The resulting modulated symbols are delivered to the antenna, where different transmit diversity schemes, which will be covered by Section 3.3.3, can be applied. Finally, Section 3.3.4 will present the organisation of the radio transmissions in the different physical channels.

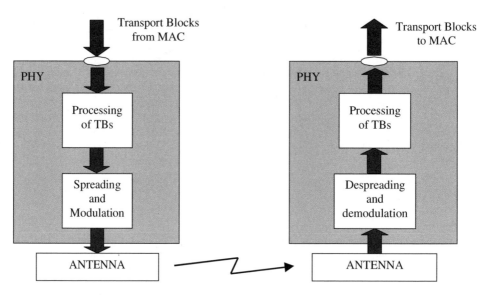

Figure 3.7   Procedures of the physical layer

3.3.1 PROCESSING OF TRANSPORT BLOCKS

In each Transmission Time Interval (TTI), the MAC delivers to the physical layer a given number of transport blocks for each of the transport channels multiplexed together according to a Transport Format Combination (TFC). The physical layer executes a set of processes to map the transport blocks onto the available physical resources (i.e. carrier frequency, code sequence and radio frame). These processes are shown in Figure 3.8 for the uplink direction [15].

The example of Figure 3.8 shows the multiplexing of two transport channels. Typically, this could correspond to a situation with a data channel (i.e. a DTCH) and a signalling control channel (i.e. a DCCH), although other situations with combinations of several DTCH are also possible and would be easily extended from this example. The time period is TTI_1 for delivering the information of transport channel 1 and TTI_2 for transport channel 2. The selected Transport Format Combination (TFC) requires the transmission of m_1 transport blocks from transport channel 1 and m_2 transport blocks from transport channel 2. Then, the following steps are carried out:

(1) CRC (Cyclic Redundancy Code) attachment. Some redundancy bits are added to each transport block in order to detect the existence of errors at the receiver side. The CRC size can be 24, 16, 12, 8 or 0 bits, depending on how the physical layer is configured. The CRC is generated from specific cyclic generator polynomials using all the bits of the transport block.

(2) Transport Block concatenation and/or segmentation. After adding CRC bits, the resulting transport blocks are serially concatenated forming a code block to be entered in the channel coding process. In the case where the resulting number of bits exceeds a maximum value Z, segmentation into several code blocks is required, they all having the same size. To this end, bits may be added if needed. The maximum code block size Z depends on the type of channel coding, being 504 bits for convolutional codes and 5114 bits for turbo-codes. In Figure 3.8, the code block size for transport channel 1 will be the minimum between Z_1 (i.e. the maximum code block size for channel 1) and the product of the number of transport blocks m_1 and the size of the transport block and the CRC (i.e. $TBsize_1 + CRC_1$)

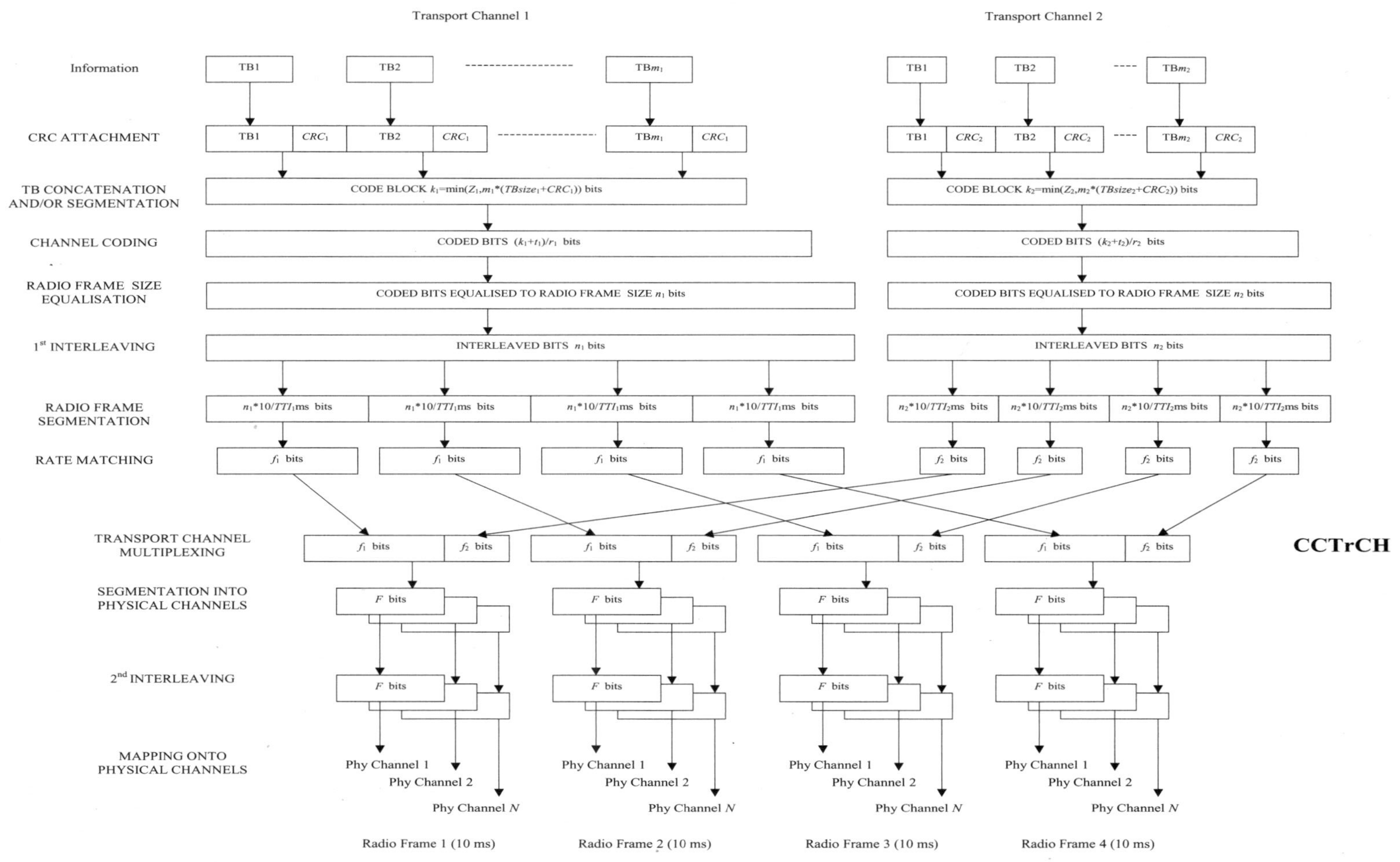

Figure 3.8 Transport channel multiplexing in the uplink direction

(3) Channel coding. With the purpose of protecting data from the effects of the radio channel, the code blocks are encoded by making use of either convolutional or turbo-code schemes. Before encoding, a number of tail bits t with value 0 are added at the end of the code block. Then, the resulting number of bits after encoding will be $(k + t)/r$, where r is the code rate and k is the code block size.

In the case of convolutional coding, the code rate may be 1/2 or 1/3, and the code constraint length is in both cases 9, thus requiring $t = 8$ tail bits. For turbo-codes, the code rate is 1/3 and the number of tail bits is $t = 4$. DCH, CPCH, DSCH and FACH transport channels may use any of the three encoding schemes, while BCH, PCH and RACH always use convolutional codes.

(4) Radio Frame size equalisation. This process consists simply of adding bits after the channel coding procedure in order to ensure that the resulting number of bits is a multiple of the number of 10 ms frames in a TTI. Depending on the TTI value, the number of frames may be 1, 2, 4 or 8.

(5) First interleaving. This process ensures that the bits are not transmitted to the radio channel in the same order that they were generated in the channel coding step. At the receiver side, the reverse deinterleaving procedure is executed before delivering the bits to the channel decoder. This ensures that, whenever a channel fading causes a burst of errors, the erroneous bits in the burst are not consecutively delivered to the channel decoder.

(6) Radio Frame segmentation. After interleaving, the flow of bits is subdivided into a number of segments and each of them will be transmitted in a 10 ms radio frame. The number of segments is equal to the number of frames in a TTI.

(7) Rate matching. This process ensures that the resulting number of bits per radio frame fits into one of the available bit rates of the physical channel, which depend on the spreading factor and on the number of parallel code sequences being used. To this end, the rate matching executes either bit puncturing or bit repetition depending on whether the bit rate should be decreased or increased. The bit repetition is the preferred operation and puncturing is only used in the following cases: where the bit repetition requires the use of additional physical channels (i.e. multicode transmission); when there are physical channel limitations (e.g. a limitation in the minimum spreading factor or in the maximum number of simultaneous codes that can be used); or when there are transmission gaps during compressed mode operation that are left to allow the terminal to make interfrequency measurements (more details about compressed mode will be given on page 79).

The maximum amount of puncturing that can be applied in the uplink direction is signalled by the network through the puncturing limit attribute, which is defined as the minimum value that can take the ratio between the number of bits after and before the puncturing process. The puncturing limit is always lower or equal to 1, and the value 1 means that no puncturing is allowed.

When several transport channels are multiplexed together, a rate matching attribute is assigned by higher layers to each transport channel. This attribute is used in the rate matching process and determines the number of repeated or punctured bits for each transport channel, so that the different requirements in terms of E_b/N_o for each one can be translated onto a single requirement for the multiplexed channel.

(8) Transport Channel multiplexing. In this step, the resulting segments from the different transport channels are multiplexed together forming the so-called Coded Composite Transport Channel (CCTrCH). In this way, each radio frame contains a segment from each transport channel.

(9) Segmentation into physical channels. This procedure is executed whenever the resulting number of bits in one radio frame does not fit into a single physical channel (i.e. a spreading code sequence). In this case, the bit flow is segmented into blocks of equal size and each of them will be mapped into a different physical channel.

(10) Second interleaving. This interleaving procedure is done separately for each of the segments that will be mapped into different physical channels.

After all these procedures, the resulting bit flows are mapped onto the corresponding uplink physical channels (e.g. DPDCH, PRACH or PCPCH) and sent to the spreading and modulation procedures, which will be described in Section 3.3.2.

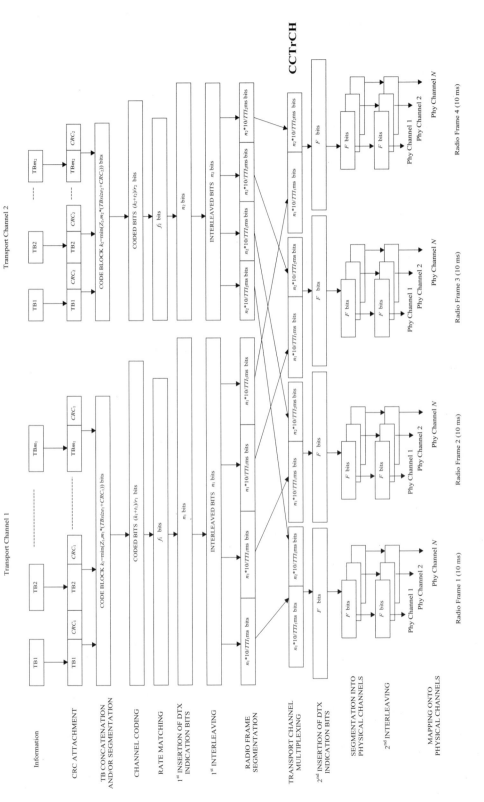

Figure 3.9 Transport channel multiplexing in the downlink direction

In the case of the downlink direction the procedures that are executed to map the transport blocks onto physical channels are shown in Figure 3.9 [15]. One of the main differences in the downlink direction compared to the uplink is the insertion of DTX (Discontinuous Transmission) indication bits, whose purpose is to indicate the instants when the transmission should be turned off for this transport channel. The reason for this difference is that transmissions through dedicated channels in the downlink direction make use of discontinuous transmission to change the channel bit rate without modifying the spreading factor, while in the uplink direction transmission is continuous and changes in the bit rate are achieved by modifying the spreading factor of the code sequence being used. The insertion of DTX indication bits is done at the end of each segment and before both interleaving procedures, after which the DTX bits will be distributed among all the slots of a radio frame.

The use of DTX in the downlink direction allows for two possibilities when multiplexing several transport channels: 'fixed' and 'flexible' positions, as depicted in Figure 3.10, which shows an example with two channels and four different possible TFCs. In the first case, the starting positions in the frame of the multiplexed transport channels are always the same no matter the selected TFC. This simplifies the blind TFC detection in the mobile terminal. In turn, when flexible positions are used, as in the uplink case, the starting position of each transport channel may vary from frame to frame depending on the selected TFC.

When fixed positions are used (see the examples in Figure 3.10a), the number of bits of the physical channel allocated to each transport channel is fixed whatever the selected TFC (in the example in Figure 3.10a there are $B1$ bits for transport channel 1 and $B2$ bits for transport channel 2). These numbers of bits are computed taking into account the maximum TF and the rate matching attribute of every multiplexed transport channel. Once the fixed number of bits allocated per transport channel has been computed, the parameters to execute rate matching are determined for each transport channel. The required number of bits to repeat or puncture is such that the number of bits of the fixed allocation (i.e. $B1$ or $B2$ in the example) equals the resulting number of bits associated with the maximum TF. Consequently, no DTX is used with the maximum TF (see the example in Figure 3.10a where the maximum transport format is TF2). For lower TFs, the rate matching parameters of the channel remain

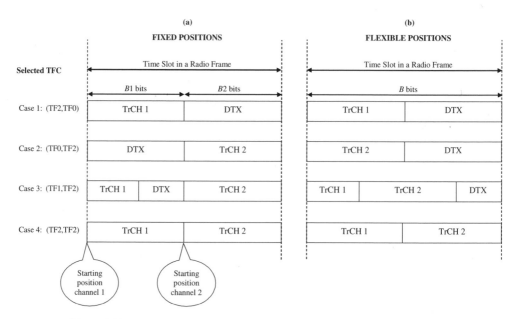

Figure 3.10 Downlink transport channel multiplexing with flexible and fixed positions

the same which means that the ratio of punctured or repeated bits is approximately equal and that DTX is applied to fill the fixed bit allocation. Note that with fixed positions the rate matching process of a given channel depends on its TF and not on the TFC. Therefore the DTX is independent for each channel and is done in the 1st DTX insertion process before channel multiplexing, as depicted in Figure 3.9.

However, when flexible positions are used (see the example in Figure 3.10b), the number of bits allocated to each transport channel is variable and depends on the selected TFC. It is computed dynamically during the rate matching process taking into account the specific TFs that form the TFC combination and the rate matching attribute of each channel. The rate matching is configured so that when the highest TFC is used no DTX is required (see the example in Figure 3.10b, where the highest TFC is TF2,TF2). For lower TFCs, the number of DTX bits to add depends on the number of repeated or punctured bits for each channel. Because of this, the DTX is done in the 2nd insertion process after channel multiplexing, as shown in Figure 3.9. In general, the rate matching with flexible positions is more efficient and less puncturing is required because it operates on a TFC basis, while the use of fixed positions operates on a TF basis [16].

Just to illustrate the overall transport block processing procedures, let us consider an example in which transport channel 1 corresponds to the interactive service characterised in Table 3.1 that is multiplexed together with the signalling radio bearer detailed in Table 3.2. Let's assume that TFC = (TF2,TF1) is selected in a given TTI (i.e. $m_1 = 2$ transport blocks are sent for the interactive service and $m_2 = 1$ for the signalling).

The number of bits after each of the procedures is computed in Table 3.3 for both transport channels. After CRC attachment and TB concatenation, the code block sizes are $k_1 = 688$ bits and $k_2 = 164$ bits. In both cases, the sizes are below the maximum values for the corresponding channel coding schemes and therefore no segmentation is required. After channel coding, the number of bits per TTI are 2076 and 516 for TrCH1 and TrCH2 respectively. Note that no additional bits are required to equalise the radio frame size because the above values can be segmented into 2 frames of 1038 bits for TrCH1 and 4 frames of 129 bits for TrCH2.

The rate matching procedure should adjust the total number of bits per frame to one of the possible values of a DPDCH physical channel in the uplink direction. The closest value is 1200 bits, as will be shown in Section 3.3.4.1, corresponding to a single physical channel with spreading factor 32. As a result, $1200 - 1038 - 129 = 33$ bits should be repeated. When the rate matching attribute is equal for both transport channels, this leads to repeating 29 bits from TrCH1 and 4 bits from TrCH2. So, finally, a segment of 1067 bits from TrCH1 will be transmitted in each frame together with a segment of 133 bits from TrCH2.

Table 3.3   Example of channel multiplexing in the uplink

	TrCH1	TrCH2
TTI	20 ms	40 ms
TFC	TF2 = 2 TBs	TF1 = 1 TB
TB size	336 bits	148 bits
CRC attachment	(336 + 16) = 352 bits	(148 + 16) = 164 bits
TB concatenation	2*(336 + 16) = 688 bits	1*(148 + 16) = 164 bits
Channel coding	Turbo-code $r = 1/3$	Convolutional code $r = 1/3$
	3*(688 + 4) = 2076 bits	3*(164 + 8) = 516 bits
Radio Frame size equalisation	2076*10/20 = 1038 bits/frame	526*10/40 = 129 bits/frame
Radio Frame segmentation	1038 bits/frame	129 bits/frame
Rate Matching	1038 + 29 = 1067 bits/frame	129 + 4 = 133 bits/frame
CCTrCH	1067 + 133 = 1200 bits/frame = 120 kb/s	

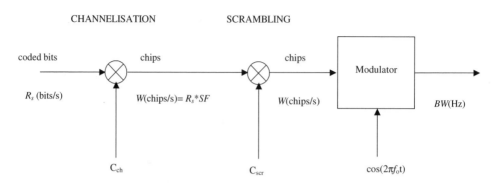

Figure 3.11 Generic structure of the WCDMA spreading and modulation procedures

3.3.2 SPREADING AND MODULATION

The spreading process is applied over all the bits that are transmitted through the physical channel in order to spread the bandwidth of the transmitted signal. This is the key procedure in the WCDMA access that allows the distinguishing of transmissions from each physical channel operating at the same frequency by means of different code sequences. In UTRAN FDD, the spreading process is done in two steps, the channelisation and scrambling procedures, so two types of code sequences are used [13].

Figure 3.11 presents a general scheme of the spreading and modulation processes. As can be observed, the coded bits that have been mapped onto the physical channel are initially multiplied by the channel-isation code C_{ch}. The multiplication assumes that the values $\{0,1\}$ of the bits are mapped to levels $\{+1,-1\}$, respectively. The code sequence has a length of SF chips and, as a result, SF chips are obtained from each bit. Then, the chip rate W after the channelisation procedure is SF times the original bit rate. In UTRAN FDD, the chip rate is equal to 3.84 Mchips/s, which means that the chip period of the channelisation codes is always the same (i.e. 260.4 ns) no matter what spreading factor is being used.

After the channelisation procedure, the obtained sequence is multiplied by the scrambling code C_{scr}. This procedure does not modify the bandwidth of the signal, since the chip rate of the scrambling code is equal to the chip rate of the sequence after channelisation. Finally, the chips resulting from the scrambling procedure are modulated and transmitted to the antenna at the corresponding carrier frequency f_o. The final bandwidth of the signal at antenna BW will depend on the chip rate W, the modulation scheme and the pulse shaping roll-off.

The purpose of the channelisation process is to separate physical channels that are generated at the same transmitter. Thus, in the uplink direction, the channelisation codes allow the distinguishing of the different physical channels belonging to the same user. In turn, in the downlink direction, the channel-isation codes separate signals from different users served by the same node B.

In turn, the scrambling procedure allows the separation of the physical channels generated at differ-ent transmitters. Then, in the uplink direction, it allows the separation of signals coming from different users while in the downlink direction it allows the separation of signals coming from different nodes B. Note that, provided that two signals make use of different scrambling codes, they can be separated even if both make use of the same channelisation code.

3.3.2.1 Channelisation Codes

The channelisation codes used in UMTS belong to the family of OVSF (Orthogonal Variable Spreading Factor) codes. They are organised according to a code tree structure, as depicted in Figure 3.12, depending on the code length (i.e. the spreading factor SF). $C_{ch,SF,i}$ ($i = 0,\ldots, SF - 1$) represents the ith code with spreading factor SF. Note that there is a single code with $SF = 1$, two codes with $SF = 2$, four with $SF = 4$, and so on. The tree generation rule states that, for each code $\{x\}$ with spreading factor SF,

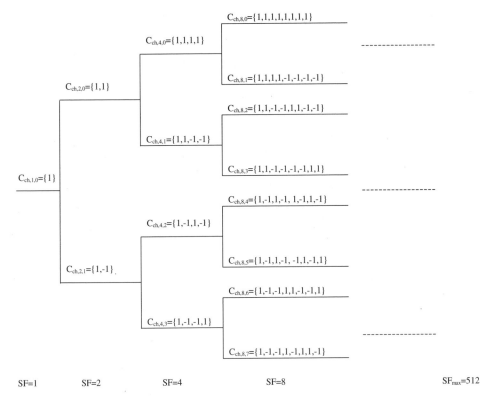

Figure 3.12 OVSF code tree

two descendant codes with spreading factor $2*SF$ are obtained: the first one, in the upper branch, is generated simply by repeating the predecessor code (i.e. $\{x,x\}$) and the second one, in the lower branch, is generated by repeating and inverting the predecessor code (i.e. $\{x,-x\}$). Although this generation procedure can be extended indefinitely, in practice the allowed values of spreading factor for UTRAN FDD range from 4 to 256 in the uplink and from 4 to 512 in the downlink direction.

Taking into account the generation procedure, the two descendant codes that are obtained from a given code are orthogonal, but they do not keep orthogonality with their predecessor (e.g. codes $C_{ch,4,0}$ and $C_{ch,4,1}$ are orthogonal, but they are not orthogonal with $C_{ch,2,0}$). As a result, a pair of codes will be orthogonal only if they are located in different branches of the code tree. This consideration poses a restriction on the way OVSF codes must be allocated, since whenever a code is allocated to a given transmission neither its predecessors nor its descendants can be used simultaneously.

Due to the restrictions in terms of orthogonality, when trying to allocate OVSF codes for n simultaneous transmissions, not all the combinations of spreading factors are feasible. In particular, a combination is feasible provided that it retains the following inequality, known as Kraft's inequality [17]:

$$\sum_{j=1}^{n} \frac{1}{SF_j} \le 1 \tag{3.1}$$

where SF_j is the spreading factor of the code allocated to the jth transmission. As an example, if $n = 5$, it would not be possible to allocate all the transmissions with $SF = 4$, but it would be possible to allocate e.g. three with $SF = 4$ and two with $SF = 8$.

Another important characteristic of OVSF codes is that orthogonality is only guaranteed provided that the code sequences are perfectly synchronised at chip level at the receiver side. Furthermore, the crossed correlation between OVSF codes may reach very high values in some situations depending on the relative delay between code sequences. As an example, note that $C_{ch,4,1}$ and $C_{ch,4,3}$ are orthogonal, but if $C_{ch,4,3}$ is received one chip later, it turns into the same sequence as $C_{ch,4,1}$, so transmissions with the two codes cannot be distinguished. This is the reason why OVSF codes are only used to separate physical channels that have been generated in the same transmitter.

3.3.2.2 Scrambling codes

Scrambling codes are used after the channelisation process to make signals that originate in different transmitters distinguishable, so they allow the separation of signals from different users in the uplink and from different nodes B in the downlink. Scrambling codes in UTRAN FDD are complex codes, so the scrambling procedure consists of a complex multiplication symbol by symbol, as depicted in Figure 3.13. At the receiver side, the received sequence is de-scrambled by multiplying it with the complex conjugate of the scrambling code sequence. It is worth mentioning that the channelisation step is done separately for phase I and quadrature Q components, and the resulting chips from both components are combined to form complex symbols prior to the scrambling process.

Note that before scrambling, the I and Q components might have different amplitudes (for example, because they carry different physical channels with different spreading factors). After modulation, this would result in signals with a high crest factor (i.e. the peak-to-average power ratio) thus leading to poor efficiency of the power amplifiers. Thanks to complex scrambling, the amplitudes of the I and Q components are equalised and the phase rotations between consecutive chips are limited to 90°. As a result, the crest factor of the modulated signal is reduced [18].

In the downlink direction, a total of 8912 scrambling codes are used. They are generated through the combination of two real sequences into a complex sequence. Each real sequence is a segment of 38 400 chips obtained from a Gold sequence. Then, the scrambling code fits exactly one 10 ms radio frame and is repeated periodically frame to frame. As a difference from OVSF codes, Gold codes are not orthogonal but present low levels of both cross-correlation and auto-correlation, which makes them suitable for distinguishing transmissions even if they are not synchronised chip to chip.

The 8912 downlink scrambling codes are divided into 512 sets, each one having one primary code and 15 secondary codes. Every node B must have one and only one primary scrambling code, which carries the CPICH and the P-CCPCH physical channels. If necessary, additional secondary scrambling codes from the set to which the primary code belongs can also be allocated to node B. In this case, physical channels from users belonging to the same cell are distinguishing by both the scrambling and the OVSF

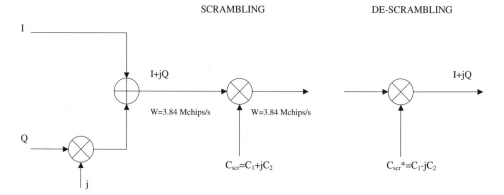

Figure 3.13 Scrambling and de-scrambling procedures

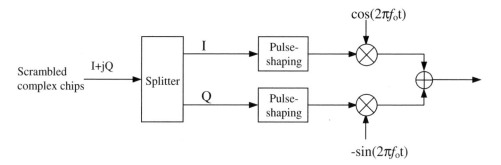

Figure 3.14   Modulation scheme in UTRAN FDD

code. This solution avoids the situation where the lack of OVSF codes limits the cell capacity, but on the other hand it represents an increase of the interference, since scrambling codes are not orthogonal. As a result, it is more usual to have cells with only a single scrambling code, at least in the initial deployment phase of UMTS.

In the uplink case, two types of scrambling code sequences are defined: long and short codes. Long scrambling codes are, as in the downlink case, generated by the combination of two real 38 400 chips segments of Gold codes into a complex sequence. In turn, short scrambling codes are generated by combining two codes from the family of periodically extended S(2) codes into a complex sequence. In this case, the code length is 256 chips, which are repeated periodically until reaching the 38 400 chips required in order to scramble a radio frame. There are 2^{24} long and 2^{24} short scrambling code sequences that can be used in the uplink and that are assigned by higher layers on a connection by connection basis.

3.3.2.3 Modulation

The modulation process converts the base band scrambled chips into physical signals that can be transmitted through the antenna at the corresponding carrier frequency. The modulation scheme is QPSK [13], as shown in Figure 3.14. Also, for the HS-PDSCH channel, 16-QAM modulation can be used.

The transmit pulse shaping filter is a root-raised cosine with roll-off $\alpha = 0.22$ in the frequency domain [11][12], which results in a total bandwidth at the antenna output of 3.84 Mchips/s*$(1 + \alpha) = 4.68$ MHz.

For the modulated signal, it is a requirement that the Adjacent Channel Leakage power Ratio (ACLR), which measures the ratio between the spectrum at the allocated channel and at the adjacent channel, should be at least 33 dB for a separation of ± 5 MHz and 43 dB for ± 10 MHz. In turn, the frequency stability should be within ± 0.1 ppm for the mobile terminal and within ± 0.05 ppm for node B [11][12].

After modulation, signals are amplified until reaching the desired transmitted power level, which depends on the power control mechanism. For the mobile terminal, different power classes are defined, ranging from 21 to 33 dBm, and the minimum power level is -50 dBm. The transmitted power must be adjusted in steps of 1, 2 or 3 dB at the mobile terminal, while the base station should be able to adjust the power level in steps of 1 dB mandatory and 0.5 dB optional [11][12].

3.3.3 DOWNLINK TRANSMIT DIVERSITY SCHEMES

The flow of symbols that result from the modulation process described in the above section are finally power amplified and sent to the antenna for transmission. In the downlink direction, it is possible to apply different diversity schemes with two transmit antennas. Essentially, transmit diversity schemes are classified into closed loop and open loop, depending on whether feedback from the UE is required or not, respectively. These two schemes cannot be used simultaneously on the same radio link. The support of downlink transmit diversity is mandatory in the UE and optional for UTRAN [7][19].

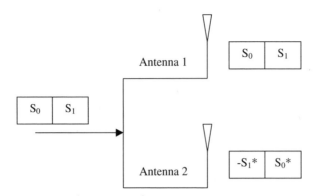

Figure 3.15 STTD downlink transmit diversity with QPSK modulation

With respect to the open loop downlink transmit diversity, it makes use of a Space Time block coding based Transmit Diversity (STTD). The main feature of this method is that, by encoding the symbols transmitted in the two antennas, it is possible for the receiver at the UE to decode separately the two signals and achieve with only one antenna a similar performance to when two antennas were used in the receiver.

For the case of QPSK modulation, the STTD scheme is shown in Figure 3.15, where the consecutive complex symbols S_0 and S_1 are transmitted. In antenna 1, the same symbols are transmitted, while in antenna 2 the sequence of transmitted symbols is $-S_1^*$ and S_0^*. As a result, the symbol sequences transmitted in the two antennas are orthogonal. The STTD scheme for the case of 16-QAM modulation used in HS-PDSCH would be the same, but in this case each symbol would be 4 bits long.

The STTD scheme can be applied in all the downlink physical channels except the synchronisation channel (SCH), where another type of open loop downlink transmit diversity is used. It is the so-called Time Switched Transmit Diversity (TSTD), and consists of sending the physical channels alternatively through antenna 1 or antenna 2 on a time slot to time slot basis.

When an associated uplink channel is available (i.e. in dedicated physical channels as well as in PDSCH, HS-PDSCH and the DL control channel of the CPCH), it is possible to apply the closed loop mode transmit diversity. In this case, as depicted in Figure 3.16, the scrambled and modulated symbols

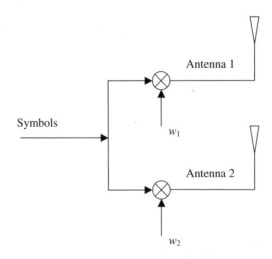

Figure 3.16 Closed loop mode transmit diversity scheme

are sent to the two antennas after weighting them with two complex values w_1 and w_2. In the uplink direction, the UE indicates the information on the power and phase values that should be assigned to each weight in order to maximise the downlink received power. This information is obtained from the estimation of the channel corresponding to each transmit antenna. The differentiation between both channels is possible because different symbol sequences are used by the two antennas when transmitting the CPICH channel.

Two different modes of closed loop transmit diversity are defined depending on how the information is fed back from the UE. In mode 1, the UE simply sends phase rotations for w_2, while w_1 from antenna 1 remains constant. In mode 2, however, it sends information regarding the setting of both weights in terms of amplitudes and phase difference.

3.3.4 ORGANISATION OF THE PHYSICAL CHANNELS

As shown in Figures 3.8 and 3.9, after the processing of the transport blocks in the uplink and downlink directions, a set of F coded bits is mapped onto specific physical channels. In this mapping process, the data bits are combined with other physical layer control bits and organised in temporal structures conceived to support physical procedures such as synchronisation or power control. This organisation is different depending on the characteristics and requirements of each specific physical channel, as it will be described in the following section for the main physical channels [7].

3.3.4.1 Dedicated Physical Channel

The information from DCH transport channels is mapped onto the Dedicated Physical Data Channel (DPDCH), which is multiplexed together with the Dedicated Physical Control Channel (DPCCH) that carries physical layer control information. Such multiplexing is done differently in the uplink and downlink cases.

Uplink The organisation of the dedicated physical channel in the uplink direction is shown in Figure 3.17. The 10 ms radio frame is subdivided in 15 time slots, each one having 2560 chips. Here, the concept of time slot is not at all the same as in TDMA systems, since in UTRAN FDD there is no user multiplexing from slot to slot in the dedicated physical channels. On the contrary, each time slot corresponds to a power control cycle, which means that the power amplifier should be able to modify the transmitted power on a time slot basis according to the commands that are received in the downlink direction. This ensures a power control rate of 1500 Hz.

The DPDCH and DPCCH are multiplexed in the I and Q components, respectively, which guarantees the existence of a continuous transmission even when there are no data to be sent and avoids audible interference in audio equipments close to the terminal, which would be present during discontinuous transmission instants if DPDCH and DPCCH were time multiplexed.

The DPCCH uses a constant spreading factor of 256, which is equivalent to 10 bits per slot. Several slot formats are defined depending on the amount of bits devoted to each of the existing fields, as shown in Table 3.4 [7]. Some modifications to these formats exist where compressed mode is used, as is explained on page 79.

The DPCCH contains four fields in each time slot:

- Pilot bits transmit a specific bit sequence different from slot to slot and are used to perform channel estimation for coherent detection.
- TFCI (Transport Format Combination Indicator) bits. This field informs the receiver about the selected TFC corresponding to the CCTrCH channel that is being transmitted in the DPDCH. It is an optional field, and when it is not present the receiver should apply blind transport format detection unless the physical channel corresponds to a fixed bit rate service, since in this case the transport format is always the same.

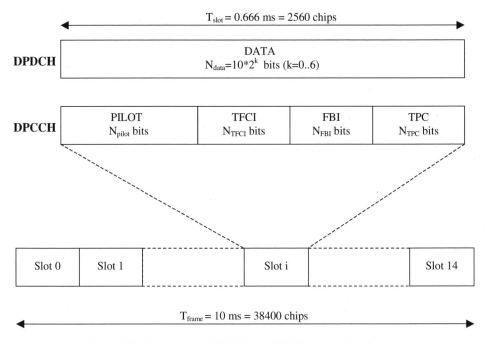

Figure 3.17 Organisation of DPDCH and DPCCH in the uplink direction

- FBI (Feedback Information) bits. This field is used to support two optional downlink transmission techniques that require some feedback from the UE in the uplink direction. They are the closed loop mode transmit diversity (in which the UE passes information about the appropriate weights to be applied at the diversity antennas in the downlink) and the SSDT (Site Selection Diversity Transmission), in which a UE in soft handover indicates the cell to be used for transmission in the downlink [19].
- TPC (Transmit Power Control) bits. This field is used to indicate whether the downlink direction should increase or reduce the transmission power during power control procedure.

With respect to the DPDCH, a total of seven time slot formats are possible, depending on the spreading factor being used, which ranges from 4 to 256. Taking into account the fixed chip rate $W = 3.84$ Mchips/s, this corresponds to a variation in the number of bits per slot from 640 to 10 or, equivalently, a channel bit rate from 960 to 15 kb/s, as shown in Table 3.5.

Table 3.4 Time slot formats for the uplink DPCCH without compressed mode

Format	N_{pilot} (bits)	N_{TFCI} (bits)	N_{FBI} (bits)	N_{TPC} (bits)
0	6	2	0	2
1	8	0	0	2
2	5	2	1	2
3	7	0	1	2
4	6	0	2	2
5	5	2	2	1

Table 3.5 Slot formats for the uplink DPDCH

SF	N_{data} (bits)	Bits per frame	Channel bit rate (kb/s)
4	640	9600	960
8	320	4800	480
16	160	2400	240
32	80	1200	120
64	40	600	60
128	20	300	30
256	10	150	15

Transmissions at variable bit rate can be obtained in the DPDCH channel simply by modifying the slot format or equivalently the spreading factor, whenever the transport format is changed at the MAC layer. The physical channel allows the performance of this operation from frame to frame, although in practice the time between variations will be given by the TTI of the corresponding transport channel, which is an integer number of frames.

In order to retain the quality requirement in terms of (E_b/N_o) when variable bit rate transmission is used, the *SIR* target should be modified depending on the bit rate due to the relationship that exists between both parameters $(E_b/N_o) = (W/R_b)*SIR$. As a result, if the bit rate R_b is increased by a factor of 2, the *SIR* requirement and consequently the transmitted power level must be increased by the same magnitude. This situation introduces the fact that the power control should be consistent with the selected transport format combination and vary the power accordingly. This is illustrated schematically in Figure 3.18, where four TTIs of a given uplink transport channel are shown. While the power of the DPCCH remains the same, the power of the DPDCH is adjusted depending on the selected transport format. Note that for the case TF0, no discontinuous transmission is applied since the DPCCH channel must be continuously transmitted.

The maximum uplink bit rate of 960 kb/s can be increased by means of multiple DPDCHs (i.e. multi-code transmission) for a single user. In this case, up to 6 DPDCH can be used, with $SF = 4$ in all of them. This results in a set of channel bit rates from 960 up to 5760 kb/s. In any case, only one DPCCH channel is transmitted.

Figure 3.19 shows the spreading process when a single DPDCH is used [13]. In this case, the channelisation code of the DPDCH depends on the spreading factor being used and is denoted as $C_{ch,SF,SF/4}$. In all the cases, it is located in the branch of the OVSF code tree that starts in code $C_{ch,4,1}$. In turn, the DPCCH always uses the channelisation code $C_{ch,256,0}$, located at the end of the upper branch of the OVSF code tree. The coefficients β_c and β_d are used to establish the relative amplitude level from DPDCH with respect to DPCCH. They are between 0 and 1, where 0 means that the corresponding

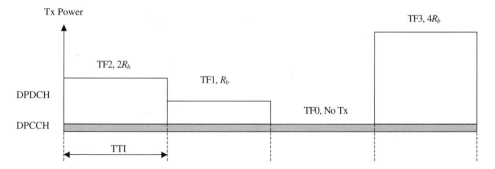

Figure 3.18 Power control and variable bit rate transmission in the uplink direction

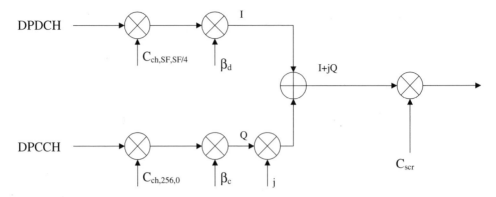

Figure 3.19 Channelisation and scrambling for a single uplink DPDCH

channel is switched off, and at every instant at least one of the two should take the value 1. The values are quantised with 4 bits, leading to 16 levels. They are changed from frame to frame depending on the selected TFC for the corresponding transport channel.

Figure 3.20 shows the channelisation and scrambling schemes when multicode transmission is applied. In this case, a maximum of three DPDCH with $SF = 4$ are transmitted together in either the I or Q components [13]. Note that the same channelisation OVSF codes are used for the DPDCH channels in the I and Q components. This is not a problem since in this case the orthogonality is assured because of the orthogonality of the I and Q components in the QPSK modulation. The allocated OVSF codes occupy the three lower branches of $SF = 4$, as depicted in Figure 3.12. Notice that the upper code $C_{ch,4,0}$ cannot be used in the Q component since it is located in the same branch as the code $C_{ch,256,0}$ used for DPCCH. On the other hand, note that the same coefficient β_d is used for all six DPDCH channels.

Downlink Figure 3.21 presents the structure of the downlink dedicated physical channel (DPCH). As in the uplink case, it is organised into 10 ms radio frames divided into 15 time slots, each one corresponding to a power control cycle [7]. However, and as a difference from the uplink case, the data and physical control channels (i.e. DPDCH and DPCCH) are time multiplexed within each time slot. Note that if code multiplexing were used, each DPCH would require two channelisation codes, and thus be less efficient from the code utilisation point of view.

In each time slot, there are two data fields containing the DPDCH and three for the DPCCH whose contents are similar to those in the uplink, i.e. the TPC contains uplink power control commands, the TFCI notifies the transport format combination being used and the pilot bits are used to perform channel estimation. As in the uplink case, the power of the DPDCH and DPCCH fields is not the same. In particular, three values (PO1, PO2 and PO3) define the power offset with respect to the DPDCH for TFCI, TPC and pilot bits, respectively (see Figure 3.22).

Another difference compared to the uplink case is that in downlink both the I and Q components transmit bits from the DPDCH and the DPCCH. Specifically, the bits in a given slot are grouped into QPSK symbols, so that odd bits are transmitted through the I component and even bits through the Q component. Then, for a spreading factor $SF = 512/2^k$ ($k = 0, 1, \ldots, 7$ corresponding to SF from 4 to 512) the number of bits per slot is $10*2^k$.

Table 3.6 presents the available time slots formats for the downlink dedicated physical channel [7]. These formats are modified where compressed mode is used. The selection of which format to use is done by means of higher layer signalling, and it can be reconfigured during a connection lifetime. The maximum channel bit rate that can be achieved with a single physical channel is 1872 kb/s, but it is possible to increase this bit rate by using multi-code transmission with different physical channels in parallel, all of them having the same spreading factor. In this case, the DPCCH fields are only transmitted

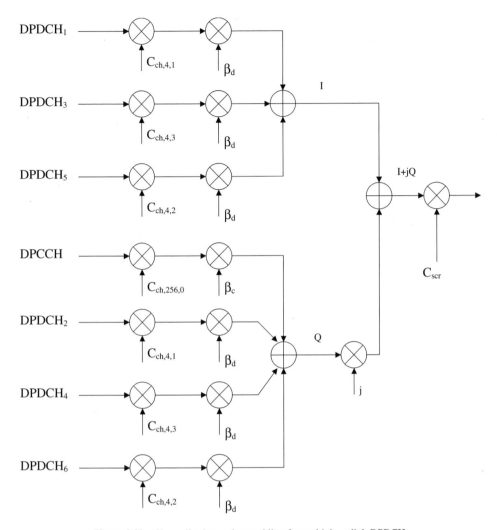

Figure 3.20 Channelisation and scrambling for multiple uplink DPDCHs

with the first physical channel. In the rest, discontinuous transmission is applied during these fields as shown in Figure 3.22.

The allocated channelisation code for the downlink dedicated physical channel is selected among the OVSF codes with the spreading factor that corresponds to the highest bit rate of the TFCS of the specific transport channel. Then, the allocated spreading factor remains constant unless the physical channel is reconfigured by means of higher layer signalling and by making use of the corresponding RRC procedure. As explained in Section 3.3.1, the variable bit rate for different transport formats is implemented with discontinuous transmission, switching off the transmission during the DPDCH fields in certain time slots. Note that, if variable spreading factor were used in the downlink, the positions of the DPCCH fields would change with the spreading factor being used. Consequently, it would be more difficult for the receiver to detect the corresponding TFC.

Figure 3.23 shows the channelisation and scrambling process for the downlink dedicated physical channel. The QPSK symbols are serial to parallel converted, thus obtaining the I and Q components, and the channelisation step is done separately to both components by multiplying them by the same

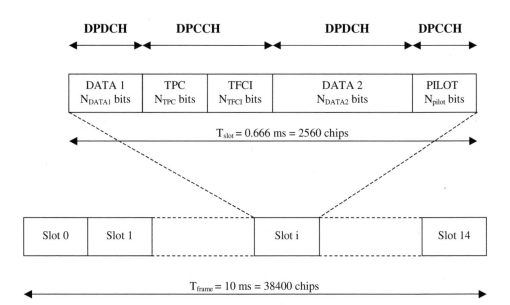

Figure 3.21   Organisation of DPDCH and DPCCH in the downlink direction

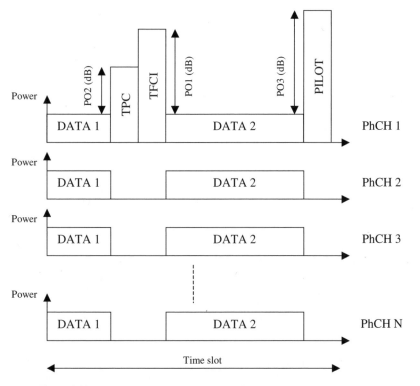

Figure 3.22   Transmission with multiple downlink dedicated physical channels

Table 3.6 Time slot formats for the downlink dedicated physical channel without compressed mode

Format	SF (bits)	N_{TPC} (bits)	N_{TFCI} (bits)	N_{pilot} (bits)	N_{data1} (bits)	N_{data2} (bits)	DPDCH channel bit rate (kb/s)
0	512	2	0	4	0	4	6
1	512	2	2	4	0	2	3
2	256	2	0	2	2	14	24
3	256	2	2	2	2	12	21
4	256	2	0	4	2	12	21
5	256	2	2	4	2	10	18
6	256	2	0	8	2	8	15
7	256	2	2	8	2	6	12
8	128	2	0	4	6	28	51
9	128	2	2	4	6	26	48
10	128	2	0	8	6	24	45
11	128	2	2	8	6	22	42
12	64	4	8	8	12	48	90
13	32	4	8	8	28	112	210
14	16	8	8	16	56	232	432
15	8	8	8	16	120	488	912
16	4	8	8	16	248	1000	1872

OVSF code. Afterwards the complex scrambling process is done and the resulting flow is amplified to the required power level before combining it with the rest of channels in the cell, which usually make use of the same scrambling code and are separated by means of the channelisation codes.

With respect to the timing relation between uplink and downlink dedicated physical channels, it is established that the uplink frame must start 1024 chips after the reception of the first path of the downlink frame.

Compressed Mode One important difference between CDMA and TDMA techniques is that in the former case the mobile transmits and receives continuously during all the frames and slots, while in TDMA the terminal is only active during one slot in a frame. Then, during the rest of the slots the receiver can be tuned to other frequencies and perform measurements. With CDMA, however, it is not possible to make interfrequency measurements in the mobile terminal unless the receiver structure is duplicated. Interfrequency measurements include other UTRAN cells operating at different frequencies

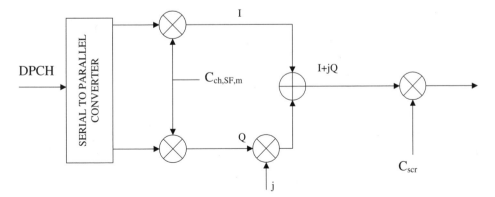

Figure 3.23 Channelisation and scrambling for the downlink dedicated physical channel

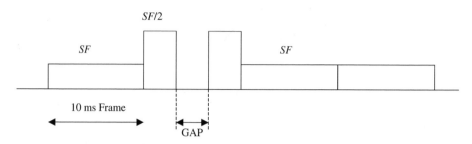

Figure 3.24 Discontinuous transmission during compressed mode

as well as GSM cells in order to enable the possibility of handover and cell reselection between UMTS and GSM systems. The coexistence between both systems has been taken into account in the UMTS radio interface specifications by means of *compressed mode transmission*, which is schematically presented in Figure 3.24. This allows making interfrequency measurements with a single receiver simply by switching off the transmission during certain slots in the downlink direction and also, in some cases, in the uplink. Compressed mode in the uplink direction is required when making measurements from GSM1800, since in this case the downlink GSM frequency is too close to the uplink UMTS frequency to perform simultaneous transmission and reception.

Compressed frames can occur periodically or on demand, depending on the type of required interfrequency measurements. The maximum length of the transmission gap in one frame is 7 slots, and it can be located either at the middle of a frame or between two frames. When using compressed mode, there are special slot formats for both the uplink and the downlink direction in order to compensate for the absence of transmission in data and control channels.

There exist three alternatives to introduce transmission gaps in compressed mode without loss of information [15]:

- Spreading factor reduction. This method is based on transmitting slots of compressed frames with half the spreading factor of the non-compressed frame. In this way the instantaneous bit rate is doubled in certain slots, which means other idle slots can be left without reducing the average bit rate along the frame. This situation is depicted in Figure 3.24. However, the reduction in the spreading factor is done at the expense of a higher power in the slots outside the transmission gap, which may cause higher interference patterns to other terminals. In terms of channelisation code allocation, the OVSF code of the immediate preceding layer is used during compressed frames (i.e. for the downlink case, if the code $C_{ch,SF,m}$ is used in normal frames, the code $C_{ch,SF,\lfloor m/2 \rfloor}$ is used in compressed frames).
- Puncturing. Taking into account the redundancy that is added in the coding process, it is possible to increase the bit rate simply by puncturing some of the data bits as done in the rate matching procedure that was presented in Section 3.3.1. This method is only followed in the downlink direction.
- Higher layer scheduling. With this method, the responsibility relays on the MAC layer, which selects appropriate transport format combinations during compressed frames requiring the transmission of a lower number of bits.

3.3.4.2 Common Pilot Channel (CPICH)

The Common Pilot Channel (CPICH) is a physical channel that does not carry data from higher layers, so it is not mapped to any transport channel. On the contrary, it transmits a pre-defined bit sequence that is used for channel estimation. It constitutes the channel estimation reference for the rest of the downlink common control channels, and it can also be the reference for the downlink dedicated and shared physical channels [7]. The frame structure is shown in Figure 3.25.

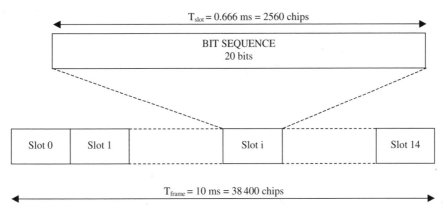

$T_{slot} = 0.666$ ms $= 2560$ chips

BIT SEQUENCE
20 bits

| Slot 0 | Slot 1 | | Slot i | | Slot 14 |

$T_{frame} = 10$ ms $= 38\,400$ chips

Figure 3.25 Frame structure of the CPICH channel

The CPICH has a fixed bit rate of 30 kb/s, corresponding to a spreading factor 256, with 20 bits per time slot. When no downlink transmit diversity is applied, the CPICH simply transmits a sequence of all 0s. However, with transmit diversity, a different bit sequence is transmitted in the second antenna: the sequence 0011110000111000011 is transmitted in even slots and the inverted sequence in odd slots. In this way the mobile terminal has the capability to discriminate between the channel in one or the other antenna and to act accordingly in closed loop mode transmit diversity procedure [19].

There are two types of CPICH channels, namely the primary and the secondary CPICH. The primary CPICH is always transmitted with the channelisation code $C_{ch,256,0}$ and with the primary scrambling code (see Section 3.3.2.2), so only a single primary CPICH can exist in each cell. The primary CPICH is the phase reference for all the downlink common control channels and possibly for dedicated and shared channels. Also, in specific situations (e.g. operations with narrow antenna beams), it is possible to have other secondary CPICH channels that are transmitted with arbitrary channelisation codes of $SF = 256$ and with either the primary or secondary scrambling codes. A secondary CPICH can only be the phase reference for downlink dedicated or shared physical channels.

The transmission power of the primary CPICH is broadcast in the BCH channel, so that it is possible for a mobile terminal to measure the path loss with respect to a given cell simply by measuring the received power of the CPICH and comparing it with the transmitted power. Furthermore, the CPICH is the reference channel used for cell selection and handover procedures. Specifically, the mobile terminal measures the received power of the CPICH channel from different cells (called the Monitored Set) and the one with the highest level is selected as the serving cell. The decision to add or remove cells from the Active Set (i.e. the set of cells to which the mobile is simultaneously connected during soft handover) is taken depending on the ratio between the received CPICH level of the best cell and each individual cell. This feature of the pilot channel provides a way to balance the load between adjacent cells simply by modifying the transmitted power of the CPICH, which as a matter of fact modifies the coverage area of the cell. Then, by increasing the pilot power, the cell will capture users from adjacent cells while, by decreasing the pilot power, users connected to the cell will tend to handover to the adjacent cells.

It is worth mentioning that, in order to be able to measure the CPICH of a given cell, the mobile needs to know the primary scrambling code in advance. The determination of the primary scrambling code is done in the synchronisation procedure, as will be explained in Section 3.3.4.4.

3.3.4.3 Common Control Physical Channels

Primary CCPCH (P-CCPCH)   The broadcast transport channel is mapped onto the Primary Common Control Physical Channel, whose frame structure is depicted in Figure 3.26. Like the dedicated physical

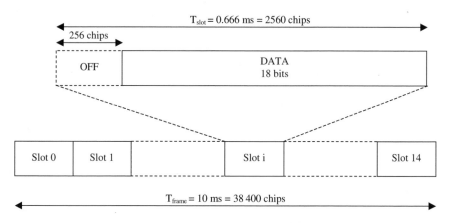

Figure 3.26   Frame structure of the Primary CCPCH

channel, it is organised into 10 ms frames each one with 15 time slots. At the beginning of each slot there is an interval of 256 chips (i.e. 0.0666 ms) in which the transmission of the P-CCPCH is switched off [7]. This interval corresponds with the instants in which the synchronisation channel SCH is transmitted, as will be shown in Section 3.3.4.4.

The P-CCPCH is a downlink fixed bit rate channel with a spreading factor of 256, and the corresponding channelisation code is $C_{ch,256,1}$. When more than one scrambling code is allocated in the cell, the P-CCPCH is always transmitted with the primary scrambling code (see Section 3.3.2.2). A total of 18 bits are transmitted in each slot, which yields a raw channel bit rate of 27 kb/s.

No physical layer control information is sent in the P-CCPCH channel, since no power control is applied (i.e. no TPC bits are needed), the transport format is always the same (i.e. no TFCI indication is present) and, finally, there are no pilot bits because of the presence of the CPICH channel, which is transmitted in parallel and that allows for channel estimation.

Note that, in order to be able to decode the broadcast information, the mobile terminal needs to know the primary scrambling code. This can be done with the aid of the SCH (see in Section 3.3.4.4).

Secondary CCPCH (S-CCPCH)   The downlink FACH and PCH transport channels are mapped onto the Secondary Common Control Physical Channel, whose frame structure is depicted in Figure 3.27. In each time slot there exist three fields: the data field that carries the coded bits from the transport

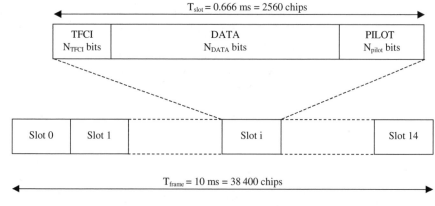

Figure 3.27   Frame structure of the Secondary CCPCH

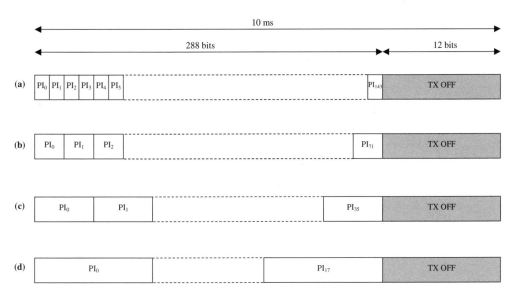

Figure 3.28   Structure of the four configurations of the PICH channel: (a) with 144 PIs, (b) with 72 PIs, (c) with 36 PIs, (d) with 18 PIs

channel; the TFCI bits, indicating the selected TFC; and finally the pilot bits, devoted to channel estimation. There are no TPC bits since no closed loop power control is applied with this channel. As in the downlink dedicated physical channel, different time slot formats exist depending on the number of data bits, related with the channel bit rate, and on whether the TFCI and pilot bits are present or not.

The S-CCPCH supports multiple transport format combinations, which turn into multiple channel bit rates. As a result, the spreading factor ranges from 256 down to 4 in powers of 2. Different data rates are possible for the same spreading factor depending on the number of bits allocated to the data field. Specifically , the minimum coded data rate is 15 kb/s (corresponding to $SF = 256$ and $N_{data} = 10$ bits) and the maximum is 1908 kb/s (corresponding to $SF = 4$ with $N_{data} = 1272$ bits).

The channelisation code for a given S-CCPCH channel is not fixed but depends on the specific spreading factor and on the current OVSF code allocation. On the other hand, when multiple scrambling codes are present in a given cell, one S-CCPCH can be allocated either in the primary or in a secondary scrambling code. If it carries the PCH channel, it must be allocated in the primary scrambling code.

Those S-CCPCH channels that carry the PCH transport channel are associated with a Paging Indicator Channel (PICH). The PICH is a constant bit rate channel with $SF = 256$ and scrambled with the primary scrambling code of the cell that includes 288 bits per radio frame, leaving 12 bits at the end of the radio frame during which transmission is switched off. Depending on the number of bits per Paging Indicator (PI), the PICH may carry 18, 36, 72 or 144 indicators per frame (see Figure 3.28). Whenever a paging indicator is set to 1, this means that the mobiles associated with this indicator should read the corresponding S-CCPCH frame. In this way, the terminals simply need to monitor their indicator in the PICH channel and decode the paging messages of the S-CCPCH channel only when needed. The associated S-CCPCH is transmitted 7680 chips (i.e. 2 ms) after the end of the transmission of the PICH frame.

3.3.4.4 Synchronisation Channel (SCH)

The Synchronisation Channel (SCH) is the first physical channel that a mobile terminal detects when it is switched on. By means of this detection the terminal discovers how many networks are available and

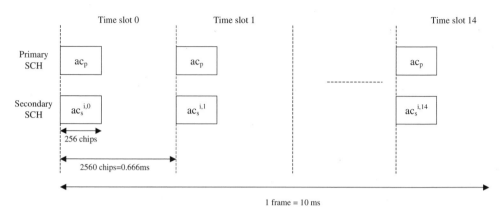

Figure 3.29   Organisation of the synchronisation channel

achieves the frame and slot synchronisation with each of them. Furthermore, the synchronisation channel eases the process of detecting the primary scrambling code of the cell, prior to be able to decode the corresponding BCH transport channel.

As depicted in Figure 3.29, the synchronisation channel consists of a primary SCH and a secondary SCH organised into 10 ms radio frames, subdivided into 15 time slots [7]. Both channels are transmitted simultaneously, once per time slot, during the initial 256 chip period of the slot in which the P-CCPCH channel that carries the BCH is switched off (see Figure 3.26).

The synchronisation procedure operates in the following steps [19], as depicted schematically in Figure 3.30:

(a) Time slot synchronisation. The primary SCH transmits a generalised hierarchical Golay sequence c_p, which is common for all the cells and therefore it is known to the mobile terminal a priori. The sequence is multiplied by a symbol a (see Figure 3.29), whose value depends on whether or not STTD is applied in the P-CCPCH channel, being $+1$ if STTD is used and -1 if it is not used. Sequence c_p has 256 chips and identical real and imaginary parts. Its autocorrelation function presents a peak for $t = 0$, so that it is easy to detect it by means of the correlation of the received signal with the expected c_p sequence. The output of the correlator will then contain in one time slot period as many peaks as cells are detected (see Figure 3.31), in addition to the different replicas of the same cell due to multi-path. The separation between peaks of the same cell indicates the time slot duration. Consequently, after detecting the primary SCH, the mobile terminal has acquired both the chip level and the slot level synchronisation with each one of the detectable cells.

(b) Frame synchronisation. In order to achieve synchronisation at the frame level, the mobile terminal makes use of the secondary SCH. There exist 16 secondary synchronisation codes $c_{s,1}$, $c_{s,2}$, $c_{s,3}$, ..., $c_{s,16}$ and, like c_p, their autocorrelation functions present peaks for $t = 0$ in order to be detected through correlation. A total of 64 sequences are constructed with the 16 codes. Each sequence

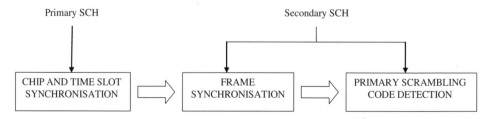

Figure 3.30   Steps of the synchronisation procedure

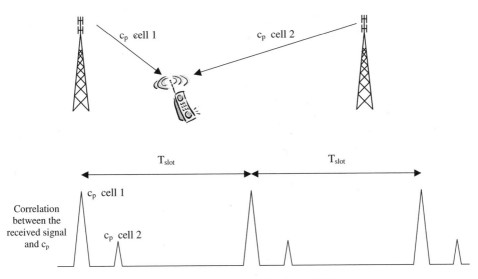

Figure 3.31 Detection of the primary SCH

V_i $(i = 0, 1, \ldots, 63)$ is composed of 15 codes $c_s^{i,0}$, $c_s^{i,1}$, $c_s^{i,2}$, $\ldots$, $c_s^{i,14}$, where $c_s^{i,j}$ is the code transmitted during time slot j $(j = 0, 1, \ldots, 14)$ and belongs to one of the 16 secondary synchronisation codes (see Figure 3.29). As an example, one of the 64 available sequences is given by:

$$V_0 = \{c_s^{0,0}, c_s^{0,1}, \ldots, c_s^{0,14}\} = \{c_{s,1}, c_{s,1}, c_{s,2}, c_{s,8}, c_{s,9}, c_{s,10}, c_{s,15}, c_{s,8}, c_{s,10}, c_{s,16}, c_{s,2}, c_{s,7}, c_{s,15}, c_{s,7}, c_{s,16}\}$$

The mobile terminal must correlate the received secondary SCH with the 16 possible synchronisation codes in order to detect the transmitted sequence V_i in a given frame. The 64 sequences V_i have the important property that none of them can be generated by means of a cyclical rotation of another one. As a result, once the terminal has detected the specific sequence the position of the first time slot in the frame is known and the frame level synchronisation has been acquired.

Continuing with the previous example with sequence V_0, let us assume that the mobile starts decoding the secondary SCH in an unknown time slot and detects the following sequence along 15 consecutive time slots:

$$R = \{c_{s,8}, c_{s,9}, c_{s,10}, c_{s,15}, c_{s,8}, c_{s,10}, c_{s,16}, c_{s,2}, c_{s,7}, c_{s,15}, c_{s,7}, c_{s,16}, c_{s,1}, c_{s,1}, c_{s,2}\}$$

Note that R is a cyclical rotation of V_0 starting at slot $j = 3$. Consequently, the unknown time slot where the mobile started the reception of the secondary SCH must be time slot 3 and the new frame starts then after 12 time slots (i.e. with code sequence $c_{s,1}$), which concludes the synchronisation at the frame level.

(c) Primary scrambling code detection. The detection of the sequence V_i used in the secondary SCH is also needed for detecting the primary scrambling code of the cell. There exists a total of $512 = 64*8$ primary scrambling codes, as was explained in Section 3.3.2.2, so each one of the 64 sequences V_i is mapped with a set of 8 primary scrambling codes. As a result, once the terminal has identified the sequence it simply has to correlate the received signal at the P-CCPCH (in the channelisation code $C_{ch,256,1}$) with each of the 8 possible primary scrambling codes and select the one with the highest correlation. After this procedure, the terminal will be able to decode the BCH transport channel and receive all the information required to attach to the network.

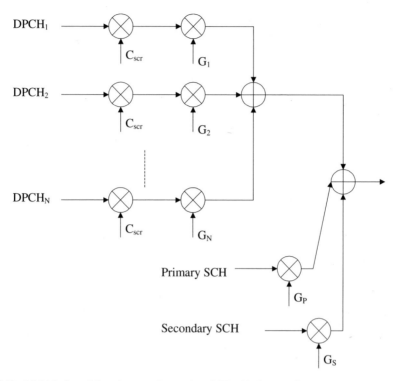

Figure 3.32 Multiplexing of the primary and secondary SCH with the rest of the downlink physical channels

Neither the primary nor the secondary SCH are scrambled. They are simply multiplexed with the rest of the scrambled downlink channels, as depicted in Figure 3.32, and weighted to achieve the required power level. The SCH channels can be transmitted using TSTD (Time Switched Transmit Diversity) with two antennas. In this case, during even slots both primary and secondary SCH are transmitted with antenna 1 and during odd slots both channels are transmitted with antenna 2.

3.3.4.5 Physical Random Access Channel (PRACH)

The Physical Random Access Channel carries the RACH transport channel used for the transmission of short packets and initial messages in the uplink direction. It is the channel used by the terminal to start the communication with the network. Transmissions on the PRACH are based on the S-ALOHA/CDMA protocol and it operates in conjunction with the downlink AICH (Acquisition Indicator Channel).

The frame structure of both the PRACH and the AICH is shown in Figure 3.33. They are organised into 20 ms periods divided into 15 access slots (AS) of 1.333 ms, so that each period corresponds to two 10 ms radio frames and each access slot to two slots of the radio frame [7]. The AICH is time aligned with the P-CCPCH channel, so the first access slot AS0 starts at the same time as the first time slot of the P-CCPCH. In turn, the access slots of the PRACH are advanced 7680 or 12 800 chips (i.e. 1.5 or 2.5 access slots), with respect to the AICH. The specific value 7680 or 12 800 depends on the network configuration and particularly on the value of the *AICH_transmission_timing* parameter, broadcast by the cell.

An Access Slot defines the instants in which the transmission through the PRACH can start. Transmissions are done in two steps, namely access phase and message transmission, as described by the

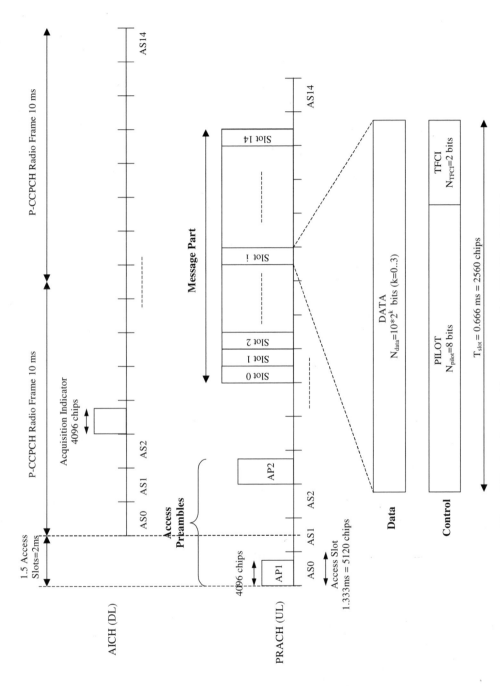

Figure 3.33 Organisation and frame structure of the PRACH and AICH channels

Physical Random Access Procedure. In the first step, and whenever a mobile needs to start the transmission, an access preamble is sent during a randomly selected Access Slot. The power of this access preamble is set taking into account an open loop power control scheme depending on the measured path loss and the total uplink interference in the cell. The path loss is computed as the difference between the transmitted and the received CPICH powers. Both the transmitted CPICH power and the uplink interference are broadcast by the cell. Then, the preamble initial power is given by:

$$\text{InitialPower} = \text{CPICH Tx power} - \text{CPICH Rx power} + \text{UL interference} + \text{Constant\_Value} \qquad (3.2)$$

Constant_Value is broadcast by the network and ranges between -35 and $-10\,\text{dB}$.

The access preamble is a 4096 chip complex sequence constructed as the product of a signature S_n and a scrambling code C_{scr}, as denoted by [13]:

$$C_{pre,n}(k) = S_n(k) * C_{scr}(k) * e^{j\left(\frac{\pi}{4} + \frac{\pi}{2}k\right)} \qquad k = 0, 1, 2, \ldots, 4096 \qquad (3.3)$$

Note that the duration of the access preamble is shorter than the duration of the access slot in order to ensure that the preamble is always received within the time limits of the access slot taking into consideration the propagation delay.

With respect to the scrambling code $C_{scr}(k)$ of the access preamble, it is a real sequence obtained by taking the real part from one of the complex long scrambling sequences used in the uplink direction, as explained in Section 3.3.2.2. A total of 8192 scrambling code sequences exist that can be used in the PRACH channel. These sequences are divided into 512 groups of 16 codes and each group has a one-to-one correspondence with the downlink primary scrambling code of the cell. Consequently, a maximum of 16 scrambling codes (i.e. 16 PRACH channels) can be allocated to a cell. The specific codes being used are broadcast by the cell.

On the other hand, there exist a total of 16 signatures, corresponding to the 16 OVSF codes of $SF = 16$. The 4096 chips of the signature are achieved by means of 256 repetitions of the corresponding OVSF code. The signature of the access preamble is randomly selected by the MAC layer of the mobile terminal. Since the different signatures are orthogonal (provided that they are perfectly time aligned), it is possible to distinguish between transmissions of terminals that have selected different signatures depending on their relative propagation delays. However, if two users have selected the same signature, their transmissions will mutually interfere and they will be indistinguishable.

After the transmission of the access preamble in a given access slot, the mobile terminal listens to the same access slot of the AICH that carries the Acquisition Indicators (AI) for each of the signatures. The AI can take three values: 0 if no transmission has been detected in the corresponding signature; $+1$ if an access preamble has been detected in the signature thus indicating that the access phase has been completed successfully and that the message transmission can start; and -1 indicating a negative acknowledgement (e.g. if a preamble has been detected in one signature but the PRACH channel is already occupied by another user). The AICH channel uses a channelisation code with $SF = 256$ and it is always scrambled with the primary scrambling code. It only occupies 4096 chips of the total 5192 chips of the access slot.

When the AI takes the value -1, the mobile terminal must interrupt the access procedure and inform the corresponding MAC layer. Transmission is postponed by a random number of 10 ms intervals. However, if $AI = 0$ the terminal will continue with the access phase by transmitting a second access preamble in another access slot. The minimum separation between consecutive access preambles is set to either 3 or 4 access slots depending on the value of the *AICH_transmission_timing* parameter. The power of the second access preamble is increased with respect to the power of the initial preamble in a value ranging between 1 and 8 dB, which is broadcast by the network. This power ramping procedure continues with additional access preambles until receiving either a positive or a negative AI or until reaching the maximum number of preambles allowed, which again is configured by the network.

When the AI takes the value $+1$, the transmission of the message part starts 3 or 4 access slots after the last preamble (again depending on the *AICH_transmission_timing* parameter). The power of the message part is computed by adding to the power of the last preamble a magnitude between -5 and 10 dB that is broadcast by the network. This power is kept constant for the duration of the total message, since no closed loop power control procedure is conceived for the PRACH channel.

The message part may last either 10 or 20 ms depending on PRACH configuration and it is composed of a data part and a physical control part, like the uplink dedicated physical channel, as depicted in Figure 3.33. It is subdivided into 15 time slots just like the radio frame. The data part is transmitted in the I component and the spreading factor can be 32, 64, 128 or 256, so the number of bits per slot can be 80, 40, 20 or 10, respectively. In turn, the control part is transmitted in the Q component and the spreading factor is fixed to 256, corresponding to 10 bits per time slot. The control part only contains two fields: the pilot bits required for channel estimation and the TFCI indicating the transport format being used in the data part. The power ratio between the data and control parts is controlled in the same way as the power ratio between uplink DPDCH and DPCCH.

The spreading of the message part is done depending on the signature selected in the last preamble. As has been mentioned, each preamble signature S_n ($n = 0, 1, \ldots, 15$) corresponds to one OVSF code with $SF = 16$ (i.e. $S_n = C_{ch,16,n}$). Then, starting from this OVSF code, and as depicted in Figure 3.34, the channelisation code of the data part is selected depending on the allocated SF following the upper sub-tree, thus allocating the code $C_{ch,SF,SF*n/16}$. In turn, the channelisation code of the control part is selected by following the lower sub-tree, thus allocating the code $C_{ch,256,16n+15}$.

The same long scrambling code used for the access preamble is used for the scrambling code of the message part, but in this case both the real and imaginary parts of the complex code sequence are utilised.

Due to the random nature of this access mechanism, collisions may occur in the access phase resulting in a collision in the message part. This will happen whenever two terminals select the same signature in the same access slot and a single preamble is detected because of, for example, a capture effect.

In the PRACH access mechanism, it is also possible to establish some prioritisation among users by using Access Service Classes (ASC). Each ASC is formed by a set of access slots that are allowed to be used only by users belonging to this ASC. In this way, there may be users that enjoy a higher number of transmission opportunities than others depending on the ASC to which they belong. Furthermore, each

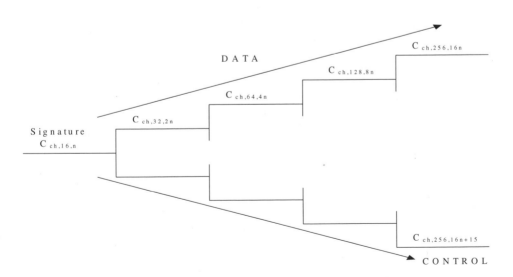

Figure 3.34 Channelisation code allocation for the message part of the PRACH

ASC contains a persistence probability p_i to be applied before sending an access preamble. This means that a random number between 0 and 1 is drawn and the preamble transmission is only carried out if the number is below p_i, otherwise, the terminal waits 10 ms before attempting again. By setting different persistence probabilities to the different ASC, it is then possible to control the load of each service class in the PRACH channel.

3.3.4.6 Physical Common Packet Channel (PCPCH)

The PCPCH is the physical channel that supports the CPCH transport channel devoted to the transmission of long sporadic messages without requiring the set-up of a dedicated channel. Transmission is based on the contention protocol DSMA/CD (Digital Sense Multiple Access with Collision Detection) and fast acquisition indication [7]. It can be seen as an extension of the PRACH channel with the inclusion of a closed loop power control mechanism and a collision detection procedure in order to avoid two users acquiring and transmitting simultaneously in the message part. Therefore, the frame structure of the PCPCH is essentially the same of the PRACH, organised in 20 ms periods each one subdivided into 15 access slots, as depicted in Figure 3.35. The uplink frame structure is advanced 1.5 or 2.5 access slots with respect to the downlink frame structure, depending on the value of the T_{cpch} parameter, which is broadcast by the network and plays the same role as the *AICH_transmission_timing* parameter for the PRACH.

Each PCPCH is identified by a scrambling code. There are 32 678 available codes that are divided into 512 groups with 64 codes. Each group has a one-to-one correspondence with the downlink primary scrambling code of the cell, so that each cell may have a maximum of 64 PCPCH channels.

There are two operation modes for the CPCH depending on whether Channel Assignment (CA) is active or not. If CA is active, it is the network that allocates a given PCPCH to the terminal, depending on the required minimum spreading factor. If CA is not active, however, it is the mobile terminal that selects the PCPCH, and in this case the maximum number of PCPCH channels per cell is limited to 16.

Transmissions over the CPCH are regulated by means of the CPCH Status Indicator Channel (CSICH), which passes information about the availability of the existing PCPCHs of the cell. This allows identification of those channels that are currently in use by other mobile terminals, so that transmissions in those channels are inhibited. When CA is active, the CSICH also passes information about the minimum available spreading factor in the PCPCH channels or, equivalently, the maximum bit rate that can be achieved. The CSICH is transmitted with a fixed $SF = 256$ occupying the last 1024 chips of each Access Slot in the downlink, with the same channelisation and scrambling codes as the AP-AICH (i.e. the equivalent channel in the PCPCH operation to the AICH of the PRACH), which occupies the first 4096 chips.

Then, when a terminal needs to start the transmission in the PCPCH, it must first read the CSICH and detect the channel and/or minimum spreading factor availability. When no PCPCHs are available or if the minimum spreading factor does not allow the transmission at the required bit rate, the procedure is aborted and the terminal waits for some time before attempting the transmission again. If there are PCPCHs available, however, and if the minimum spreading factor is acceptable, it initiates the Access Preamble phase, as in the PRACH case, but with the difference that each one of the 16 signatures has a correspondence either with a PCPCH channel (if CA is not active) or with a maximum bit rate (if CA is active).

In the case where CA is not active, the terminal randomly selects an available PCPCH and a signature and access slot corresponding to the selected PCPCH and transmits the Access Preamble. In turn, if CA is active, the signature and access slot are selected depending on the bit rate requirements, but no PCPCH is selected. The signatures are constructed in the same way as the signatures of the PRACH but with a different scrambling code. The initial power of the preamble is set by means of an open loop power control procedure with the same parameters as in the PRACH case. Similarly, the terminal monitors the AP-AICH downlink channel that contains the Acquisition Indicators (AI) for each of the signatures of each access slot. As long as $AI = 0$ for the selected signature, the mobile will reattempt to access a

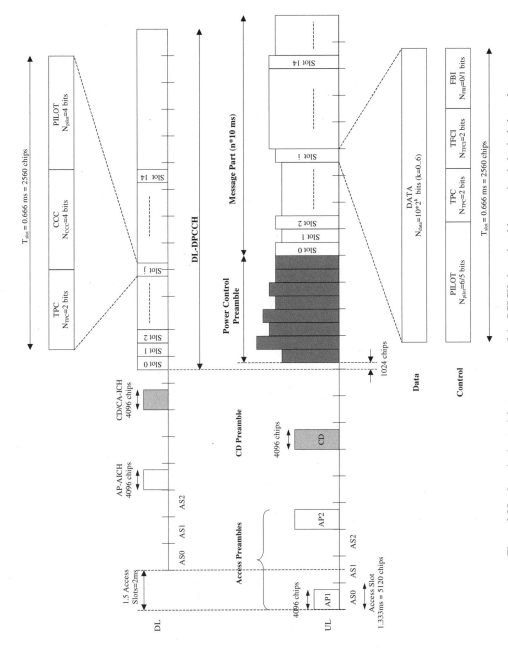

Figure 3.35 Organisation and frame structure of the PCPCH channel and its associated physical channels

minimum of 3 or 4 access slots (depending on the T_{cpch} parameter) after the last preamble and will increase progressively the power of the transmitted preambles. The process will be repeated until reaching the maximum allowed number of preambles or until detecting $AI \neq 0$ for the selected signature. If, during this phase, the CSICH indicates that the selected channel or bit rate is unavailable, the procedure will be aborted. The same happens if $AI = -1$.

When a positive acknowledgement ($AI = 1$) is detected, the mobile starts the Collision Detection (CD) phase (see Figure 3.35). The purpose of this phase is to detect the possibility that two or more users selected the same signature in the AP phase and transmitted simultaneously the message part, which would result in corruption of the information. In this phase, the mobile transmits a CD preamble after selecting one of the available signatures for the CD phase. The set of signatures for the CD and the AP preambles are the same, but the scrambling code is different (in some cases it could be the same, but then it is necessary to separate some signatures for the AP and some others for the CD preambles). The power of the CD preamble is the same of the last AP preamble.

After transmitting the preamble, the terminal listens to the Collision Detection/Channel Assignment–Indicator Channel (CD/CA-ICH). This channel is constructed in a similar way to the AP-AICH. If CA is not active, it simply contains CD indicators for each signature. If the $CD = +1$ for the selected signature in the preamble, this means that the terminal has succeeded in the access and can start the message transmission with the selected PCPCH. However, if $CD = 0$ this means that the terminal must abort the access. When two or more mobiles have selected the same signature in the AP phase, the collision will have been avoided provided that they select different signatures in the CD phase, so that the collision probability is highly reduced. When CA is active, this channel also contains the Channel Assignment indicators that point to the PCPCH that will finally be allocated to the terminal.

If the CD procedure has succeeded, the message transmission may start. In this case, a closed loop power mechanism is used, by allocating a downlink DPCCH that includes uplink power control commands, pilot bits and CPCH Control Commands (CCC), which are used to indicate certain situations, such as the start of the message transmission or the requirement to stop the transmission in emergency situations. This channel is organised into 10 ms radio frames divided in 15 slots, as depicted in Figure 3.35.

As in the PRACH case, the message part consists of a data and a control part, with the difference that the spreading factor of the data part may range from 4 to 256. Furthermore, TPC and FBI bits are also included in the control part, with a fixed $SF = 256$. The control part is spread with channelisation code $C_{ch,256,0}$ and the data part with $C_{ch,SF,SF/4}$. The message is organised into 10 ms radio frames subdivided into 15 time slots, each one corresponding to a power control period. The maximum number of frames is broadcast by the network and it can be up to 64.

Prior to the transmission of the message part, a power control preamble with duration of 8 slots can be used, depending on the information broadcast by the network. During these slots, only the control channel is transmitted in order to stabilise the closed loop power control procedure, so that inaccuracies resulting from the open loop power control of the AP and CD phases are avoided during the transmission of the message. The initial power of the power control preamble is set as in the PRACH case, adding just a certain value broadcast by the network to the transmission power of the CD preamble. The transmission of the power control preamble starts 3 or 4 access slots after the CD preamble, depending on the T_{cpch} parameter. Furthermore, the transmission in the uplink starts 1024 chips after the reception of the downlink DPCCH.

3.3.4.7 Physical Downlink Shared Channel (PDSCH)

The Physical Downlink Shared Channel (PDSCH) is used to carry the data of the DSCH transport channel. Its frame structure is shown in Figure 3.36. Like the downlink dedicated physical channel, it is organised into 10 ms radio frames divided into 15 time slots. The main difference with respect to the dedicated channel is that the PDSCH only contains data and it does not transmit any physical control information. The reason is that a mobile making use of the DSCH must always have an associated DCH

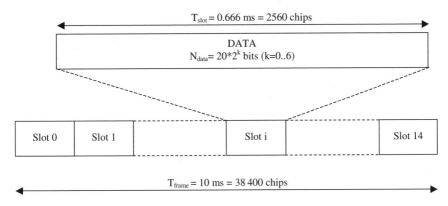

$T_{slot} = 0.666$ ms $= 2560$ chips

DATA
$N_{data} = 20*2^k$ bits (k=0..6)

| Slot 0 | Slot 1 | | Slot i | | Slot 14 |

$T_{frame} = 10$ ms $= 38\,400$ chips

Figure 3.36 Frame structure of the PDSCH

channel. Then, the power control and the channel estimation can be done by means of the control fields of the associated downlink dedicated physical channel [7].

Another important difference between dedicated channels and the PDSCH is that the latter makes use of variable spreading factor to achieve variable transmission bit rate, while discontinuous transmission is used in dedicated physical channels. The spreading factor of the PDSCH ranges from 4 to 256 in powers of 2, corresponding to a number of bits per radio frame ranging from 19 200 to 300.

Figure 3.37 presents an example of how the PDSCH channel can be allocated in the OVSF code tree. It occupies a sub-tree starting from the PDSCH root code, which can be a code with $SF = 4$ or higher. The codes below this sub-tree are allocated to DSCH users on a frame by frame basis depending on a packet scheduling algorithm. Consequently, different simultaneous transmissions at different bit rates are possible for different mobile terminals, provided that there are enough OVSF codes in the PDSCH sub-tree. It is also possible to allocate multiple parallel PDSCHs to the same terminal in a given frame, with the restriction that all of them must have the same spreading factor and must be frame synchronised. The PDSCH channel may be scrambled either with the primary or with a secondary scrambling code.

With respect to the timing relationship between the PDSCH and the associated DPCH, the start of the PDSCH frame should be between 3 and 18 time slots after the end of the associated DPCH frame. Thanks to this time separation, the allocation of transmissions in a PDSCH can be indicated through the TFCI of the associated dedicated channel. The TFCI informs about both the channelisation code of the allocated PDSCH and the transport format being used.

3.3.4.8 High Speed-Physical Downlink Shared Channel (HS-PDSCH)

The HS-PDSCH is used to carry the HS-DSCH transport channel, which significantly increases the bit rate of the DSCH by introducing the possibility of transmissions with 16-QAM depending on channel conditions, while at the same time modifying the frame structure to allow a faster packet scheduling mechanism. Transmissions are organised according to the frame structure shown in Figure 3.38. A 2 ms sub-frame is specifically defined for HS-PDSCH sub-divided into three time slots, so that different users can be time multiplexed on a sub-frame basis. The time slot duration is the same as in dedicated channels, so that the 10 ms radio frame used by the other physical channels corresponds to five HS-PDSCH sub-frames [7].

Each time slot of the HS-PDSCH contains only data bits and operates with a fixed $SF = 16$. Then, the number of bits depends on the modulation scheme, so that 320 bits per slot are transmitted with QPSK and 640 bits can be transmitted with 16-QAM modulation, corresponding to grouping the data bits into 4-bit modulation symbols. The higher speed achieved with 16-QAM modulation is at the expense of a higher signal to interference ratio needed to keep the same bit error rate, so that link adaptation

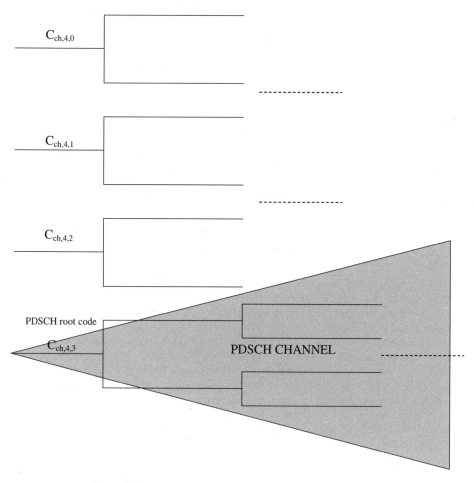

$C_{ch,4,0}$

$C_{ch,4,1}$

$C_{ch,4,2}$

PDSCH root code

$C_{ch,4,3}$

PDSCH CHANNEL

Figure 3.37 Example of PDSCH allocation in the OVSF code tree

mechanisms are necessary in order to decide on the most suitable modulation scheme to be used, depending on channel and interference conditions as well as distance to the serving cell. Multiple parallel HS-PDSCH channels can be allocated to the same user, thus increasing the bit rate.

The signalling associated with the HS-PDSCH, which includes the allocation of this channel to the specific terminals, together with the transport block size being used, is carried by the High Speed Shared Control Channel (HS-SCCH), which follows the same sub-frame structure but advanced 2 time slots with respect to the HS-PDSCH. It is transmitted with $SF = 128$, corresponding to 40 bits per time slot. Also the HS-SCCH includes all relevant physical layer information regarding the HS-PDSCH. Multiple parallel HS-SCCHs can be associated with a single HS-PDSCH.

A terminal that operates with HSDPA must monitor the corresponding HS-SCCH channel in order to know if there is control information intended for it. If this is the case, it should receive the HS-PDSCH channel in the next sub-frame. After reception of the data, the terminal must acknowledge in the uplink direction the received packet. This is done through the uplink High Speed Downlink Physical Control Channel (HS-DPCCH), whose frame structure is depicted in Figure 3.39. It is organised into 5 sub-frames per radio frame, and in each sub-frame two data fields are transmitted: the HARQ-ACK, which contains the acknowledgement of the received packet and occupies one time slot, and the Channel

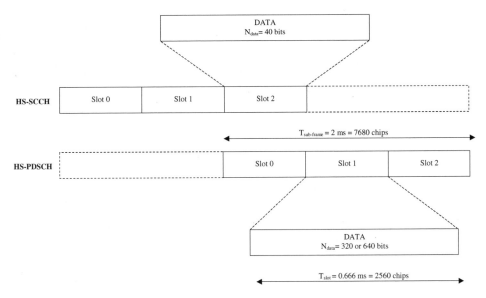

Figure 3.38   Organisation and frame structure of the HS-PDSCH and HS-SCCH

Quality Indicator (CQI), which occupies the other two slots of the radio frame. If the acknowledgement is negative, the downlink will proceed to the retransmission according to an Hybrid Automatic Repeat Request (HARQ) strategy.

The CQI is reported by the terminal and contains a measurement of the quality that is being perceived in the HS-PDSCH transmissions. This should be used by the link adaptation algorithm of the packet scheduler to decide the appropriate modulation scheme for this mobile. The CQI determines the transmitted power, which is kept constant during a sub-frame.

The uplink HS-DPCCH is code multiplexed with the rest of the dedicated channel DPDCHs of the user. Consequently, the channelisation code of the uplink HS-DPCCH depends on the number of simultaneous DPDCHs (see Figure 3.20). Particularly, when there are 2, 4 or 6 DPDCH channels, it is allocated in the I component with channelisation code $C_{ch,256,1}$. In turn, when there is only 1 DPDCH, it is allocated in the Q component with channelisation code $C_{ch,256,64}$ and when there are 3 or 5 DPDCH channels it is allocated in the Q component with channelisation code $C_{ch,256,32}$ [13].

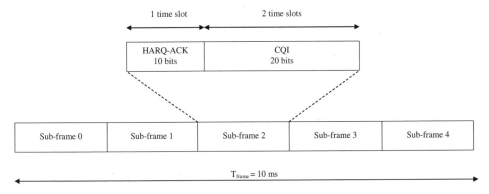

Figure 3.39   Frame structure of the uplink HS-DPCCH

3.4 LAYER 2 PROTOCOLS

Layer 2 offers services of information transmission to layer 3 in the form of Radio Bearers for data services and Signalling Radio Bearers for control information originated either in the Radio Resource Control protocol or in the Non Access Stratum (see Figure 3.3). With respect to the flow of control information, it goes through the RLC and MAC layers, while in the case of data information, depending on the specific service, there exist two additional sub-layers, namely the PDCP and the BMC. In this section the relevant characteristics of the protocols in layer 2 are described.

3.4.1 MEDIUM ACCESS CONTROL (MAC) PROTOCOL

The Medium Access Control protocol exists in the lowest sub-layer of layer 2 protocol architecture of the radio interface [21]. It exists in the UE, the node B and the RNC entities. The MAC provides data transfer services to the Radio Link Control sub-layer by means of logical channels and it is serviced by the physical layer by means of the transport channels, so one of the main functionalities of the MAC is the mapping between logical and transport channels.

The MAC layer at either the UTRAN or the UE receives MAC SDUs (Service Data Units) from the RLC and it is responsible for transferring them to the corresponding peer MAC entity at the other side. A MAC SDU is the minimum amount of information that can be transferred between the two sub-layers in a logical channel. This transfer service is done in unacknowledged mode, which means that the delivery to the other side is not guaranteed, so the RLC layer must have mechanisms to detect errors and losses of SDUs as well as perform retransmissions. Furthermore, the MAC protocol does not execute any type of segmentation of the MAC SDUs.

For each MAC SDU, a MAC header is added, whose length and contents depend on the specific transport channels. The MAC header includes the C/T field that identifies the specific logical channel when several logical channels are multiplexed onto the same transport channel, as well as fields to identify the specific UE in the case of common transport channels like RACH or FACH. It is also possible that the MAC header is empty. Typically, this would be the case of the transfer of user information through a DCH transport channel that is not multiplexed with any other channel at the MAC layer, and in which no UE identification is required.

The combination of a MAC SDU and a MAC header is a MAC PDU (Protocol Data Unit), which corresponds to a Transport Block transferred to the physical layer through the corresponding transport channel. In each TTI (Transmission Time Interval) the MAC layer selects a suitable Transport Format (TF) or a Transport Format Combination (TFC), depending on the instantaneous source rate, the service characteristics and the Transport Format Combination Set (TFCS). Each TF is related to a given instantaneous bit rate. Once the selection is done, the MAC layer delivers a set of Transport Blocks to the physical layer, denoted as the Transport Block Set. The transport blocks must be delivered in the same order in which the corresponding MAC SDUs were delivered by the RLC layer.

When several transport channels are multiplexed onto the same physical channel, the selection of the appropriate TFC must take into account the priorities of the different data flows associated with each transport channel. This can be understood as a dynamic scheduling for the different flows of a given user, which would allow the provision of different QoS to each flow. As an example, this MAC scheduling algorithm would allocate to the most demanding services transport formats associated with higher bit rates while allocating the lower transport formats to low bit rate services or to ones with lower priority. It is worth mentioning that this functionality is implementation dependent: the specifications only establish that the TFC selection should be done within the TFCS that is indicated by the network in each moment, and the specific algorithm that carries out the selection in the uplink direction may change from terminal to terminal depending on the manufacturer.

The transmission in uplink common transport channels (i.e. RACH and CPCH) is also controlled by the MAC protocol. Specifically, it is responsible for selecting the Access Service Class that determines the available slots and signatures as well as for applying the persistence probability before attempting the transmission (see Section 3.3.4.5). When the procedure fails at the physical layer (i.e. a NACK is received in the access), the MAC protocol decides the random instant in which the access will be attempted again. By controlling the Access Service Classes available to each user the network is able to establish priorities between users transmitting in these channels.

In the case of common and shared transport channels in the downlink direction (e.g. FACH, DSCH or HS-DSCH), where several users are multiplexed together, the MAC layer executes a scheduling algorithm to decide the user to which the channel is allocated in each transmission interval. Again, this algorithm is implementation dependent and should take into account the priorities and requirements of each of the services and users. In this way it becomes a key component in the radio resource management process to ensure the required level of QoS while simultaneously maintaining high utilisation of the scarce radio resources.

The introduction of the HSDPA access has created the inclusion of new functionalities to the MAC layer of UTRAN, since any packet access is necessarily based on a suitable Medium Access Control protocol. These functionalities are carried out in the Node B and include the scheduling algorithm to decide the appropriate allocation of the channel to each user as well as the Hybrid ARQ functionality, which combines the different retransmissions of a transport block to reduce the error probability. The scheduling algorithm, apart from including the requirements of each service, should take also into account the channel quality indicators to decide the appropriate selection of the modulation scheme according to a certain link adaptation algorithm.

The MAC layer is also responsible for performing the ciphering algorithm for RLC data transmitted in transparent mode. This ensures that the data are delivered only to authorised users. The ciphering is executed on a TTI basis, taking as a ciphering unit the concatenation of all the MAC SDUs transmitted in a given TTI. The configuration of the ciphering algorithm is done by the upper layers: the block cipher algorithm KASUMI is used. More details about the ciphering algorithm can be found in Reference 22.

The previous functionalities of the MAC layer were essentially intended for the service of data transfer offered to the RLC layer. Communication between the MAC and RLC layers is done by means of the primitives MAC-DATA-Req, MAC-DATA-Ind – intended to request to the MAC or to indicate to the RLC the transfer of MAC SDUs – and MAC-STATUS-Ind and MAC-STATUS-Rsp – mainly intended to indicate to the RLC the rate at which it may transfer data to the MAC.

Furthermore, the MAC layer also provides services to the RRC layer, which is responsible for reconfiguring the MAC functions and parameters. Specifically, the RRC may change the TFCS to limit the TFC selection at the MAC layer, which is equivalent to limiting the maximum transmission bit rate. This can be done in the case of congestion situations, in which the interference associated with high bit rate transmissions may not be tolerated. Another important procedure is the Transport Channel Type Switching, which is requested by the RRC and executed by the MAC layer. This procedure consists of changing the transport channel type that a given service is using depending on the source and the network status at a given moment. An example of this situation would be the change between the use of RACH/ FACH and DCH channels for interactive services (e.g. web browsing). Since these services normally exhibit long inactivity periods, it is usual to operate them in RACH/FACH mode during the low activity periods, so that low resource consumption in terms of power and OVSF codes is ensured with respect to occupying a dedicated channel. However, during activity periods, transmission is switched to DCH, which is able to ensure a higher quality.

Communication between MAC and RRC regarding the above procedures is carried out by means of the primitive CMAC-CONFIG-Req.

The interaction between the MAC and RRC layers also requires that the MAC layer send some measurements so that the RRC may take the appropriate decisions with respect to the configuration of the MAC parameters. These measurements include traffic volume monitoring and quality indications. Based on the reported traffic volume measurements, which include the buffer occupation, the RRC may take the decision to execute a transport channel type switching. The primitives used by the RRC to ask for measurements to the MAC layer are the CMAC-MEASUREMENT-Req, and the corresponding

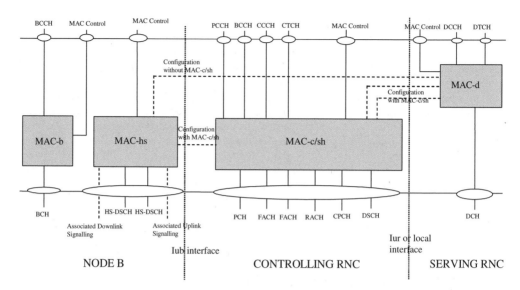

Figure 3.40   Architecture of the MAC layer entities at the UTRAN side

result is given in CMAC-MEASUREMENT-Ind. The CMAC-Status-Ind primitive is also used by the MAC to notify the RRC about the status information.

The architecture of the MAC layer consists of several entities, which execute the functions described above, as shown in Figure 3.40 for the case of the UTRAN side. These entities depend on the specific transport channels that are handled, and are listed below:

- MAC-b. This entity exists in the Node B and the UE and handles the transfer of data from the BCCH to the BCH transport channel, except when the BCCH is mapped onto the FACH. In this case, it is handled by the MAC-c/sh entity.
- MAC-c/sh. This entity handles the common and shared channels. In the UTRAN side, this entity is located in the controlling RNC of a set of nodes B.
- MAC-d. This entity handles the transmission through the DCH channels. In the UTRAN side, this entity is located in the serving RNC of each UE. When the UE also uses common or shared channels, there is a connection between the MAC-d and the MAC-c/sh entities. In the most usual situation, in which the serving RNC and the controlling RNC are physically the same node, the connection is done through a local interface. However, when they are two different nodes, the connection is done through the Iur interface.
- MAC-hs. This entity was included with the introduction of the HSDPA functionality and handles the transmission through the HS-DSCH. In the UTRAN side, this entity is located in the Node B. The MAC-hs is also connected to the MAC-d and/or to the MAC-c/sh entities where a UE also uses dedicated and/or common channels.

3.4.2 RADIO LINK CONTROL (RLC) PROTOCOL

The Radio Link Control sub-layer is located in both the UE and the RNC immediately above the MAC sub-layer according to the radio interface protocol architecture. In the control plane, it provides services directly to layer 3, while in the user plane it may also provide services to the PDCP and BMC sub-layers [23]. In turn, it receives information transfer services from the MAC layer by means of the logical channels, as described in Section 3.4.1. The RLC protocol receives from the upper layer RLC SDUs

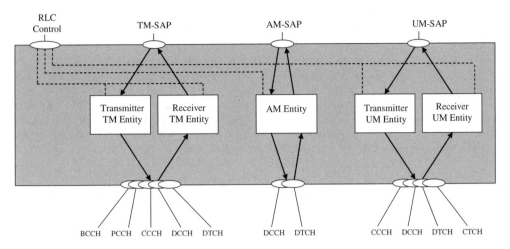

Figure 3.41   RLC sub-layer architecture

(Service Data Units) and transmits RLC PDUs (Protocol Data Units) to the MAC layer (note that the RLC PDUs are the same as the MAC SDUs).

Essentially, the RLC provides three types of data transfer services corresponding to the three modes of operation: Transparent Mode (TM); Unacknowledged Mode (UM); and Acknowledged Mode (AM) [23]. Each mode is associated with a different Service Access Point (SAP) for upper layers, denoted as TM-SAP, UM-SAP and AM-SAP, respectively, and with different RLC entities, as depicted in Figure 3.41. Different types of RLC PDUs are defined depending on the considered mode.

3.4.2.1 Transparent Mode

The transparent data transfer service simply consists of sending the RLC SDUs received from the upper sub-layer through the TM-SAP without adding any type of overhead (i.e. any RLC header). The length of a RLC SDU must be a multiple of the minimum amount of information that can be delivered to the lower layer, corresponding to the TMD PDU (Transparent Mode Data PDU), so that segmentation of RLC SDUs in an integer number of TMD PDUs is possible. All the TMD PDUs containing the same RLC SDU are transmitted in the same Transmission Time Interval.

At the receiver side, the different TMD PDUs are reassembled in order to deliver the original RLC SDU. The values of the lengths for the RLC SDU and TMD PDU are specified in the definition of the corresponding Radio Bearer or Signalling Radio Bearer. When errors are detected in a RLC SDU at the receiver entity, it can be either discarded or marked erroneous, depending on how the RLC is configured, but no recovery mechanism by means of retransmissions is provided by the Transparent Mode.

When transmission is not possible for a period of time, and in order to avoid buffer overflow, some RLC SDUs may be discarded by the RLC layer at the transmitter side, without explicit signalling of this situation to the RLC entity at the receiver side. The upper layers of the transmitter side are notified about this situation.

The transparent mode can be used by any of the logical channels except the CTCH, and it is the only mode that can be used by the BCCH and PCCH for the transfer of broadcast and paging messages. It is the mode typically used by the streaming service class in the case of dedicated channels.

3.4.2.2 Unacknowledged Mode

In Unacknowledged Mode (UM), the RLC transmits the RLC SDUs received through the RLC-SAP, and, as in the TM, the correct delivery to the receiver side is not guaranteed. The segmentation of RLC

SDUs into UMD PDUs (Unacknowledged Mode Data PDUs) is possible, and additional padding bits may be added in order to ensure an integer number of RLC PDUs of valid length. Similarly, several RLC SDUs may be concatenated if its length is lower than that of the RLC PDU. Each UMD PDU contains a RLC header. This header contains the sequence number of the PDU as well as a length indicator for each of the RLC SDUs that are included in the PDU. Thanks to the information contained in this header, concatenation and padding functions, which were not allowed in TM, are possible with UM.

At the receiver side, the different UMD PDUs are reassembled and the resulting RLC SDU is delivered immediately to the upper layer. Thanks to the sequence number included in the RLC header, it is possible to detect corrupted RLC SDUs because of the lack of some UMD PDUs. Depending on the RLC configuration, erroneous RLC SDUs can be discarded or marked and delivered to the upper layers. Note that in UM, the existence of erroneous UMD PDUs can be detected either because the MAC layer has indicated this (i.e. because the physical layer has detected errors through the CRC that is added to each transport block) or because a complete PDU is lost and the received sequence numbers are not consecutive. With TM, however, errors are only detected when the MAC layer indicates this situation, since no sequence numbers are available. In any case, the unacknowledged mode does not provide retransmission mechanisms for corrupted UMD PDUs, so that delivery at the peer entity is not guaranteed.

At the transmitter side, some RLC SDUs may be discarded when transmission is not possible for a given period of time. As in the TM, no signalling to the receiver entity exists and the upper layer is informed when a SDU is discarded.

The UMD PDUs transmitted in UM are ciphered in the RLC layer before delivering them to the MAC layer. For TM, the ciphering is performed in the MAC layer, as stated in Section 3.4.1.

UM can be used by CCCH, CTCH, DTCH and DCCH logical channels, and it is normally used by some RRC control procedures, in which there exist specific RRC acknowledgement messages, so that acknowledgement at the RLC layer is not necessary. Voice over IP services may also use this mode.

3.4.2.3 Acknowledged Mode

Acknowledged Mode (AM) is used exclusively by dedicated logical channels, either DCCH or DTCH, and it extends the functionalities of UM by incorporating a retransmission mechanism to recover those AMD PDUs (Acknowledged Mode Data PDUs) that are received in error. Consequently, it is required that both transmitter and receiver entities are active at both UE and UTRAN sites, while in UM and TM it was only necessary to have one transmitter entity at one side and one receiver entity at the other side (see Figure 3.41). As a matter of fact, in AM, transmitter and receiver entities at one side must operate jointly to update the transmitter buffer according to the received acknowledgements. Therefore, both transmitter and received entities are considered as a single AM entity.

As in UM, transmission in AM executes the functions of concatenation and segmentation of RLC SDUs into AMD PDUs, padding, RLC SDU discard and ciphering. But thanks to the retransmissions, a certain quality level can be ensured since delivery of RLC SDUs to the peer entity can be guaranteed to a certain extent. Note that, due to retransmissions, it is possible that some RLC SDUs are not delivered in the same order they were generated. In this case, the RLC can be configured to force in-sequence delivery to maintain the original delivery order. This is at the expense, however, of a certain delay before delivering the RLC SDU to the upper layer, which will depend on the number of retransmissions that are required for the preceding RLC SDUs.

The RLC protocol in AM defines several types of control PDUs which contain data, apart from the AMD PDUs. The first bit in the RLC header specifies whether a PDU contains data or control, so that the next bits of the header can be decoded differently for data and control PDUs. The STATUS-PDU is used to exchange information about the successfully received and the missing AMD PDUs, so that the transmitter buffer is updated and retransmissions are performed according to the received STATUS-PDUs from the receiver side. Depending on how the RLC is configured, the STATUS-PDU can be transmitted whenever a missing PDU is detected or also on a periodic basis. Similarly, it is possible for the transmitter side to request a STATUS-PDU by making use of the Polling bit included in the AMD PDU header. In the case of bi-directional services, in which both sides transmit AMD PDUs, it is also possible to piggyback a STATUS-PDU at the end of an AMD PDU.

The RLC protocol in AM defines the maximum number of retransmissions that can be allowed for a given AMD PDU. When this maximum is reached, depending on the configuration the protocol may discard the corresponding RLC SDU that contains the AMD PDU. This requires informing the receiver side by means of a STATUS PDU in order to update accordingly the reception window with the expected PDUs. In turn, another possibility is that the RLC SDU is not discarded, and in this case the RLC reset procedure is started between the transmission and reception entities. This procedure involves the transmission of a RESET PDU and the reception of the corresponding RESET ACK PDU.

The setting of the parameter that specifies the maximum number of retransmissions is important in order to establish a trade-off between the quality in terms of delay and in terms of packet losses. For background services, in which the delay restrictions are not critical, this parameter can be set higher to ensure data integrity than for interactive services, in which data transfer is expected to be carried out under certain time constraints. Note that when a RLC SDU is lost, it will be the responsibility of the upper layers (e.g. TCP protocol) to recover it in order to preserve data integrity.

RLC operating in AM also provides flow control functionality. In this way, the receiver side may control the rate at which the transmitter sends the AMD PDUs by adjusting the transmitter window size by means of a STATUS PDU.

3.4.2.4 Communication between RLC and Upper Layers

Communication between the RLC and the upper layers for the transmission of RLC SDUs is done by means of the RLC-TM-DATA, RLC-UM-DATA and RLC-AM-DATA primitives, for each of the three modes. Each primitive can be of type Req, for requesting transmission, Ind, for delivering the received RLC SDUs, and Conf, to indicate discarded RLC SDUs in the three modes and to confirm the reception of the RLC SDU at the peer entity in AM.

The RLC layer is configured by the upper layers, which can establish, re-establish, stop, continue or modify the RLC by means of the CRLC-CONFIG-Req primitive. This allows the control of the QoS of the provided service, for example, by setting the maximum number of retransmissions to a suitable value. In UM and AM, it is also possible to suspend temporarily the RLC operation, and in this case the RLC is not allowed to send sequence numbers higher than a given value. The RLC is suspended by CRLC-SUSPEND-Req and CRLC-SUSPEND-Conf, and operation is resumed by CRLC-RESUME-Req. CRLC-STATUS-Ind is used by the RLC layer to indicate its status to the upper layers.

3.4.3 PACKET DATA CONVERGENCE PROTOCOL (PDCP)

The Packet Data Convergence Protocol only exists in the user plane and is specifically for Packet Switched services. Its main functionality is to improve the efficiency in the radio transmission by means of executing header compression of the IP data packets coming from upper layers. It receives PDCP SDUs from upper layers and delivers to the RLC sub-layer different PDCP PDUs. The types of PDCP PDUs are [24]:

- PDCP Data PDU, which contains a PDCP SDU, either compressed or uncompressed, or header compression related control signalling. The type of information contained is indicated in a one-byte header.
- PDCP-No-Header PDU, which contains an uncompressed PDCP SDU without adding any type of header.
- PDCP SeqNum PDU, which, apart from containing a compressed or uncompressed PDCP SDU, also includes a PDCP SDU sequence number.

The specific header compression protocol depends on the particular upper layer protocols that form the incoming PDCP SDU and on the configuration of the PDCP sub-layer by upper layers. For TCP/IP packets, the IP header compression mechanism described in IETF RFC 2507 [25] is used, which is essentially based on providing a variable length TCP/IP header whose contents are specified by the bits of the first byte. The amount of information in the header is reduced by sending only the changes from one packet to the next. In this way, it is possible to reduce the TCP/IP header from the usual value of 40 bytes down to 4 or 5 bytes. Note that when such a TCP/IP compressed packet is lost due to errors in the lower layers, the incremental condition of the header compression mechanism makes it no longer possible to decode the headers of the subsequent packets. Consequently, in such cases it is necessary to send sporadic uncompressed TCP/IP segments. Different types of compressed headers are accepted by the protocol. The type of compression is indicated in the PDCP Data PDU header.

For RTP/UDP/IP packets, in which certain time constraints must be met, the Robust Header Compression (ROHC) method defined in IETF RFC 3095 is used [26]. This protocol supports segmentation of packets and can transmit information packets from several contexts, distinguished by the CID (Context Identifier), which is included either in the PDCP header or in the upper layer packet. The ROHC protocol involves some signalling parameters that must be exchanged between the compressor and the decompressor entities at the transmitter and the receiver sides.

The PDCP protocol may operate with any of the three RLC modes (i.e. transparent, unacknowledged and acknowledged, as described in Section 3.4.2), and the selection will depend on the specific service characteristics. When operating with acknowledged RLC mode configured with in-sequence delivery, the PDCP with ROHC protocol also provides support for a lossless SRNS relocation procedure by means of PDCP sequence numbering.

3.4.4 BROADCAST/MULTICAST CONTROL (BMC) PROTOCOL

Like the PDCP, the Broadcast/Multicast Control protocol only exists in the user plane, and provides a broadcast/multicast transmission service operating in RLC unacknowledged mode [27].

The BMC entity at the UTRAN side is located in the RNC and its functions include the storage of Cell Broadcast Messages, which are transmitted in a given cell to certain UEs that support the SMS Cell Broadcast Service. These messages, which carry information that depends on the geographical area, are transmitted by means of the CTCH logical channel and the mobile should be able to receive them in idle mode as well as in the rest of RRC states (see Section 3.5.2). The stored messages are scheduled by the BMC in order to decide the appropriate instant for their transmission. The specific scheduling algorithm is implementation dependent. At the UE side, the BMC entity delivers the received cell broadcast messages to the upper layers.

Traffic volume monitoring is done by the BMC – by making predictions on the expected amount of capacity in terms of bit rate that is needed for the transmission of cell broadcast messages – and indicates it to the RRC entity, so that it can act accordingly.

Each BMC entity serves only those broadcast messages that should be transmitted in a specific cell, so there should exist in the RNC some additional functions that resolve the geographical area in which each message should be transmitted in order to identify the corresponding cell.

3.5 RADIO RESOURCE CONTROL (RRC) PROTOCOL

According to the radio interface protocol architecture that was presented in Figure 3.3, the Radio Resource Control (RRC) protocol only exists in the control plane and occupies the lowest sub-layer of layer 3 while, at the same time, being the highest layer of the Radio Access Network (i.e. the UTRAN part) protocol stack [20]. Although layer 3 is partitioned in other sub-layers above RRC, they belong to the Non Access Stratum (NAS) and are specified between the UE and the Core Network parts of the UMTS architecture. Some examples of control protocols above the RRC include Session Management

(SM), Connection Management (CM), Mobility Management (MM) or GPRS Mobility Management (GMM).

The RRC protocol provides the service of transferring signalling information to the NAS upper layer entities. Besides, RRC handles the control plane signalling between the UE and the UTRAN through Signalling Radio Bearers. This includes procedures specific of the Access Stratum (AS) that allow the appropriate configuration of the lower layers in both control and user planes taking into account the network status. Therefore, these signalling procedures provide the support for the execution of the Radio Resource Management strategies.

3.5.1 ARCHITECTURE

Figure 3.42 shows the architecture of the RRC sub-layer as well as the interrelations between RRC and the upper and lower sub-layers [20]. The RRC entities are located in both the UE and the RNC. With respect to the interactions with the upper layers, the Access Stratum offers services to the Non Access Stratum (NAS) protocols through three different Service Access Points; the General Control (GC) SAP, which provides an information broadcast service for all the mobiles in a given geographical area; the Notification (Nt) SAP, which provides paging and notification broadcast services addressed to specific mobiles in a given geographical area; and the Dedicated Control (DC) SAP, which provides services for the establishment and release of dedicated connections and for the transfer of signalling messages through these connections.

The messages that come from these three SAPs are collected by the RFE (Routing Function Entity), which transfers them to the corresponding SAPs of the RRC. Similarly, the RFE at the peer side receives the messages that have passed through the lower layer protocols and transfers them to the upper NAS protocols.

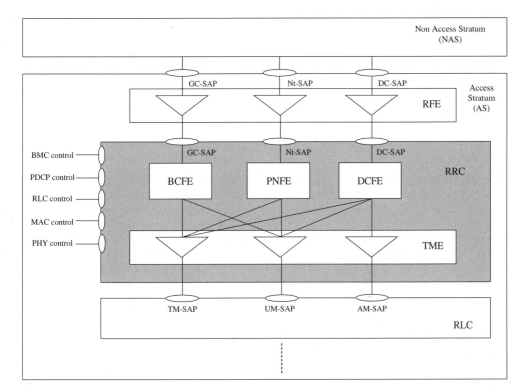

Figure 3.42 RRC architecture and interrelations with other layers

The three types of SAPs are directly related with the three entities that form the RRC sublayer, as depicted in Figure 3.42. The Broadcast Control Function Entity (BCFE) handles the delivery of broadcast messages that may come from the Core Network through the GC-SAP or that may originate in the UTRAN. These messages can be delivered by using either the Transparent or the Unacknowledged Mode at the RLC sub-layer, although normally the Transparent Mode is used. The Paging Notification Function Entity (PNFE) is used to deliver the paging messages coming from the Nt-SAP to those mobiles that do not yet have an established RRC connection (i.e. they are in idle mode). As in the previous case, they can be transferred either in Transparent or in Unacknowledged RLC modes. Finally, the Dedicated Control Function Entity (DCFE) is used to handle the signalling procedures of specific mobiles that have already established an RRC connection and to transfer the messages created in the Core Network that come through the DC-SAP. The DCFE can use any of the three modes of the RLC sub-layer. It is worth mentioning that paging messages to mobiles with an established RRC connection are handled by the DCFE and not by the PNFE. For the transfer of signalling generated at the NAS, the RRC layer discriminates between two priority classes, namely high and low priority. All the signalling connections for a given mobile are mapped on a single RRC connection.

The messages transferred from the previous entities, and which originated either at the NAS or at the RRC layer itself, are collected by the Transfer Mode Entity (TME), which simply maps them to the appropriate SAP from the RLC sub-layer. Figure 3.42 also depicts the control service access points that are used by the RRC to interact with layer 2 and layer 1 protocols (i.e. BMC, PDCP, RLC, MAC and PHY) in order to configure them and to receive the measurements that are used to decide the best configuration parameters.

The RRC messages are transferred by the lower layers through Signalling Radio Bearers (SRBs), which specify the characteristics of the logical, transport and physical channels being used. Different SRBs are specified depending on the type of messages that are transferred. According to the numbering given in 3GPP specifications, SRB#0 is used for the transfer of messages through the CCCH logical channel, which include the initial message that is sent by the mobile when establishing a RRC connection. It uses the RACH transport channel in the uplink and the FACH in the downlink. SRB#1 and SRB#2 transfer dedicated messages originating in the RRC layer through the DCCH. They use Unacknowledged and Acknowledged Mode, respectively. Finally, SRB#3 and SRB#4 transfer messages through the DCCH in Acknowledged mode that are generated at the NAS layer, with high and low priority, respectively. Additional SRBs up to 32 may be also used for the transfer of messages with Transparent Mode (e.g. through DCCH, PCCH or BCCH).

3.5.2 RRC STATES

The operation of the RRC protocol is intrinsically coupled with the different states in which a mobile can be during its lifetime. This is because, depending on these states, not all the functionalities may be available for this mobile and therefore there will be variation in the type of information that it is ready to send or receive. The defined RRC states are depicted in Figure 3.43, together with the transitions between them and the events that lead to such transitions [20]. The figure considers the general case of a UE that supports UTRAN as well as GSM/GPRS functionalities. The different states are essentially related with the activity of the mobile and the radio resources that it may have allocated depending on this activity. Note that in the new services existing in 3G systems that involve bursty transmissions with long inactivity periods, it is necessary to define a larger set of states than in 2G systems like GSM, where the service is mainly active throughout the connection lifetime.

In both GSM/GPRS and UTRAN, there are essentially two main operation modes, namely the Idle and Connected modes [28][29]. The idle mode is characterised by the absence of an RRC connection with the network, so that the mobile does not have any radio resources allocated, and therefore it cannot start any type of user data transfer in the dedicated or common channels. When a terminal is switched on, it selects an appropriate cell and camps on it by detecting the synchronisation and broadcast channels, as explained in Section 3.3.4.4. The mobile is then in the idle mode. The only transmission that is allowed for a

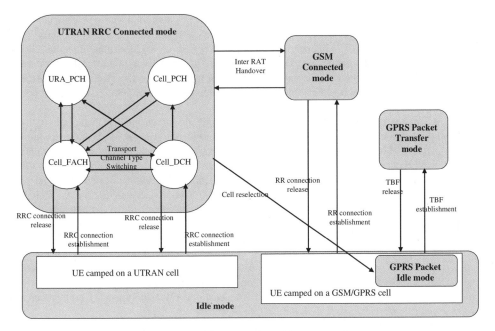

Figure 3.43 RRC states and transitions among them

terminal in idle mode is the Initial Message that is sent through the RACH transport channel to initiate the establishment of an RRC connection (for example, in order to make the initial attach procedure or to start a new service). Similarly, the establishment of an RRC connection can also be triggered by the network through a paging message coming from the NAS, for example, in the case of an incoming call. Since no RRC connection is available, the terminal can only be identified by Non Access Stratum identifiers, like the IMSI (International Mobile Subscriber Identity) or the TMSI (Temporary Mobile Subscriber Identity), which belong to the Mobility Management (MM) functions. During idle mode, the network knows the position of the terminal on a registration area basis, where each registration area includes a certain number of cells. As long as it remains in the same registration area, it is not necessary for the terminal to start an RRC connection. However, when the mobile enters a new registration area, it must establish an RRC connection to indicate the new area to the corresponding MM entity.

In connected mode, the terminal has established an RRC connection with the network, so that it has some radio resources allocated and therefore can start a data transfer service. A specific identity is assigned to each mobile in the RRC connected mode. This is the RNTI (Radio Network Temporary Identity) and can be used as a UE identifier whenever the terminal transmits in the common channels like RACH or CPCH. In GSM, the connected mode corresponds to the situation in which the terminal has a dedicated channel, either Traffic Channel (TCH) or Stand-alone Dedicated Control Channel (SDCCH), while in GPRS, in packet transfer mode (that is the equivalent to the connected mode) the terminal has a PDCH (Packet Data Channel) allocated for the transmission of a number of radio blocks denoted as Temporary Block Flow (TBF). Note that these allocations do not necessarily mean that the terminal is transmitting user data, since it can be simply transmitting signalling data (for example, attach, location area update or session establishment). In any case, in GSM and in GPRS, the mobile can only be in connected or in idle mode, while in UTRAN the RRC connected mode includes different states. This is required in order to cope with the different traffic characteristics of the higher number of provided services as well as to be able to provide different levels of QoS differentiation. In particular, four different RRC states are defined in connected mode:

- Cell_DCH state. This corresponds to the situation in which a dedicated transport channel both in uplink and downlink directions is allocated to the terminal, so that it can transmit either signalling or data information. A DSCH channel may also be allocated in the downlink direction. During Cell_DCH state, the network knows the mobile position at a cell level, and a handover procedure is required to change from one cell to another or to add or remove new cells to the active set in case of soft handover. This state would be equivalent to the connected mode of GSM or the packet transfer mode in GPRS.
- Cell_FACH state. A terminal in this state has no dedicated transport channel, so the terminal can only transmit through the RACH or CPCH in the uplink and receive through the FACH in the downlink. In any case, a DCCH logical channel is available for the transfer of signalling information and a DTCH logical channel for the transfer of data may also be available. As in the Cell_DCH state, the mobile location is known by the network at a cell level. The mobile may reach the Cell_FACH state either from the idle mode by means of the establishment of an RRC connection or from the Cell_DCH states. Transitions from Cell_FACH to Cell_DCH are normally subject to the activity detected for a given traffic source, typically for services with long inactivity periods (for example, web browsing where long reading times occur between consecutive downloading periods). In such a case, the mobile would be in Cell_DCH during activity periods, i.e. when the terminal makes an efficient use of the dedicated channel, and it would be switched to the Cell_FACH state during inactivity periods, in which the maintenance of a dedicated channel would not be efficient since it could block other users that would like to start a connection in the cell. The Transport Channel Type Switching procedure is used to change between Cell_FACH and Cell_DCH states.
- Cell_PCH. During this state, the terminal cannot transmit in the uplink direction in neither the dedicated nor in the common transport channels. Neither DTCH nor DCCH logical channels are allocated to the terminal, so it simply decodes the information from the broadcast and paging channels, and can also receive BMC messages if it supports the Cell Broadcast Services. The mobile location is known at the cell level. This requires that, when the terminal enters a new cell, a cell reselection procedure is initiated by moving to the Cell_FACH state and transmitting through the RACH. After the cell reselection procedure, the terminal switches back to Cell_PCH if no more activity is detected.
- URA_PCH. Essentially, this state is the same as the Cell_PCH, with the difference that the mobile position is known on a registration area basis, so that the terminal does not require informing the network when it performs a cell reselection procedure between cells of the same registration area. In turn, when the terminal reselects to a cell of a different registration area, it moves to the Cell_FACH and transmits through the RACH to inform about this change.

The mobile may enter the Cell_PCH and the URA_PCH states either from the Cell_DCH or the Cell_FACH states after the release of the dedicated logical channels DCCH or DTCH.

3.5.3 RRC FUNCTIONS AND PROCEDURES

The RRC layer handles all the control signalling related to the radio access between a terminal and the UTRAN. As a result, its functions cover all the procedures that a mobile must carry out depending on its RRC state in order to ensure an efficient radio transmission according to the mobile requirements. This includes, among others: the broadcasting of system messages so that the mobile can know the network and cell configuration; the paging messages to notify incoming calls; the establishment and release of the appropriate radio resources to enable transmission; the corresponding configuration and reconfiguration of the lower layers in terms of the parameters of radio bearers, transport channels and physical channels; the control of measurement reporting; and the required mobility functions to keep the RRC connection when the mobile changes from cell to cell. Therefore, RRC procedures can be seen as the mechanisms provided by UTRAN for the support of the Radio Resource Management algorithms in the sense that the decisions taken by these algorithms will be executed by specific RRC signalling procedures.

In the following, a summary of the main RRC signalling procedures is provided. For a detailed specification of the corresponding messages involved in each of these as well as a complete list of procedures, the reader is referred to Reference 20.

3.5.3.1 RRC Connection Management Procedures

This set of procedures is intended to cover the RRC functions that support the establishment, maintenance and release of RRC connections. They include the following:

- Broadcast of system information. This procedure allows the broadcasting of system information messages from the UTRAN to the terminals of a given cell. These messages are organised into System Information Blocks (SIBs), ranging from 1 to 18, which are sent periodically on the BCCH. There exists a Master Information Block that includes the reference information to decode the rest of the SIBs and which is the first block that the mobile should read when it selects a new cell. Additionally, it contains a tag that allows the detection of changes in certain SIBs whose information does not change frequently, so that the terminals do not need to decode continuously all the broadcast messages. In the case of important changes in the SIBs, the network may notify these to the mobiles in Cell_PCH and URA_PCH states through a *Paging Type 1* message in the PCH channel, while the mobiles in Cell_FACH are informed through the FACH channel by means of the *System Information Change Indication* message.
- Paging. This procedure is used to transmit paging information to selected UEs that are in idle mode or in Cell_PCH and URA_PCH states. It makes use of the *Paging Type 1* message. The reasons for a paging message can be the establishment of a network originated call or session set-up, the request to trigger a cell update procedure, the change to Cell_FACH state because of downlink packet data activity, the request to start the release of an RRC connection and the request to read updated system information in the broadcast channel. A similar procedure exists for paging mobiles in Cell_DCH and Cell_FACH states, but in this case a *Paging Type 2* message is sent through the DCCH.
- RRC connection establishment and release procedures. Terminals in idle mode that require the initiation of a signalling connection make use of the RRC connection establishment procedure. The procedure starts with a *RRC Connection Request* message mapped to the CCCH logical channel and transmitted through the RACH. The mobile identifies itself by means of NAS identifiers like the IMSI or the TMSI and it includes the establishment cause. There exist several causes including the registration, the establishment of originating calls for each of the four possible service classes (conversational, streaming, interactive and background) or the transfer of higher layer signalling. Upon receipt of this message, the network may either accept or reject the request by means of a *RRC Connection Setup* or a *RRC Connection Reject* message, respectively, which is mapped to the CCCH logical channel and the FACH transport channel. In the case of acceptance, the *RRC Connection Setup* message includes the RNTI for the mobile and the indication about whether to pass to Cell_DCH or to Cell_FACH. It also includes the characterisation of the allocated dedicated radio channel in terms of code sequence and TFCS in both the uplink and downlink direction, when the user is moved to the Cell_DCH. In any case, the mobile terminal is now in connected mode and there is a DCCH logical channel allocated to it that includes SRB#1, SRB#2, SRB#3 and optionally SRB#4 (see Section 3.5.1). The procedure completes when the mobile sends the *RRC Connection Setup Complete* message through DCCH and either DCH or RACH transport channel, thus acknowledging the correct reception and configuration of the allocated channel. Only one RRC connection may exist for a given mobile.

  The RRC connection release procedure is started by the network to release an existing RRC connection for a mobile in Cell_DCH or Cell_FACH states. Depending on the logical channels that are active for the corresponding terminal, the procedure may be carried out either in the CCCH or the DCCH logical channels. In the DCCH case, it consists of two messages, the *RRC Connection Release* and the corresponding *RRC Connection Release Complete* sent by the terminal, while in the CCCH case the procedure is the same but no confirmation is sent by the terminal.

- Transmission of UE capability information. This procedure is used by the mobile to indicate to the network about their capabilities corresponding to the different layers of the radio interface (e.g. support of PDSCH, HS-DSCH, PCPCH, RLC AM buffer sizes, supported PDCP compression protocols, etc.). The procedure can be started by the mobile by sending the *UE Capability Information* message, which is acknowledged by the network with the *UE Capability Information Confirm*, or requested by the network by means of the *UE Capability Enquiry* message.

- Establishment and release of signalling connections between the UE and the Core Network and direct transfer of signalling messages. These procedures are intended to establish and release NAS signalling connections between the terminal and the different core network domains (i.e. CS and PS domains). This allows the direct transfer of signalling messages between the upper layer entities of mobiles that have previously established a RRC connection. The term 'direct transfer' refers to the transmission of signalling NAS messages through the RRC layer either in the uplink or in the downlink direction. These messages are delivered through the DC-SAP and handled at the DCFE, and therefore they require a DCCH logical channel established between the UE and the RNC.

  The establishment of the signalling connection is done by means of the Initial Direct Transfer procedure, which is initiated by the NAS of the UE. In this case, the RRC layer of the UE sends an *Initial Direct Transfer* message to the peer entity at the SRNC that includes the message denoted as *Initial UE Message* (which belongs to the RANAP protocol defined between RNC and CN and that contains a NAS message, see Reference 30 for details) and some information about the core network domain (i.e. CS or PS) to which the NAS message should be delivered. Once the signalling connection has been established with the Initial Direct Transfer message, subsequent NAS messages corresponding to this connection are transmitted with the *Uplink Direct Transfer* and *Downlink Direct Transfer* messages between RRC entities. Some examples of NAS messages that can be exchanged could be, for example, a CM Service Request in order to start a call, a MM Location Updating Request, etc.

  The release of a signalling connection is notified by the network to the corresponding mobile by means of the *Signalling Connection Release* message. The RRC at the UE side will then pass the message to the upper layers that will be responsible for releasing the signalling connection. When this has been completed and indicated by the upper layers, the RRC entity of the UE will transmit a *Signalling Connection Release Indication* message. Upon reception of this message, the SRNC will pass to the upper layers the indication to release the signalling connection in the core network side. Note that, after releasing the signalling connection, the RRC connection may also be released if the terminal has no more signalling connections.

- Security mode control. This procedure is used to start the ciphering of the communication between a terminal and the UTRAN or to restart the ciphering with a new configuration. It is also used to start the integrity protection of the signalling messages or to modify its configuration. The integrity protection consists of a checksum that is inserted in the majority of RRC signalling messages that are transferred between RRC peer entities (with some exceptions, like the *RRC Connection Request* during the RRC connection establishment, in which integrity is not yet configured), and that ensures that the messages are correctly delivered.

- Counter check procedure. This procedure is initiated by the network to check that the amount of data sent in the uplink and downlink direction during a RRC connection is identical in both UE and UTRAN sides, which allows the detection of possible intrusions.

- Inter RAT handover information transfer. This procedure is used by UEs operating in a Radio Access Technology (RAT) different from UTRAN, for example, GSM, in order to transfer certain information about user capabilities (for example, supported frequency bands) to prepare for a handover to UTRAN. In this procedure, the RRC entity of the UE transmits an *Inter RAT Handover Info* message to the target RNC.

3.5.3.2 Radio Bearer Control Procedures

This set of procedures contains mechanisms to cope with the management of the different radio bearers that are allocated to the terminal [20][28]. They allow defining the configuration of the lower layers of the radio interface so that transfer of both data and signalling is possible. The procedures are:

- Radio bearer establishment and release. The radio bearer establishment is a procedure initiated by the upper layers of the network side in order to request the allocation of radio resources to a mobile terminal that previously has established a RRC connection. As described in Section 3.5.3.1, the establishment of a RRC connection involves the establishment of different SRBs by means of the allocation of the required radio resources to allow the transfer of signalling messages between the UE and the network. At a given instant during the RRC connection, the upper layer signalling messages exchanged by the UE and the Core Network may request the initiation of a user service belonging to a certain service class and with different requirements (e.g. a circuit switched call by means of a *CC Setup* message or a packet transfer by means of a *SM Activate PDP Context Request*). This user service requires the extension of the radio resource allocation to the corresponding user taking into consideration the service requirements. Then, after the acceptance of the new service by the admission control, the RRC of the SRNC will receive from the core network (i.e. from the MSC for CS services or from the SGSN for PS services) the order to allocate the corresponding radio resources to the terminal. This will initiate the establishment of a radio bearer through the corresponding RRC procedure, which starts with the transmission of a *Radio Bearer Setup* message from the RRC at the SRNC to the peer entity at the UE side. This message includes all the parameters to configure the RLC/MAC/PHY layers according to the transport and physical channels that are being assigned for both the uplink and downlink direction (e.g. transport channel type, code sequence, TFCS, RLC mode, etc.). Note that depending on the service nature (i.e. CS or PS) and the specific service requirements, this procedure may or may not involve the establishment of a dedicated channel. Similarly, and in the case when previous physical dedicated channels are already allocated to the user, the procedure may involve the modification of the physical channel characteristics. When the transport and physical channels allocated in the radio bearer are successfully established, the RRC at the UE side will issue a *Radio Bearer Setup Complete* message. In the case of failure, it will issue a *Radio Bearer Setup Failure* message.

  Figure 3.44 tries to summarise in a schematic way the sequence of the RRC procedures that are required to establish a service between a UE and the Core Network. RRC messages are shown between UE and SRNC with the corresponding logical channel while RANAP messages are shown between the SRNC and the CN.

  At service termination, the release procedure will be initiated by the network with the *Radio Bearer Release* message, whose response in the case of a successful release of the allocated radio resources will be the *Radio Bearer Release Complete*.

- Reconfiguration procedures. At any instant during a service lifetime, the network may decide on a modification of the resources that are allocated to this service. Normally this procedure would be the result of specific Radio Resource Management decisions depending on the network status at a given moment. There are three reconfiguration procedures depending on the degree of reconfiguration that is required. The lowest degree of reconfiguration is achieved with the Physical Channel Reconfiguration, which modifies the physical channels that are assigned to a given terminal (e.g. the code sequences, the carrier frequency or the maximum allowed uplink power) and can be executed during a handover procedure. This procedure involves replacing the current physical resources by the new ones, but it does not imply changes in the transport channel. In turn, the Transport Channel Reconfiguration is used to modify the characteristics of the currently assigned transport channel, so that it can be used, for instance, for a transport channel type switching (i.e. moving from a DCH to a RACH/FACH channel) or for a change in the TFCS that can be used by the mobile. The transport channel reconfiguration procedure may also involve the reconfiguration of the physical channels. Finally, the

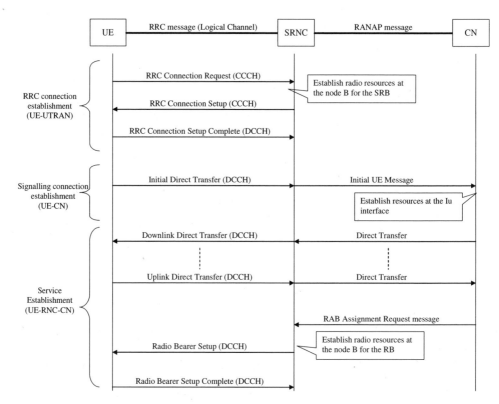

Figure 3.44 Summary of the RRC procedures required to establish a service

highest degree of reconfiguration is achieved by the Radio Bearer Reconfiguration procedure, which apart from changing the transport and physical channels, may also modify the RLC and PDCP layers or the MAC multiplexing of different logical channels. The three reconfiguration procedures are initiated by the RRC layer of the UTRAN with the corresponding *Radio Bearer/Transport Channel/Physical Channel Reconfiguration* message and acknowledged by the peer RRC layer at the UE side with the *Radio Bearer/Transport Channel/Physical Channel Reconfiguration Complete* message.

A special case of transport channel reconfiguration occurs when the network simply modifies the TFCS that can be used by the UE in the uplink direction, in order to reduce or increase the maximum instantaneous bit rate and consequently control the existing interference. In this case, the network sends a *Transport Format Combination Control* message that includes the new subset of transport format combinations that the MAC layer can select.

3.5.3.3 RRC Connection Mobility Procedures

This set of procedures handles the functions required to reconfigure the resources that are allocated to the terminals depending on their geographical location and on their RRC state. The procedures are:

- UTRAN mobility information. This procedure is used to allocate either a new RNTI or new mobility related information to the mobile. It is initiated by the network with the *UTRAN Mobility Information* message and confirmed by the mobile with the *UTRAN Mobility Information Confirm* message.

- Cell and URA update. The Cell Update procedure is mainly used by mobiles in Cell FACH and Cell_PCH states in order to indicate to the network that they have entered in a new cell. Similarly, the URA update is used by mobiles in the URA_PCH state to indicate the entrance in a new registration area. This could lead to a change in the RNTI assigned to the terminal. The procedure is triggered by the mobile with the *Cell/URA Update* message and confirmed by the network with the *Cell/URA Update Confirm* message. If the latter includes the new RNTI, then a *UTRAN Mobility Information Confirm* message is required from the mobile. The confirmation from the network may also require a physical or a transport channel reconfiguration.
- Active Set update. This procedure is triggered by the network for mobiles in Cell_DCH state indicating the requirement to modify the Active Set – i.e. the cells to which the mobile is simultaneously connected in soft handover – by adding or removing cells. The decision is taken by the network depending on the measurements reported by the mobile. The procedure is initiated by the *Active Set Update* message and confirmed with the *Active Set Update Complete* message issued by the mobile.
- Hard Handover. This procedure is used to change the allocation of resources of one mobile in Cell_DCH state from one UTRAN carrier frequency to another, thus executing an inter-frequency handover. It is executed by means of any of the reconfiguration procedures explained in Section 3.5.3.2 (i.e. physical channel, transport channel or radio bearer reconfiguration).
- Inter-RAT handover and cell reselection procedures. These procedures are used to execute handovers between another RAT (e.g. GSM/GPRS) and UTRAN for users in Cell_DCH state or to execute cell reselections involving different RATs for users in Cell_FACH, Cell_PCH or URA_PCH states.

3.5.3.4 Measurement Procedures

These procedures allow the mobile terminals to provide the network with different measurement reports that will be used by the radio resource management strategies to take the appropriate decisions that maintain the required QoS for the accepted mobiles.

The network configures the measurements that should be provided by the mobile terminal by indicating the objects to be measured (i.e. the cells, the transport channels and the physical channels), the criteria to be used (i.e. periodic reporting or event-triggered reporting when certain events are detected at the UE) and the RLC mode to be used (i.e. acknowledged or unacknowledged). This configuration is done by means of the *Measurement Control* message. Measurements are required only by terminals in Cell_DCH and Cell_FACH states, although in some cases such as traffic volume monitoring, terminals in Cell_PCH may also send measurement reports.

The Measurement Reports provided by the terminals include several types of measurements, which are classified into the following groups [20]:

- Intra-frequency measurements. These correspond to downlink physical channels in the cells with the same frequency as the cells from the Active Set. The measured cells belong to the Monitored Set, which is broadcast in the cell where the mobile is allocated. These measurements include [9][31]:
  - Ec/No of the primary CPICH channel, which is equivalent to the ratio between the power of the pilot channel and the total received power at the antenna connector.
  - Downlink path loss, which can be measured as the difference between the transmitted and the received CPICH power. The transmitted CPICH power is broadcast by the network.
  - Downlink RSCP (Received Signal Code Power) for the primary CPICH, corresponding to the power measured at the code of the primary CPICH.
  - Measured time difference between P-CCPCH frames of the different cells.
- Inter-frequency measurements. These are done over downlink physical channels of cells with a different frequency to that of the cells in the active set. The measurements included in this group are essentially the same as in the intra-frequency measurements.

- Inter-RAT measurements. These are done over cells from other RATs, like GSM/GPRS, and will be required to decide the execution of inter-RAT handovers. The measured quantities for GSM cells include the GSM carrier RSSI and the observed time difference with respect to the GSM cell.
- Traffic volume measurements. These include uplink measurements of the RLC buffer occupancy, providing instantaneous and average values as well as the measured variance.
- Quality measurements. These report downlink quality parameters, such as the transport block error rate for specific transport channels.
- UE internal measurements. These measurements include the UE transmission power, the UE received RSSI and the observed difference between reception and transmission times (i.e. the difference between the start of the uplink DPCCH/DPDCH transmission and the reception of the first path of the downlink DPCH).

3.5.3.5 General Procedures

The specification of the RRC protocol also defines a set of general procedures and some actions to be executed when an information element is missing in the received messages. Some of the defined general procedures include [20]:

- Selection of initial UE identity to be used in the RRC Connection Establishment procedure.
- Actions when entering idle mode form connected mode.
- Open loop power control upon the establishment of an uplink DPCCH.
- Actions to be taken when the mobile is in an 'out of service area' and when it re-enters an 'in service area' and to detect these situations.
- Criteria to detect radio link failure and actions to be taken when it occurs.
- Open loop power control for PRACH and PCPCH transmission, as described in Section 3.3.4.5.
- Maintenance of Hyper Frame Numbers used in ciphering and integrity protection procedures.
- Establishment of Access Service Classes to be used in PRACH transmission.
- PRACH selection.
- Selection of the RACH TTI.
- PLMN selection.

3.6 EXAMPLES OF RADIO ACCESS BEARERS

The above sections have detailed the different layers of the UTRAN FDD radio interface in terms of protocols and functions. The configuration of these layers must be done taking into account the specific service requirements that are indicated during service establishment in the signalling exchanged at the NAS. The result of this signalling will be the requirement for the access network to establish a Radio Bearer between the RNC and the UE that, jointly with the Iu bearer between the RNC and the CN, creates a Radio Access Bearer, as explained in Section 3.1 (see Figure 3.2). The establishment of the resources that form the Radio Bearer is done by means of the RRC protocol, as explained in Section 3.5.3.

The UMTS architecture has been defined in a very flexible way, which in fact provides multiple possibilities for radio bearers and will allow the provision of very different services. However, in practice, this enormous flexibility revealed to be a problem due to incompatibility between terminals and networks and so it was decided to fully specify a set of radio bearers with which the initial mobiles in the market must be compliant [10], apart from others that could appear in the future.

The objective of this section is to provide some examples of radio bearer definitions in order to illustrate the concepts given throughout the chapter and to give the reader a practical view of how the radio interface is configured in specific situations for different services.

Table 3.7 Characteristics of the 3.4 kb/s SRB in the uplink

Higher layer	RAB/signalling RB User of Radio Bearer	SRB#1 RRC	SRB#2 RRC	SRB#3 NAS_DT High priority	SRB#4 NAS_DT Low priority
RLC	Logical channel type	DCCH	DCCH	DCCH	DCCH
	RLC mode	UM	AM	AM	AM
	Payload sizes (bits)	136	128	128	128
	Max. data rate (b/s)	3400	3200	3200	3200
	AMD/UMD PDU header (bits)	8	16	16	16
MAC	MAC header (bits)	4	4	4	4
	MAC multiplexing		4 logical channel multiplexing		
Layer 1	TrCH type		DCH		
	TB size (bits)		148		
	TFS				
	TF0 (bits)		0×148		
	TF1 (bits)		1×148		
	TTI (ms)		40		
	Coding type		Convolutional 1/3		
	CRC (bits)		16		
	Max. number of bits/TTI before rate matching		516		
	Max. number of bits/radio frame before rate matching		129		
	RM attribute		155–185		

3.6.1 SIGNALLING RADIO BEARER 3.4 kb/s THROUGH DCH

The characteristics of this Signalling Radio Bearer (SRB) are shown in Tables 3.7 and 3.8 for the uplink and downlink directions, respectively. It is established between RRC entities to support the transfer of signalling information that can be generated either at the RRC or the NAS. Therefore, the SRB includes a total of four SRB, denoted as SRB#1, SRB#2, SRB#3 and SRB#4 (as explained in Section 3.5.3.1), each

Table 3.8 Characteristics of the 3.4 kb/s SRB in the downlink

Higher layer	RAB/signalling RB User of Radio Bearer	SRB#1 RRC	SRB#2 RRC	SRB#3 NAS_DT High priority	SRB#4 NAS_DT Low priority
RLC	Logical channel type	DCCH	DCCH	DCCH	DCCH
	RLC mode	UM	AM	AM	AM
	Payload sizes (bits)	136	128	128	128
	Max. data rate (b/s)	3400	3200	3200	3200
	AMD/UMD PDU header (bits)	8	16	16	16
MAC	MAC header (bits)	4	4	4	4
	MAC multiplexing		4 logical channel multiplexing		
Layer 1	TrCH type		DCH		
	TB size (bits)		148		
	TFS				
	TF0 (bits)		0×148		
	TF1 (bits)		1×148		
	TTI (ms)		40		
	Coding type		Convolutional 1/3		
	CRC (bits)		16		
	Max. number of bits/TTI before rate matching		516		
	RM attribute		155–230		

Table 3.9   Parameters of the uplink physical channel for the 3.4 kb/s SRB

DPCH Uplink	Min. spreading factor	256
	Max. number of DPDCH data bits/radio frame	150
	Puncturing Limit	1

of them mapped onto a DCCH logical channel. The SRB#1 and SRB#2 are used to transfer messages from RRC in unacknowledged and acknowledged RLC modes, respectively, while the SRB#3 and SRB#4 are used for the Direct Transfer (DT) of NAS messages with high and low priority, respectively, also using the acknowledged RLC mode. Note in Tables 3.7 and 3.8 that the use of AM requires eight more bits in the RLC header, and therefore the payload, i.e. the amount of information from the higher layers that can be transferred, contains eight bits less. As a result, the rate of 3.4 kb/s is only achieved for SRB#1 in UM, while the rest of SRB in AM achieve 3.2 kb/s.

The MAC layer executes the multiplexing of the above four logical channels and therefore it requires 4 bits of the MAC header (i.e. the C/T field) to indicate the logical channel to which the corresponding MAC-SDU should be delivered. In any case, and for the four logical channels, the sum of the payload from higher layers, the RLC header and the MAC header lead to 148 bits, which corresponds to the size of the transport blocks delivered by the MAC layer to the physical layer through a DCH.

At the physical layer, the TTI is 40 ms, and the MAC layer can select two transport formats of the Transport Format Set, corresponding to no transmission (TF0) or transmission of a single 148 bits transport block (TF1). When selecting TF1, 16 CRC bits are added to the transport block and convolutional code with a rate of 1/3 applied, so taking into account that 8 tail bits are required in the coding (see Section 3.3.1) this leads to 516 bits $= (148 + 16 + 8)*3$. Since the TTI consists of four radio frames, the number of bits that should be transmitted per radio frame is 129.

The characteristics of the dedicated physical channels in the uplink and downlink are shown in Tables 3.9 and 3.10, respectively. In the uplink direction, the 129 bits are adjusted by means of rate matching to the value of 150 bits that is transmitted in the physical DPDCH channel, by using spreading factor $SF = 256$. Note that the puncturing limit is 1, which means that no puncturing is allowed for this DCH. In turn, in the downlink direction the slot format 4 is used (see Table 3.6) so the spreading factor is also 256 and the rate matching must extend the 129 bits up to 210 bits. Note that, while the uplink rate matching must add 21 bits, in the downlink it must add 81 bits. The rate matching (RM) attribute would be used when the DCH was multiplexed with other DCHs, allowing the addition of a different number of bits depending on the characteristics of the channels that are multiplexed. Furthermore, there is no need to apply DTX in the downlink direction because there are always 210 bits at the downlink physical channel if the selected transport format is different from TF0.

Table 3.10   Parameters of the downlink physical channel for the 3.4 kb/s SRB

DPCH Downlink	DTX position	N/A
	Minimum spreading factor	256
	DPCCH	
	Number of TFCI bits/slot	0
	Number of TPC bits/slot	2
	Number of Pilot bits/slot	4
	DPDCH	
	Number of data bits/slot	14
	Number of data bits/frame	210

3.6.2 RAB FOR A 64/384 kb/s INTERACTIVE SERVICE AND 3.4 kb/s SIGNALLING

This RAB defines the configuration of the radio interface parameters for a service that has a maximum bit rate of 64 kb/s in the uplink direction and 384 kb/s in the downlink. In the CN, this service is provided through the PS domain, and therefore the Iu bearer service will be established through the Iu_PS interface. In this example, the 64/384 kb/s Radio Bearer will be multiplexed with the 3.4 kb/s SRB explained in Section 3.6.1. Furthermore, this RAB assumes that a DCH transport channel is used, although this service could also be provided with other RABs using DSCH.

The characteristics of the uplink RB are shown in Table 3.11. At the RLC layer, it uses AM mode, since an interactive PS service tolerates some delay margin to allow for possible retransmissions. This mode requires a RLC header of 16 bits to indicate, among other things, the sequence number of the transmitted PDUs. The payload from upper layers, corresponding to the data traffic, is 320 bits. Therefore, AMD PDUs with a total of 336 bits are delivered to the MAC layer by means of a DTCH logical channel. At the MAC layer, no MAC header is required, because no multiplexing of different logical channels is done and furthermore the user identification is implicit due to the use of a DCH transport channel. Consequently, the MAC PDUs delivered to the physical layer (i.e. the transport blocks) contain 336 bits.

The Transport Format Set (TFS) is composed of five different transport formats, and the maximum bit rate of 64 kb/s is achieved with TF4, in which four transport blocks, each of them containing 320 payload bits from the upper layers, are transmitted in one TTI of 20 ms. For the other transport formats, the bit rates are 0 (TF0), 16 kb/s (TF1), 32 kb/s (TF2) and 48 kb/s (TF3). When the DCH containing the 64/384 kb/s RB is multiplexed with the DCH containing the 3.4 kb/s SRB of Section 3.6.1, the five transport formats of the RB are combined with the two transport formats of the SRB leading to a TFCS with ten TFC, shown in Table 3.12.

The corresponding physical layer parameters are presented in Table 3.13, and Table 3.14 shows the computation of the spreading factor for every TFC in the TFCS, reflecting that the minimum spreading factor to be used is 16. The transport block processing requires adding 16 CRC bits and applying the rate

Table 3.11 Characteristics of the 64/384 kb/s RB in the uplink

Higher layer	RAB/Signalling RB	RAB
RLC	Logical channel type	DTCH
	RLC mode	AM
	Payload size (bits)	320
	Max. data rate (b/s)	64000
	AMD PDU header (bits)	16
MAC	MAC header (bits)	0
	MAC multiplexing	N/A
Layer 1	TrCH type	DCH
	TB size (bits)	336
	TFS	
	TF0 (bits)	0×336
	TF1 (bits)	1×336
	TF2 (bits)	2×336
	TF3 (bits)	3×336
	TF4 (bits)	4×336
	TTI (ms)	20
	Coding type	Turbo Code (1/3)
	CRC (bits)	16
	Max. number of bits/TTI after channel coding	4236
	Uplink: Max. number of bits/radio frame before rate matching	2118
	RM attribute	130–170

Table 3.12 Uplink TFCS for the 64/384 kb/s RB + 3.4kb/s SRB

TFCS size	10
TFCS	(64 kb/s RB, 3.4 kb/s SRB) =
	(TF0, TF0), (TF1, TF0), (TF2, TF0), (TF3, TF0), (TF4, TF0),
	(TF0, TF1), (TF1, TF1), (TF2, TF1), (TF3, TF1), (TF4, TF1)

Table 3.13 Parameters of the uplink physical channel for the 64/384 kb/s RB + 3.4 kb/s SRB

DPCH	Min. spreading factor	16
Uplink	Max. number of DPDCH data bits/radio frame	2400
	Puncturing Limit	0.96

1/3 turbo code that requires four tail bits over the concatenation of transport blocks that are transmitted in the TTI. For example, in the case of using TF4, this leads to a total of $3*(4*(336+16)+4) = 4236$ bits per TTI, or equivalently 2118 bits per radio frame. Note that, in all the combinations, rate matching is done by repeating bits instead of puncturing, because the bit repetition does not require using an additional physical channel in any of the combinations. The specific number of bits repeated in the puncturing process for each channel depends on the specific value that takes the rate matching attribute for the channel.

Table 3.15 shows the characteristics of the radio bearer in the downlink direction. In this case, the TTI is 10 ms, and there are a total of six transport formats, that combined with the two transport formats of the SRB lead to a TFCS with 12 combinations, as shown in Table 3.16. In this case, the spreading factor takes the fixed value of eight for all of the combinations. Note that for the combination with the maximum bit rate (TF5,TF1), the total number of bits per radio frame would include 12 684 bits from the RB and 129 bits from the SRB, resulting in 12 813 bits. However, as indicated in Table 3.17, the spreading factor is eight, and the selected slot format 15 (see Table 3.6) only allows the transmission of 9120 bits per radio frame. Therefore, puncturing is necessary in the rate matching procedure. Note that bit repeating could have been done if the selected spreading factor was four, however, this would have imposed a strict limitation in the number of available code sequences, since only three users could have been allocated in the OVSF code tree.

Table 3.14 Computation of the uplink spreading factor for every TFC

TFCS		Bits/TTI before RM		Bits/frame before RM				
RB64	SRB	RB64	SRB	RB64	SRB	Total	Bits/frame after RM	SF
TF0	TF0	0	0	0	0	0	0	N/A
TF1	TF0	1068	0	534	0	534	600	64
TF2	TF0	2124	0	1062	0	1062	1200	32
TF3	TF0	3180	0	1590	0	1590	2400	16
TF4	TF0	4236	0	2118	0	2118	2400	16
TF0	TF1	0	516	0	129	129	150	256
TF1	TF1	1068	516	534	129	663	1200	32
TF2	TF1	2124	516	1062	129	1191	1200	32
TF3	TF1	3180	516	1590	129	1719	2400	16
TF4	TF1	4236	516	2118	129	2247	2400	16

Table 3.15 Characteristics of the 64/384 kb/s RB in the downlink

Higher layer	RAB/Signalling RB	RAB
RLC	Logical channel type	DTCH
	RLC mode	AM
	Payload size (bits)	320
	Max. data rate (b/s)	384000
	AMD PDU header (bits)	16
MAC	MAC header (bits)	0
	MAC multiplexing	N/A
Layer 1	TrCH type	DCH
	TB size (bits)	336
	TFS	
	TF0 (bits)	0×336
	TF1 (bits)	1×336
	TF2 (bits)	2×336
	TF3 (bits)	4×336
	TF4 (bits)	8×336
	TF5 (bits)	12×336
	TTI (ms)	10
	Coding type	Turbo Code (1/3)
	CRC (bits)	16
	Max. number of bits/TTI after channel coding	12684
	RM attribute	110–150

Table 3.16 Downlink TFCS for the 64/384 kb/s RB + 3.4 kb/s SRB

TFCS size	12
TFCS	(384 kb/s RB, 3.4 kb/s SRB) =
	(TF0, TF0), (TF1, TF0), (TF2, TF0), (TF3, TF0), (TF4, TF0), (TF5, TF0)
	(TF0, TF1), (TF1, TF1), (TF2, TF1), (TF3, TF1), (TF4, TF1), (TF5, TF1)

The downlink channel multiplexing at the physical layer is done by the use of flexible positions, which means that the starting position of each transport channel is variable for each TFC. This would be a problem for the terminal if it had to perform blind TFC detection. However, this is not the case for this RAB since TFCI bits are included in the frame, as shown in Table 3.17.

Table 3.17 Parameters of the downlink physical channel for the 64/384 kb/s RB + 3.4 kb/s SRB

DPCH Downlink	DTX position	Flexible
	Spreading factor	8
	Number of DPDCH	1
	DPCCH	
	Number of TFCI bits/slot	8
	Number of TPC bits/slot	8
	Number of Pilot bits/slot	16
	DPDCH	
	Number of data bits/slot	608
	Number of data bits/frame	9120

REFERENCES

[1] 3GPP TS 23.101 'General UMTS architecture'
[2] 3GPP TS 25.401 'UTRAN overall description'
[3] 3GPP TS 23.110 'UMTS Access Stratum: services and functions'
[4] 3GPP TS 25.301 'Radio Interface Protocol Architecture'
[5] 3GPP TS 25.430 'UTRAN Iub interface: general aspects and principles'
[6] 3GPP TS 25.410 'UTRAN Iu Interface: general aspects and principles'
[7] 3GPP TS 25.211 'Physical channels and mapping of transport channels onto physical channels (FDD)'
[8] 3GPP TS 25.308 'High Speed Downlink Packet Access (HSDPA): overall description'
[9] 3GPP TS 25.302 'Services provided by the physical layer'
[10] 3GPP TS 34.108 'Common Test Environments for User Equipment (UE): conformance testing'
[11] 3GPP TS 25.101 'User Equipment (UE) radio transmission and reception (FDD)'
[12] 3GPP TS 25.104 'Base Station (BS) radio transmission and reception (FDD)'
[13] 3GPP TS 25.213 'Spreading and Modulation (FDD)'
[14] 3GPP TS 25.201 'Physical layer – General description'
[15] 3GPP TS 25.212 'Multiplexing and channel coding (FDD)'
[16] R. Tanner, J. Woodard, *WCDMA Requirements and Practical Design*, John Wiley & Sons Ltd, 2004
[17] T. Minn, K.Y. Seu, 'Dynamic Assignment of Orthogonal Variable Spreading Factor Codes in W-CDMA', *IEEE Journal on Selected Areas in Communications*, pp. 1429–1440, August 2000
[18] H. Holma, A. Toskala, *WCDMA for UMTS*, John Wiley & Sons Ltd, 2nd edition, 2002
[19] 3GPP TS 25.214 'Physical layer procedures (FDD)'
[20] 3GPP TS 25.331 'Radio Resource Control (RRC): protocol specification'
[21] 3GPP TS 25.321 'Medium Access Control (MAC): protocol specification'
[22] 3GPP TS 35.202 '3G Security: specification of the 3GPP confidentiality and integrity algorithms; Document 2: KASUMI specification'
[23] 3GPP TS 25.322 'Radio Link Control (RLC): protocol specification'
[24] 3GPP TS 25.323 'Packet Data Convergence Protocol (PDCP) specification'
[25] IETF RFC 2507 'IP Header Compression'
[26] IETF RFC 3095 'RObust Header Compression (ROHC): framework and four profiles: RTP, UDP, ESP, and uncompressed'
[27] 3GPP TS 25.324 'Broadcast/Multicast Control (BMC)'
[28] 3GPP TS 25.303 'Interlayer procedures in Connected Mode'
[29] 3GPP TS 25.304 'User Equipment (UE) procedures in idle mode and procedures for cell reselection in connected mode'
[30] 3GPP TS 25.413 'UTRAN Iu interface RANAP signalling'
[31] 3GPP TS 25.215 'Physical layer – Measurements (FDD)'

4

Basics of RRM in WCDMA

In this chapter, the basics of Radio Resource Management (RRM) in WCDMA are described by following a step by step approach, with four main blocks. In the first block, the main basic concepts of RRM are introduced. In particular, Section 4.1 describes the concept of the Radio Resource Unit (RRU). Then, Section 4.2 explains how radio network planning achieves the provision of RRU along the coverage area. Section 4.3 provides a high level description of the RRM mechanisms necessary to allocate RRUs to the different users. In the second block, a generic WCDMA air interface is analytically characterised in Section 4.4 to reveal the main parameters and relationships within those parameters influencing the radio access performance. To this end, some concepts related to users and traffic are introduced. Models for both uplink and downlink are covered. In order to stress the relevance of intercell interference, a single cell model is first introduced and then multiple cells are considered. With the background provided by the first two blocks, in Section 4.5 the third block identifies the main RRM functions to be considered in a WCDMA radio access network. Some initial considerations on each RRM function are included, although particular realisations (i.e. RRM algorithm examples) and evaluation results will be covered in Chapter 5. Finally, the fourth and last part of this chapter, Section 4.6, is devoted to highlighting some intrinsic characteristics that are relevant at the RRM level, such as service heterogeneity, indoor traffic, non-homogeneous spatial traffic distributions and traffic hot spots.

4.1 RADIO RESOURCE CONCEPT

A Radio Resource Unit (RRU) is defined here by the set of basic physical transmission parameters necessary to support a signal waveform transporting end user information corresponding to a reference service. In particular:

- In FDMA, a radio resource unit is equivalent to a certain bandwidth within a given carrier frequency. For example, in TACS (Total Access Communication System), a radio resource unit is a 25 KHz portion in the 900 MHz band.
- In TDMA, a radio resource unit is equivalent to a pair consisting of a carrier frequency and a time slot. For example, in GSM, a radio resource unit is a 0.577 ms time slot period every 4.615 ms on a 200 KHz carrier in the 900 MHz, 1800 MHz or 1900 MHz bands.
- In CDMA, a radio resource unit is defined by a carrier frequency, a code sequence and a power level. The main difference arising here with respect to other techniques is that, as presented in Chapter 2, the required power level necessary to support a user connection is not fixed but depends on the interference level. For example, in UTRAN FDD a 5 MHz portion in the 2 GHz band as well as a

pair of OVSF and scrambling codes are identified to support a given service. Nevertheless, the amount of transmitted power resources will vary over time according to multiple elements of the scenario, such as propagation conditions, interference, cell load level, etc.

In addition to the main physical dimensions (frequency, time slot, code sequence and power level), there are other physical transmission elements such as the modulation scheme, channel coding scheme, etc. Clearly, depending on the exploitation of the basic dimensions in terms of the former elements, different efficiencies may follow. Nevertheless, for the conceptual definition of a radio resource unit, only the referred main transmission parameters will be retained.

It is worth noting that, in a multiservice scenario, each service may require that a different amount of radio resource units are supported. Services with higher bit rates will, consequently, require more radio resource units. These will be additional frequency bands if the access mechanism is FDMA, additional time slots in the case of TDMA or additional code sequences together with higher transmitted power levels in the CDMA case.

4.2 RADIO NETWORK PLANNING

The objective of a network operator is the deployment of a network able to support its customers with the required QoS in a target coverage area. To this end, the overall network involves several sub-problems, covering the radio network, the transmission network and the core network designs.

Focusing on the radio part, the problem is to provide enough RRUs along the service area with sufficient quality. The principles of radio network planning as well as a practical example will be described in more detail in Chapter 5.

It is worth noting that the operator's investment in radio network infrastructure [1] is proportional to the number of base stations or nodes B deployed, which, in turn, increases with:

- The number of users to be supported. The higher the number of users, the higher the network capacity must be, therefore the requirement for more base stations.
- The service area. The higher the area to provide service, the higher the number of base stations.
- The user's average transmission rate. The higher the bit rate, the higher the amount of radio resources needed.
- The desired QoS level. The better the QoS, the higher the amount of required radio resources.

Customers of 2G are used to good subjective voice service quality, with extended service coverage and good network accessibility under normal operation conditions with current 2G network deployment. Nevertheless, in the context of 3G networks, it is envisaged to provide data services at much higher bit rates as well as quantified QoS guarantees. Then, these terms would contribute to the increase of the infrastructure investment required to provide suitable 3G services.

The cost term associated with QoS provisioning plays a key role. The most basic way to guarantee QoS is by means of network overdimensioning and radio resource overprovisioning. Clearly, the challenge is to be able to provide the desired QoS level with the minimum possible resources, thus minimising the operator's investment while meeting network design requirements. Furthermore, 3G wireless networks need to support a variety of services including those that are already well-defined, as well as those that will emerge in the future. Therefore, the QoS framework for 3G air interface must be flexible and should also be practical, i.e. it should have low complexity of implementation and low volume of control signalling.

In summary, the result of the radio network planning will be the provision of RRUs along the service area by means of a certain radio network topology and a given configuration of the cell sites. It is worth noting that, as long as aspects such as service penetration and service usage vary with regard to time and space, the amount of RRUs to be provisioned also varies temporally and spatially. Consequently, the radio network planning is an evolving process. Nevertheless, the inertias associated with radio network deployment (e.g. site acquisition, civil engineering for site preparation, etc.) make this process able to respond to sustained and long-term variations of the radio network planning input parameters.

4.3 RADIO RESOURCE MANAGEMENT

As stated above, radio network rollout responds to a set of RRUs requirements, so that a given amount of RRUs are provisioned along the network service area at a certain time. Therefore, a suitable allocation of the provisioned RRUs to the different users in the network is needed, as they are requesting services and moving around. RRM functions are in charge of allocating and managing the provisioned RRUs, as reflected in Figure 4.1.

Cellular mobile communications are dynamic in nature. Dynamism arises from multiple dimensions: propagation conditions, traffic generation conditions, interference conditions, etc. Thus, the dynamic network evolution calls for a dynamic management of the radio resources, which is carried out by RRM mechanisms with an associated large number of parameters that need to be chosen, measured, analysed and optimised. Besides, RRM mechanisms may overcome to some extent the long term reactivity of radio network planning and deployment, which otherwise would prevent the network operator from accommodating sudden and transient traffic increases.

Radio Resource and QoS management functionalities are very important in the framework of 3G systems because the system relies on them to guarantee a certain target QoS, maintain the planned coverage area and offer a high capacity, objectives which tend to be contradictory (e.g. capacity may be increased at the expense of a coverage reduction; capacity may be increased at the expense of a QoS reduction, etc.). Radio network planning provides a thick tuning of these elements, while RRM will provide the fine tuning mechanisms that allow a final matching. This is reflected in Figure 4.2, where (a) represents the contradictory objectives in a radio network, which can be made compatible by (b) radio network planning and (c) eventually with RRM techniques.

In WCDMA, users transmit at the same time and frequency by means of different spreading sequences, which in most of the cases are not perfectly orthogonal. Consequently, there is a natural coupling among the different users that makes the performance of a given connection much more dependent on the behaviour of the rest of the users sharing the radio interface compared with other multiple access techniques like FDMA or TDMA. In this context, RRM functions are crucial in WCDMA because there is not a constant value for the maximum available capacity, since it is tightly coupled to the amount of interference in the air interface. Although an efficient management of radio resources may not involve an important benefit for relatively low loads, when the number of users in the system increases to a critical number, good radio resources management will be absolutely necessary.

RRM functions can be implemented in many different algorithms, and this impacts on the overall system efficiency and on the operator infrastructure cost. So, without doubt, RRM strategies will play an important role in a mature UMTS scenario. Additionally, RRM strategies are not subject to standardisation, so there can be a differentiation issue among manufacturers and operators.

In general terms, real time services have more stringent QoS requirements compared to non real time applications and, consequently, the former will require more investment by the network operator than the latter. Nevertheless, if the amount of available radio resources is too low, non real time users may experience a non-satisfactory connection, usually in terms of an excessive delay. Then, it will be necessary for the network operator to set some target QoS values for non real time applications as well.

Figure 4.1   Relationship between Radio Network Planning and Radio Resource Management

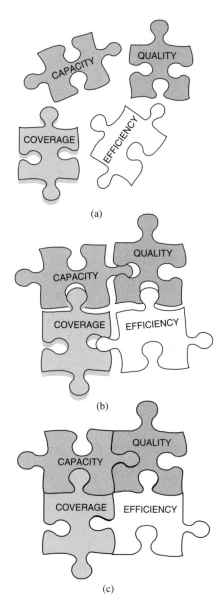

(a)

(b)

(c)

Figure 4.2 Capacity, quality, coverage and efficiency trade-offs.

QoS parameters may be classified into two different levels: network-level (such as blocking probability, dropping probability, etc.), and connection-level (such as bit error rate, maximum transmission rate, etc.).

In order to introduce the elements in the radio interface that RRM strategies will be able to manage, let's focus on the BER (Bit Error Rate) as a representative QoS parameter at connection-level, which is given as a function of the measured E_b/N_0, so that

$$BER = f\left(\frac{E_b}{N_0}\right)$$

(4.1)

where f is a monotonically decreasing function that depends on physical layer parameters like the channel coding and interleaving scheme, the channel impulse response, etc. In any case, a target BER requirement can be directly related to a target E_b/N_0 according to this function.

Further exploring the dependences on the radio interface conditions, the different terms involved in Equation (4.1) can be expressed for the ith user as a function of the total signal to interference and noise ratio according to:

$$\left(\frac{E_b(t)}{N_0(t)}\right)_i = \frac{W}{R_{b,i}(t)} \frac{S_i(t)}{N_i(t)} \tag{4.2}$$

where W is the total bandwidth, $R_{b,i}(t)$ the bit rate of the ith user and $S_i(t)$ its power at the receiver input, depending on the transmitted power $P_{T,i}(t)$ and the path loss $L_i(t)$ between the ith user and its serving cell:

$$E_{b,i}(t) = \frac{P_{T,i}(t)}{L_i(t)R_{b,i}(t)} \tag{4.3}$$

In turn, $N_i(t)$ is the total noise and interference power at the antenna input, given by:

$$N_i(t) = P_N + \sum_{\substack{j=1 \\ j \neq i}}^{n} \frac{P_{T,j}(t)}{L_j(t)} \tag{4.4}$$

where P_N is the background noise and n is the number of users simultaneously transmitting, each one with a transmitted power $P_{T,j}(t)$ and a propagation loss $L_j(t)$ with respect to the serving cell of the ith user.

One of the most important RRM tasks is to guarantee that every single connection achieves the target E_b/N_o that ensures the BER requirement. The different time varying terms involved in Equation (4.2) clearly anticipate that this task is not at all simple, since it needs to transform a highly random situation into an as predictable and controlled one as possible. It can also be anticipated that, in order to achieve the ultimate objective of providing the desired BER, it is necessary to consider several RRM functions, each of them mainly targeted to control and monitor a subset of specific parameters. The basic differentiation element among the different RRM functions will be related to the operational time scales according to the degree and nature of the variability that the intended parameters exhibit.

From the previous equations, the following parameters play a role in the quality achieved by the ith connection, so that RRM functions will mainly need to cope with and get a suitable balance between:

- The number of simultaneous users. The interference level depends on the number of users sharing the radio interface, n. As will be further stated in Section 4.4.1, the number of simultaneous users is in general random, related to the number of camping and active users respectively. Thus, there are several mechanisms in the system that will allow the control of this parameter, either in a long time scale by accepting/rejecting new calls/sessions in the system or, in a short time scale, by allowing/refraining user's transmission on the radio interface on a frame-by-frame basis.
- Bit rate. The transmission bit rate primarily depends on the traffic source behaviour, which can provide a variable number of bits per time period and, consequently, is a non predictive component for many applications. Nevertheless, there are several mechanisms in the system that will allow the control of this parameter, thus constituting another identified RRM function. Changes in the bit rate are envisaged to occur in the order of frames (i.e. in the order of tenths of milliseconds for UTRAN), whereas in a given frame the transmission rate will be constant for a given source.
- Power level. Although the number of transmitting users n and their corresponding transmission rates $R_{b,i}$ are constant in a given frame, the propagation conditions, $L_i(t)$, exhibit very short-term variations, for example, associated with Rayleigh fading. The only single transmission parameter that can provide such short term reactivity is the transmitted power level, $P_{T,i}(t)$. Power control is the RRM

function that will provide this reactivity. It is worth noting that the multiuser interference level is an increasing function with the number of users transmitting in a given moment, n. Consequently, the required transmitted power of the ith user also increases when n increases, thus reflecting the coupling existing among the different simultaneous users in a WCDMA air interface. Increases in the multiuser interference level should be overcome with increases in the transmitted power level in order to keep the E_b/N_o constant and equal to the target value.

RRM algorithms may be centralised (i.e. located at a network entity like the RNC) or distributed (i.e. located at each UE). Centralised solutions may provide better performance compared to distributed solutions because much more RRM relevant information related to all users involved in the process may be available at the RNC. Nevertheless, in the uplink direction, executing decisions taken by centralised RRM algorithms would require a higher amount of control signalling to inform each UE about the decisions. Consequently, an intermediate solution with both centralised and decentralised components is taken in the uplink of 3GPP. In particular, the instantaneous bit rate is selected at the UE in a decentralised way taking into account the maximum allowed bit rate that is set by the network in centralised way.

Figure 4.3 illustrates the role of the RRM strategies as well as the framework for their operation. As can be observed, the objectives of the RRM mechanisms are to provide the desired QoS level and ensure the target coverage while maximising the radio network capacity and efficiency. To this end, RRM techniques will mainly manage a set of radio parameters, such as the serving cell, maximum bit rate, instantaneous bit rate, transmitted power, etc., whose purpose is to allocate the necessary RRUs in the network to achieve such objectives. To set all these parameters, the support of the Radio Resource Control protocol, as explained in Chapter 3, is necessary. With RRC protocol messages, the network (i.e. RNC) is able to coordinate and manage the allocation of RRUs to every UE.

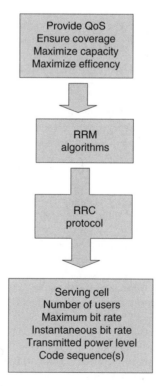

Figure 4.3 Objectives of the RRM algorithms, supporting protocol and managed parameters

4.4 AIR INTERFACE CHARACTERISATION

Since RRM strategies will aim to monitor and control the air interface, a model for the WCDMA radio access will now be introduced. It will be useful to devise the different RRM strategies needed as well as the design principles behind them in a more intuitive perspective.

At this stage, and before moving onto the above mentioned WCDMA characterisation, some definitions regarding the concept of 'user' will be described, since it is fairly common to use this term to indicate different situations.

4.4.1 CAMPING, ACTIVE AND SIMULTANEOUS USERS

It readily follows from Equation (4.4) that the number of transmitting users in a frame, n, plays a key role in the radio interface management since it mainly impacts on the interference level and must be controlled in order to keep the required transmitted power levels bounded. Nevertheless, the value n exhibits complex dependences. In order to explore these dependences, let us consider an area where a cellular system is deployed and there are M subscribers in the area generating calls (or sessions). Call arrivals are assumed to follow a Poisson distribution with an average call rate of λ calls/s per user. Let us also assume that call duration is exponentially distributed with an average value of $1/\mu$ s. Then, a continuous time birth-death Markov chain can be built [2], where the states are given by the number of users having a call in progress, usually denoted as *active users*. The birth coefficient λ_N and death coefficient μ_N in the state N (i.e. there are N users with a call in progress) are given by:

$$\lambda_N = (M - N)\lambda \tag{4.5}$$

$$\mu_N = N\mu \tag{4.6}$$

Under equilibrium conditions, it can be obtained that the probability of having N users with a call in progress is given by:

$$p_N = \frac{\binom{M}{N}\left(\frac{\lambda}{\mu}\right)^N}{\left(1+\frac{\lambda}{\mu}\right)^M} = \binom{M}{N}\frac{\rho^N}{(1+\rho)^M} \tag{4.7}$$

where it has been considered that $\rho = \lambda/\mu$.

Furthermore, during a call or session, a traffic source usually alternates active and silent periods. This behaviour is usually characterised by an activity factor α, representing the fraction of time that the source is generating traffic. Then provided that there are N users in a call, the conditional probability of n simultaneous users occupying the radio interface is given by:

$$p_{n|N} = \binom{N}{n}\alpha^n(1-\alpha)^{N-n} \tag{4.8}$$

Therefore, the probability of n users simultaneously transmitting can be computed as:

$$p_n = \sum_{N=n}^{M} p_{n|N}p_N = \binom{M}{n}\frac{(\alpha\rho)^n(1+(1-\alpha)\rho)^{M-n}}{(1+\rho)^M} \tag{4.9}$$

Although in a practical situation the methods of analysis are complex and require several hypotheses in order to become tractable, the above equations are useful in the sense that they provide insight into the basic behaviour. In particular, three different concepts and a framework for their relationships have arisen: M, the number of users camping on a cell; N, the number of users with a call in progress; and n, the number of users transmitting in a given moment.

In an interference-limited scenario like WCDMA, the relevant parameter is the number of simultaneous users n because it impacts strongly on the total interference. An excessive interference level caused by too high a n can be caused by too large a α and/or N and/or M. Consequently, the multiple variables involved in the determination of n suggest the need for multiple control levels or, in other words, multiple RRM functions involved in the control of the number of simultaneous users.

M depends on the population density in the area of the cell as well as on the service penetration. The long term variation of M indicates that this parameter is mainly controlled through a suitable radio network planning, so that the network deployment, i.e. the number of cells and their configuration, is such that the resulting number of users camping on a cell's planned coverage area is manageable with the amount of radio resources devoted to that cell.

N depends, on the one hand, on M and, on the other hand, on the call birth/death process related to the user's behaviour with respect to service usage. For a given M, the independence among the different call generation processes of the different subscribers makes N variable. Then, the admission control will be the RRM function in charge of controlling N, the number of users accepted on the cell for a connection. The main performance figure related to admission is the blocking probability (i.e. the probability that a user makes a call attempt and is not allowed to establish the connection).

Admission control would allow the absorbtion of deviations in N, due either to deviations in M with respect to the expected level or deviations in the user's call generation process from the expected behaviour (e.g. more call attempts than expected, longer calls than expected, etc.). In such situations, the resulting blocking probability would be higher than planned.

For a given N, the number of simultaneous users depends on their activity factor and on the traffic multiplexing process, so that n can be variable. In the case of deviations of n, the congestion control mechanism needs to be activated in order to preserve the performance of the different connections. The variations of n would occur in the short term, in the order of frames (i.e. in the order of tenths of milliseconds for UTRAN), so that the congestion control mechanism will be coupled at some extent with the bit rate control mechanism (note that preventing a user's transmission is equivalent to setting its transmission rate to 0 bit/s).

As an example, Figure 4.4 shows the cumulative distribution function (CDF) of the number of users in a call, N, for the case of a population of $M = 100$ camping users and $\rho = 0.5$. It can be seen that, although

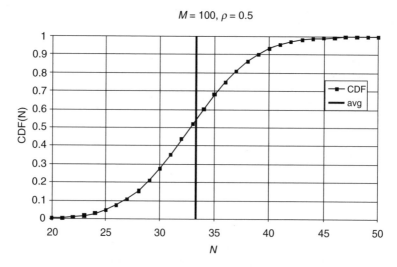

Figure 4.4 Cumulative distribution function of the number of active users N

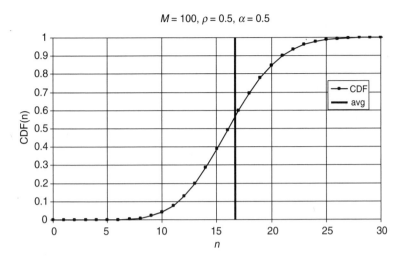

Figure 4.5 Distribution of the number of simultaneous users for activity factor $\alpha = 0.5$

the average value is 33 users, the probability of having more than 38 active users at the same time is around 10%. Similarly, the probability of having less than 27 active users at the same time is also around 10%. Clearly, uncertainties about the user's behaviour with respect to call/session generation need to be considered in the design of RRM algorithms.

As for the number of simultaneous users, retaining the case where $M = 100$ and $\rho = 0.5$, Figure 4.5 reflects the cumulative distribution function of the number of simultaneous users for an activity factor of 50%. Although the average number of simultaneous users is around 16, there is a 10% probability of having more than 21 users. The deviation in n reflects a wide variation in the interference level observed in the air interface.

Similarly, Figure 4.6 plots the probabilistic distribution of simultaneous users for the case where $M = 100$ and $\rho = 0.5$ when the activity factor is only 10%. Clearly, and by comparison with Figure 4.5, the average number of simultaneous users is much lower in this case. The same can be said of the standard deviation. Consequently, and for a proper interference control, it will be necessary to consider

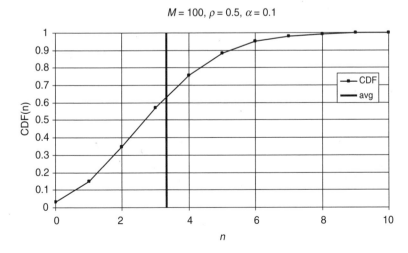

Figure 4.6 Distribution of the number of simultaneous users for activity factor $\alpha = 0.1$

both the behaviour of the user at the call/session generation process as well as the traffic generation process within a call/session.

4.4.2 UPLINK: SINGLE CELL CASE

Within a WCDMA cell, all users share the common bandwidth and each new connection increases the interference level of other connections, affecting their quality expressed in terms of a certain E_b/N_o. Let us consider an isolated cell. For n simultaneous users transmitting in a given frame, the following inequality must be satisfied for the ith user:

$$\frac{P_i(W/R_{b,i})}{P_N + [P_R - P_i]} \geq \left(\frac{E_b}{N_o}\right)_i \tag{4.10}$$

$$P_R = \sum_{i=1}^{n} P_i \tag{4.11}$$

where P_i is the ith user received power at the base station, $R_{b,i}$ is the ith user bit rate, P_N is the background noise power and $(E_b/N_o)_i$ stands for the ith user requirement. P_R is the total received own-cell power at the base station. Then, the required received power level for the ith user is:

$$P_i \geq \frac{P_N + P_R}{\dfrac{(W/R_{b,i})}{\left(\dfrac{E_b}{N_o}\right)_i} + 1} \tag{4.12}$$

Adding all n inequalities, it holds:

$$\sum_{i=1}^{n} P_i = P_R \geq \sum_{i=1}^{n} \frac{P_N + P_R}{\dfrac{(W/R_{b,i})}{\left(\dfrac{E_b}{N_o}\right)_i} + 1} \tag{4.13}$$

Then, P_R can be isolated, assuming equality, as:

$$P_R = \frac{P_N \displaystyle\sum_{i=1}^{n} \dfrac{1}{\dfrac{(W/R_{b,i})}{\left(\dfrac{E_b}{N_o}\right)_i} + 1}}{1 - \displaystyle\sum_{i=1}^{n} \dfrac{1}{\dfrac{(W/R_{b,i})}{\left(\dfrac{E_b}{N_o}\right)_i} + 1}} \tag{4.14}$$

Substituting P_R in Equation (4.12) gives:

$$P_i = \frac{P_N}{\left[\dfrac{W/R_{b,i}}{\left(\dfrac{E_b}{N_o}\right)_i} + 1\right]\left[1 - \displaystyle\sum_{i=1}^{n} \dfrac{1}{\dfrac{W/R_{b,i}}{\left(\dfrac{E_b}{N_o}\right)_i} + 1}\right]} \tag{4.15}$$

Claiming in Equation (4.15) that $P_N > 0$ leads to:

$$\sum_{i=1}^{n} \frac{1}{\dfrac{W/R_{b,i}}{\left(\dfrac{E_b}{N_o}\right)_i} + 1} < 1 \tag{4.16}$$

If all users in the cell are of the same type, the above conditions clearly indicate that the uplink WCDMA capacity is limited to:

$$n < 1 + \frac{(W/R_{b,i})}{\left(\dfrac{E_b}{N_o}\right)_i} \tag{4.17}$$

Thus, the need to control the number of simultaneous transmissions n on the air interface is clearly stressed by Equation (4.16). If Equation (4.16) does not hold, it means that it is not possible for all n simultaneous users to achieve their respective E_b/N_o targets simultaneously.

4.4.2.1 Capacity and Coverage

Capacity and coverage are closely related in WCDMA networks, and therefore both must be considered simultaneously. The coverage problem is directly related to the power availability, so the power demands deriving from the system load level should be in accordance with the planned coverage. Therefore, it must be satisfied that the required transmitted power is lower than the maximum power available at the UE side, P_{Tmax}, and high enough to be able to get the required E_b/N_o target even at the cell edge.

From Equation (4.15), and assuming a path loss $L_{p,i}$ for the ith user, its transmission power requirement is given by:

$$P_{T,i} = \frac{L_{p,i} P_N}{\left[\dfrac{W/R_{b,i}}{\left(\dfrac{E_b}{N_o}\right)_i} + 1 \right] \left[1 - \sum_{i=1}^{n} \dfrac{1}{\dfrac{W/R_{b,i}}{\left(\dfrac{E_b}{N_o}\right)_i} + 1} \right]} \tag{4.18}$$

The ith user is said to be in outage if the required transmitted power given by Equation (4.18) is higher than P_{Tmax}, which is equivalent to saying that the ith user measures at the cell site an E_b/N_o below the target value. If this situation persists, the ith user is said to be out of coverage and experiences no service availability.

The outage probability can then be computed according to:

$$\Pr\left(P_{T,i} > P_{T,\max}\right) = \sum_{n=1}^{M} \Pr\left(P_{T,i} > P_{T,\max} \mid n\right) \cdot p_n \tag{4.19}$$

$$\Pr\left(P_{T,i} > P_{T,\max}\right) = \sum_{n=1}^{M} \Pr\left(L_{p,i} > \frac{P_{T,\max}}{P_N} \left[\frac{W/R_{b,i}}{\left(\dfrac{E_b}{N_o}\right)_i} + 1 \right] \left[1 - \sum_{i=1}^{n} \frac{1}{\dfrac{W/R_{b,i}}{\left(\dfrac{E_b}{N_o}\right)_i} + 1} \right] \right) \cdot p_n \tag{4.20}$$

where p_n is the probability of having n simultaneous transmissions given by Equation (4.9).

It can be observed from Equation (4.20) that, given a number of n simultaneous users transmitting at a certain bit rate and with certain target qualities, there is a maximum path loss tolerable that makes it feasible to obtain the target quality for each connection. Consequently, the path loss distribution will determine the outage probability in a given scenario. An analytical model for computing the path loss in a cellular scenario is presented in Appendix 4.1.

To support this idea, let us consider a reference user characterised by a target $E_b/N_o = 3$ dB, a maximum transmitted power of $P_{T,max} = 21$ dBm and a transmission bit rate of $R_b = 64$ kb/s. This user has a path loss of 120 dB to the cell site and remains still. Furthermore, let us observe the air interface for a period of time, where the number of simultaneous users varies according to a certain traffic generation profile. Figure 4.7 illustrates this situation, showing the time evolution of different measurements. In the

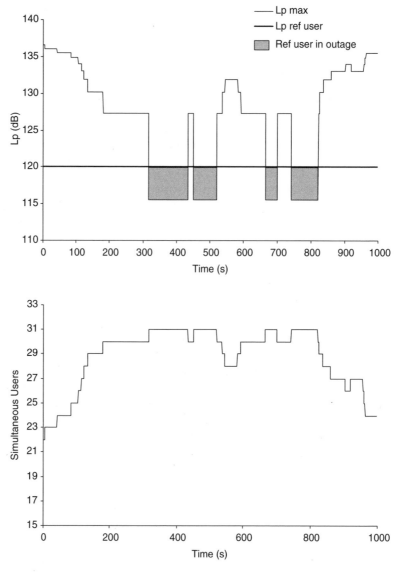

Figure 4.7   Maximum allowed path loss for a reference user depending on the number of simultaneous users n

lower part of Figure 4.7, the number of simultaneous users along time is plotted, while in the upper part the maximum tolerable path loss for the reference user as a result of the varying interference level is shown. As the number of users increases the maximum allowed path loss for the reference user decreases. In some cases (highlighted in grey in upper Figure 4.7) the result is that the reference user's path loss is higher than the maximum tolerable one. In such cases, the reference user would be in outage, so that it is not able to reach the cell site with enough power to achieve the target E_b/N_o. Therefore, it can be seen that the performance achieved depends on the cell load level or, equivalently, on the air interface interference level. This phenomenon is known as *cell breathing*, since it turns into a variable cell coverage.

Figure 4.8 presents the cumulative distribution function of the required uplink transmitted power level for different numbers of simultaneous users n. Results are obtained as the average of a sufficient large number of snapshots, where a number of simultaneous users n are randomly located within a cell, which is considered to have in this case a 500 m radius. The required transmitted power level of each as given by Equation (4.18) is captured. It can be observed that, the larger the n, the higher the required transmitted power levels. Furthermore, considering a maximum transmitted power level $P_{T,max} =$ 21 dBm, it can be observed that for $n = 10$ or 20 users, outage occurs very seldom, while for $n = 30$ the outage probability rises to about 10%. In order to assess the impact of the cell radius, Figure 4.9 shows the same result for a 1 km cell radius. It can be observed that, because of the higher propagation losses in this scenario, higher transmitted power levels are required and, consequently, higher outage probabilities are found. Figure 4.10 summarises the outage probabilities for different cell radii and different numbers of simultaneous users, showing the coupling between the radio interface occupancy and the resulting coverage.

4.4.3 UPLINK: MULTIPLE CELL CASE

In WCDMA, adjacent cells transmit at the same time and frequency band, so that a certain intercell interference χ arises. Thus, the extension of Equation (4.10) comes to be:

$$\frac{P_i\left(W/R_{b,i}\right)}{P_N + \chi + [P_R - P_i]} \geq \left(\frac{E_b}{N_o}\right)_i \tag{4.21}$$

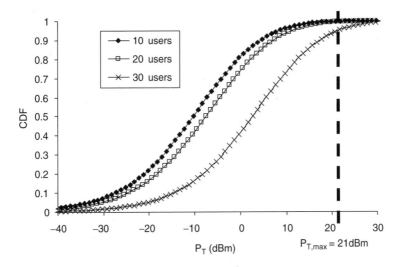

Figure 4.8 Cumulative distribution function of the transmitted power for 500 m cell radius

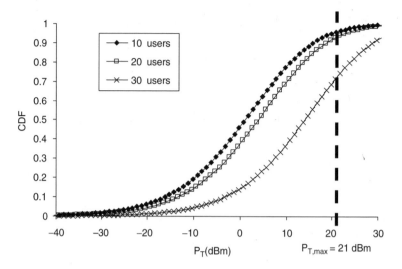

Figure 4.9 Cumulative distribution function of the transmitted power for 1 km cell radius

Setting the inequality to equality, and following a similar development as in Section 4.4.2, it can be found that the required power at the ith user receiver input is given by:

$$P_i = \frac{P_N + \chi}{\left[\dfrac{W/R_{b,i}}{\left(\dfrac{E_b}{N_o}\right)_i} + 1\right]\left[1 - \sum_{i=1}^{n}\dfrac{1}{\dfrac{W/R_{b,i}}{\left(\dfrac{E_b}{N_o}\right)_i} + 1}\right]} \tag{4.22}$$

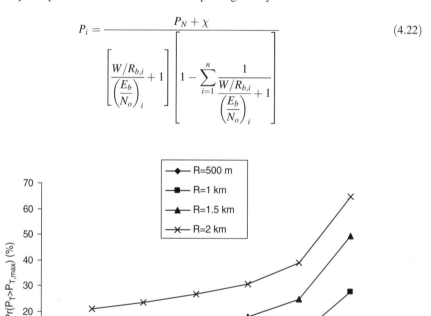

Figure 4.10 Outage probabilities for different cell radii and simultaneous users

and the required transmitted power will be given by:

$$P_{T,i} = L_{p,i} \frac{P_N + \chi}{\left[\dfrac{W/R_{b,i}}{\left(\dfrac{E_b}{N_o}\right)_i} + 1\right]\left[1 - \displaystyle\sum_{i=1}^{n} \dfrac{1}{\dfrac{W/R_{b,i}}{\left(\dfrac{E_b}{N_o}\right)_i} + 1}\right]} \tag{4.23}$$

Clearly, the required transmitted power increases with increasing intercell interference power, thus also reflecting the coupling among the different WCDMA cells because of the concurrent transmissions on the same time and frequency. Furthermore, notice that, in the case of a single cell and a constant number of users n (see Equations (4.18) or (4.23) with $\chi = 0$), the corresponding user's transmitted powers are constantly proportional to their corresponding path losses. However, in a multicellular environment the previous statement does not hold anymore, since term χ will vary continuously.

The uplink spectral efficiency of a WCDMA cell is commonly captured by the so-called uplink cell load factor, defined as [3]:

$$\eta_{UL} = 1 - \frac{P_N}{P_R + \chi + P_N} = \frac{P_R + \chi}{P_R + \chi + P_N} \tag{4.24}$$

where, similar to the single cell case, the total own-cell (i.e. intracell) power is given by:

$$P_R = \frac{(P_N + \chi) \displaystyle\sum_{i=1}^{n} \dfrac{1}{\dfrac{(W/R_{b,i})}{\left(\dfrac{E_b}{N_o}\right)_i} + 1}}{1 - \displaystyle\sum_{i=1}^{n} \dfrac{1}{\dfrac{(W/R_{b,i})}{\left(\dfrac{E_b}{N_o}\right)_i} + 1}} \tag{4.25}$$

Therefore, substituting Equation (4.25) into Equation (4.24), provides an alternative definition of the cell load factor as a function of the number of instantaneous users n:

$$\eta_{UL} = \left(1 + \frac{\chi}{P_R}\right) \sum_{i=1}^{n} \frac{1}{\dfrac{W}{\left(\dfrac{E_b}{N_o}\right)_i R_{b,i}} + 1} \tag{4.26}$$

Note that, by definition (Equation (4.24)), the cell load factor must be below unity and, consequently:

$$\left(1 + \frac{\chi}{P_R}\right) \sum_{i=1}^{n} \frac{1}{\dfrac{W/R_{b,i}}{\left(\dfrac{E_b}{N_o}\right)_i} + 1} < 1 \tag{4.27}$$

which represents a more restrictive condition than the one that arose in an isolated cell context given by Equation (4.16). Nevertheless, as long as χ varies through time, there is no fixed maximum number of simultaneous users in the cell but, rather, this maximum also becomes a function of time.

On the other hand, by combining Equations (4.23), (4.25) and (4.26), the following expression for the required mobile transmitted power can be obtained:

$$P_{T,i} = L_{p,i} \frac{(P_N + \chi + P_R)}{\frac{W}{\left(\frac{E_b}{N_o}\right)_i R_{b,i}} + 1} = L_{p,i} \frac{P_N \frac{1}{1 - \eta_{UL}}}{\frac{W}{\left(\frac{E_b}{N_o}\right)_i R_{b,i}} + 1} \tag{4.28}$$

Clearly, due to the limited available power at mobile terminals and also for efficiency reasons, Equation (4.28) reveals that the uplink cell load factor must be controlled in order to ensure the planned coverage. This control should be done according to the maximum transmission power $P_{T,max}$ and the maximum expected path loss $L_{p,max}$ depending on the desired cell coverage, thus obtaining a maximum allowable uplink load factor η_{max}. That is:

$$P_{T,max} = L_{p,max} \frac{P_N \frac{1}{1 - \eta_{max}}}{\frac{W}{\left(\frac{E_b}{N_o}\right)_i R_{b,i}} + 1} \Rightarrow \eta_{max} = 1 - \frac{P_N}{P_{T,max}} \frac{L_{p,max}}{\frac{W}{\left(\frac{E_b}{N_o}\right)_i R_{b,i}} + 1} \tag{4.29}$$

Admission control is one of the RRM strategies devoted to achieving the control of the uplink load factor below η_{max}. However, note that setting a value of η_{max} does not have a direct relationship with the number of simultaneous transmissions n, because according to Equation (4.26), even when all the users are of the same type the relationship becomes proportional to the time-varying term $1 + \chi/P_R$. Consequently, admission control becomes a challenging task. Furthermore, additional mechanisms are needed to achieve a tight control over the air interface, since the cell load factor may increase above its maximum value depending on traffic generation and on user mobility. In such cases, congestion control mechanisms are needed in order to avoid QoS degradation.

The previous equations reveal the coverage/capacity trade-off existing in WCDMA, given by the fact that the uplink load factor limits the maximum allowable path loss in a WCDMA cell according to:

$$L_{p,max} = \frac{P_{T,max}}{P_N} \left[\frac{W}{\left(\frac{E_b}{N_o}\right)_i R_{b,i}} + 1 \right] (1 - \eta_{UL}) \tag{4.30}$$

Therefore, if the load factor is high the cell coverage is reduced while if the load factor is low the cell coverage increases, thus creating a cell breathing effect. The coverage/capacity trade-off is illustrated in Figure 4.11, which presents the maximum path loss for the 64 kb/s and 384 kb/s services as a function of the cell load factor as given by Equation (4.30) with a target E_b/N_o of 3 dB. It can be observed that the higher the bit rate the lower the cell coverage because of the higher transmission power requirements.

4.4.3.1 Multiple Cell Coupling Characterisation

The characterisation introduced above, where the multicell scenario is taken into account by means of the intercell interference term χ, is already reflecting the coupling existing among the different WCDMA cells, so that changes in one cell also imply changes in the rest of the cells.

At this stage, a different perspective on cellular WCDMA is introduced by taking into account a detailed expression of the intercell interference in the cell load factor calculation. The mathematical framework captures explicitly the air interface coupling among the different cells in the scenario and the coupling among the cells is shown on a more visible form.

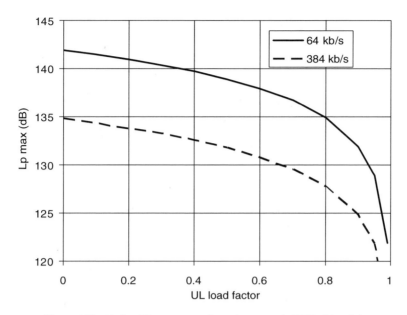

Figure 4.11   Trade-off between capacity and coverage in WCDMA uplink

Let us consider a scenario with $K + 1$ cells and focus on the central cell, chosen as the reference cell and numbered as the 0 th cell. The uplink cell load factor in this cell is given by:

$$\eta_0 = 1 - \frac{P_N}{P_{R,0} + \chi_0 + P_N} = \left(1 + \frac{\chi_0}{P_{R,0}}\right) \sum_{i_0=1}^{n_0} \frac{1}{\frac{W}{\left(\frac{E_b}{N_o}\right)_{i_0} R_{b,i_0}} + 1}$$ (4.31)

where i_0 represents the ith user in the 0th cell, with the corresponding bit rate R_{b,i_0} and requirement $(E_b/N_o)_{i_0}$, and n_0 is the total number of users transmitting in the 0th cell. $P_{R,0}$ and χ_0 are the total received intracell power and intercell interference at the 0th cell, respectively.

According to Equation (4.28), the power transmitted by the ith user of the 0th cell is given by:

$$P_{T,i_0} = L_{i_0,0} \frac{P_N \frac{1}{1 - \eta_0}}{\frac{W}{\left(\frac{E_b}{N_o}\right)_{i_0} R_{b,i_0}} + 1}$$ (4.32)

where $L_{i_0,0}$ is the path loss between the 0th cell site and the i_0 user. Then, the total received own-cell power at the 0th cell $P_{R,0}$ will be given by:

$$P_{R,0} = \sum_{i_0=1}^{n_0} \frac{P_{T,i_0}}{L_{i_0,0}} = \sum_{i_0=1}^{n_0} \frac{P_N \frac{1}{1 - \eta_0}}{\frac{W}{\left(\frac{E_b}{N_o}\right)_{i_0} R_{b,i_0}} + 1}$$ (4.33)

The intercell interference χ_0 observed at the reference cell can be expressed as:

$$\chi_0 = \sum_{j=1}^{K} \sum_{i_j=1}^{n_j} \frac{P_{T,ij}}{L_{ij,0}} \qquad (4.34)$$

where $P_{T,ij}$ is the power transmitted by the ith user in the jth cell, n_j is the total number of users transmitting in the jth cell and $L_{ij,0}$ is the path loss from the ith user in the jth cell to the reference cell site. By making use of Equation (4.32), it can be obtained that:

$$\chi_0 = \sum_{j=1}^{K} \frac{P_N}{1-\eta_j} \sum_{i_j=1}^{n_j} \frac{L_{ij,j}}{L_{ij,0}} \frac{1}{\dfrac{W}{\left(\dfrac{E_b}{N_o}\right)_{ij} R_{b,io}} + 1} \qquad (4.35)$$

At this stage, it is useful to introduce the following definitions:

$$S_{0,0}^{UL} = \sum_{i_0=1}^{n_0} \frac{1}{\dfrac{W}{\left(\dfrac{E_b}{N_o}\right)_{io} R_{b,io}} + 1} \qquad (4.36)$$

$$S_{j,0}^{UL} = \sum_{i_j=1}^{n_j} \frac{L_{ij,j}}{L_{ij,0}} \frac{1}{\dfrac{W}{\left(\dfrac{E_b}{N_o}\right)_{ij} R_{b,ij}} + 1} \qquad (4.37)$$

Note that Equation (4.36) can be interpreted as the cell load factor in the reference cell if this cell was isolated. Besides, Equation (4.37) can be interpreted as a weighted cell load factor contribution from the jth cell to the reference cell.

By substituting Equations (4.33), (4.35), (4.36) and (4.37) in to Equation (4.31), the uplink cell load factor in the 0th cell can be expressed as:

$$\eta_0 = \frac{S_{0,0}^{UL} + \dfrac{\chi_0}{P_N}}{1 + \dfrac{\chi_0}{P_N}} = \frac{S_{0,0}^{UL} + \displaystyle\sum_{j=1}^{K} \dfrac{S_{j,0}^{UL}}{1-\eta_j}}{1 + \displaystyle\sum_{j=1}^{K} \dfrac{S_{j,0}^{UL}}{1-\eta_j}} \qquad (4.38)$$

The coupling existing in a WCDMA cellular system is explicitly reflected in the above form, where the resulting cell load in a reference cell depends on all users in the scenario or, more specifically, on all other cells' respective load factors.

4.4.4 DOWNLINK: SINGLE CELL CASE

In the case of the downlink, it can also be said that all the users share the common bandwidth and each new connection increases the interference level of other connections, affecting their quality expressed in terms of a certain (E_b/N_o). For n users receiving simultaneously from a given cell, the following inequality for the ith user must be satisfied:

$$\frac{\dfrac{P_{Ti}}{L_{p,i}} \dfrac{W}{R_{b,i}}}{P_N + \rho \dfrac{P_T - P_{Ti}}{L_{p,i}}} \geq \left(\frac{E_b}{N_o}\right)_i \qquad (4.39)$$

$$P_T = P_p + \sum_{i=1}^{n} P_{Ti} \qquad (4.40)$$

P_T being the base station transmitted power, P_{Ti} being the power devoted to the ith user, $L_{p,i}$ being the path loss, $R_{b,i}$ the ith user transmission rate, W the bandwidth, P_p the power devoted to common control channels and P_N the background noise. ρ is the orthogonality factor (see page 32) since orthogonal codes are used in the downlink but some orthogonality is lost due to multi-path. ρ is between 0 and 1, where $\rho = 0$ means perfect orthogonality and $\rho = 1$ means non-orthogonal codes as in the uplink direction.

Therefore, the total transmitted power to satisfy all the users' demands should be:

$$P_T = \frac{P_p + P_N \sum_{i=1}^{n} \dfrac{L_{p,i}}{\dfrac{W}{\left(\dfrac{E_b}{N_o}\right)_i R_{b,i}} + \rho}}{1 - \sum_{i=1}^{n} \dfrac{\rho}{\dfrac{W}{\left(\dfrac{E_b}{N_o}\right)_i R_{b,i}} + \rho}} \tag{4.41}$$

and the power devoted to the ith user, P_{Ti}, is given by:

$$P_{T_i} = L_{p,i} \frac{P_N + \rho \dfrac{P_T}{L_{p,i}}}{\dfrac{W}{\left(\dfrac{E_b}{N_o}\right)_i R_{b,i}} + \rho} \tag{4.42}$$

In the downlink direction, the total power in the node B is shared among all the users. Therefore, the instantaneous location of a given user impacts on the performance of the rest of users in the cell. Note the difference with respect to the uplink, where a particular user location only impacts on its own performance.

Outage will occur in the downlink whenever P_{Tmax}, the maximum available power at the node B, is not enough to reach all the users with a suitable level to provide their corresponding target quality levels. As stated above, because all users are sharing the same resource (i.e. the base station transmitted power), the mutual dependences make the problem less tractable from the analytical point of view. Then, according to Equation (4.41), the outage probability can be formulated for the case when all users in the cell are of the same class, i.e. $R_{b,i} = R_b$ and $(E_b/N_o)_i = (E_b/N_o)$, as:

$$\Pr\left(P_T > P_{T,\max}\right) = \sum_{n=1}^{M} \Pr\left(\sum_{i=1}^{n} L_{p,i} > \frac{1}{P_N}\left[\frac{W/R_b}{\left(\frac{E_b}{N_o}\right)} + \rho\right]\left[P_{T,\max}\left(1 - n \cdot \frac{\rho}{\frac{W/R_b}{\left(\frac{E_b}{N_o}\right)} + \rho}\right) - P_p\right]\right) \cdot p_n \tag{4.43}$$

where p_n is the probability of having n simultaneous users given by Equation (4.9).

It can be seen from Equation (4.43) that, for an isolated cell and all users of the same type, the statistical distribution of the sum of corresponding path losses is the relevant parameter determining whether the node B will be able to serve all n users simultaneously or not. This indicates that the node B must effectively be able to get to all those distant users at the same time by sharing the available power. Figures 4.12 and 4.13 present the probability density function of the sum of path losses for 500 m and 1 km cell radii respectively. Similar to the uplink case, a number of snapshots have been simulated and, in each of them, n users have been scattered around the cell randomly.

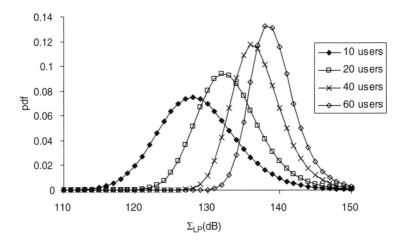

Figure 4.12 Probability distribution function of the sum of path losses term for 500 m cell radius

In order to reinforce some of the issues raised above, a few additional results are now presented. In particular, Figure 4.14 shows the node B transmitted power probability distribution when all users require a target of $E_b/N_o = 3$ dB and a transmission rate of $R_b = 64$ kb/s, the background noise level at the terminals is $P_N = -100$ dBm and the common control channels power is $P_p = 30$ dBm. The cell radius in this case is 500 m. It can be observed that as the number of users increases more transmitted power is demanded from the node B, up to the point that, in some cases the node B gets saturated (i.e. the required transmitted power to satisfy all n users requirements as given by Equation (4.40) would be above the maximum transmitted power, in this example 43 dBm). If the cell radius is extended to 1 km (see Figure 4.15), the transmitted power levels as well as the node B saturation probability also increase. Figure 4.16 summarises the node B saturation probability for different cell radii and different numbers of simultaneous users.

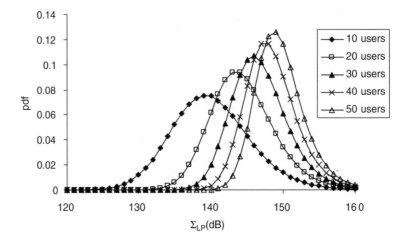

Figure 4.13 Probability distribution function of the sum of path losses term for 1 km cell radius

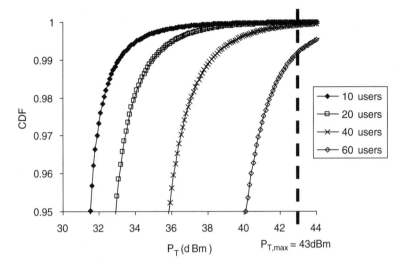

Figure 4.14   Cumulative distribution function of the node B transmitted power for 500 m cell radius

One distinguishing factor between uplink and downlink is that, while in the uplink every UE has its own power amplifier (i.e. every UE has for example, 21 dBm available) the node B transmitted power in the downlink is shared among all users simultaneously at a certain time. Therefore, the required power per user becomes very relevant in downlink, since users demanding high transmitted power levels (e.g. users far from the cell site) are consuming more radio resource units (i.e. more power) and therefore less power is available for the rest of the users. To illustrate this situation, Figures 4.17 and 4.18 present the cumulative distribution function of the transmitted power per user for a cell with radius 500 m and 1 km, respectively. From the above comments, it can be expected that limiting the maximum transmitted power per user will be a necessary radio resource management action to keep good control of the RRUs consumptions.

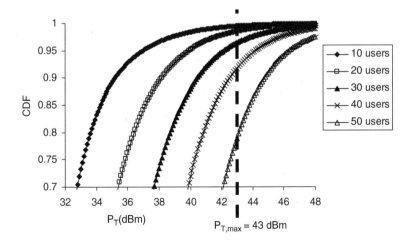

Figure 4.15   Cumulative distribution function of the node B transmitted power for 1 km cell radius

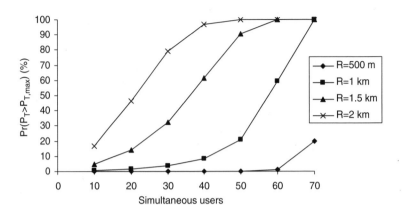

Figure 4.16   Outage probability for different cell radii and number of simultaneous users

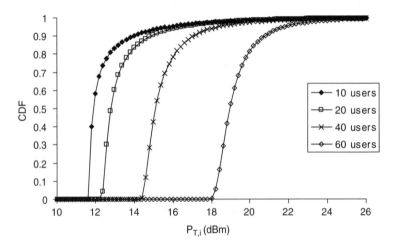

Figure 4.17   Cumulative distribution function of the transmitted power per user with 500 m cell radius

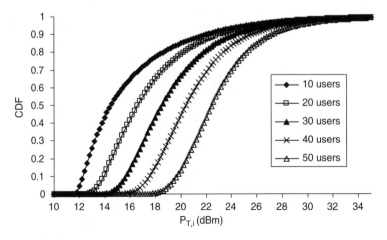

Figure 4.18   Cumulative distribution function of the transmitted power per user with 1 km cell radius

4.4.5 DOWNLINK: MULTIPLE CELL CASE

When several cells are considered in the downlink direction, the inequality in Equation (4.39) corresponding to n users receiving simultaneously from a given cell, should be modified as follows in order to reflect the ith user requirement:

$$\frac{\dfrac{P_{Ti}}{L_{p,i}}\dfrac{W}{R_{b,i}}}{P_N + \chi_i + \rho\dfrac{P_T - P_{Ti}}{L_{p,i}}} \geq \left(\frac{E_b}{N_o}\right)_i \tag{4.44}$$

where χ_i is the intercell interference observed by the ith user. Note that, different to the uplink case, downlink intercell interference is user-specific.

In this case, the total transmitted power to satisfy all the users' demands should be:

$$P_T = \frac{P_p + \displaystyle\sum_{i=1}^{n} \frac{(P_N + \chi_i)}{\dfrac{W}{\left(\dfrac{E_b}{N_o}\right)_i R_{b,i}} + \rho} L_{p,i}}{1 - \displaystyle\sum_{i=1}^{n} \frac{\rho}{\dfrac{W}{\left(\dfrac{E_b}{N_o}\right)_i R_{b,i}} + \rho}} \tag{4.45}$$

Equation (4.45) can be rearranged as:

$$P_T = \frac{P_p + P_N \displaystyle\sum_{i=1}^{n} \frac{L_{p,i}}{\dfrac{W}{\left(\dfrac{E_b}{N_o}\right)_i R_{b,i}} + \rho}}{1 - \eta_{DL}} \tag{4.46}$$

where, similar to the uplink direction, the downlink cell load factor is defined as:

$$\eta_{DL} = \sum_{i=1}^{n} \frac{\rho + \dfrac{\chi_i L_{p,i}}{P_T}}{\dfrac{W}{\left(\dfrac{E_b}{N_o}\right)_i R_{b,i}} + \rho} \tag{4.47}$$

When comparing the uplink and downlink cell load factors given by Equations (4.26) and (4.47), respectively, it can be noticed that both equations correspond to a similar expression with two differences. The first one is in the ratio between intercell and intracell power, which in the uplink is χ/P_R, (being the same for all the users) and in the downlink is user-specific and given by $\chi_i L_{p,i}/P_T$. In turn, the second difference is the orthogonality factor ρ, which in the uplink is equal to one and in the downlink is typically smaller.

With respect to the power devoted to the ith user, P_{Ti}, it is given by:

$$P_{Ti} = L_{p,i} \frac{P_N + \chi_i + \rho\dfrac{P_T}{L_{p,i}}}{\dfrac{W}{\left(\dfrac{E_b}{N_o}\right)_i R_{b,i}} + \rho} = \frac{L_{p,i}(P_N + \chi_i)}{\dfrac{W}{\left(\dfrac{E_b}{N_o}\right)_i R_{b,i}} + \rho} + \frac{\rho P_T}{\dfrac{W}{\left(\dfrac{E_b}{N_o}\right)_i R_{b,i}} + \rho} \tag{4.48}$$

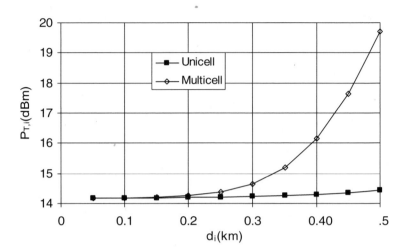

Figure 4.19 Downlink transmitted power to a reference user with cell radius 500 m

Figures 4.19 and 4.20 illustrate the transmitted power devoted to a given user i as a function of the distance d_i to its serving cell for cell radius 500 m and 1 km, respectively, according to Equation (4.48). A simple scenario in which all the cells are transmitting 33 dBm is assumed, and the reference user i is connected to the central cell and moves through the straight line that connects the central cell with one of the neighbouring cells. The macrocell propagation model presented in Appendix 5.1 of Chapter 5 is assumed, without considering any shadowing effects. A 64 kb/s service is considered, with a downlink background noise equal to -100 dBm. In both figures, the comparison with respect to the power that would be devoted to the user in a unicell scenario according to Equation (4.48) with $\chi_i = 0$ is presented, so that the impact of the intercell interference can be retained.

In the 500 m case, shown in Figure 4.19, it can be observed that the transmitted power in the unicell scenario is almost constant with respect to the distance. This is because of the low path loss existing in

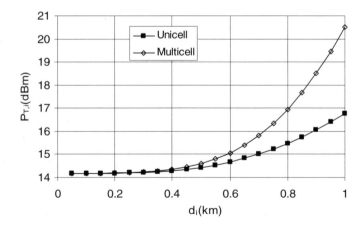

Figure 4.20 Downlink transmitted power to a reference user with cell radius 1 km

this case, which means that the first term in Equation (4.48) is negligible with respect to the second term, which does not depend on the path loss. In turn, in the multicell scenario, the increase in intercell interference experienced for higher distances increases the value of the first term in Equation (4.48) and therefore the required power at the cell edge is higher than in the unicell case.

With respect to the 1 km case, shown in Figure 4.20, the same effect is observed but here there is a power increase with the distance even in the unicell case, due to the fact that for distances higher than 500 m the first term in Equation (4.48) is not negligible. It is also worth noting that the power increase due to the intercell interference is somewhat lower in the 1 km case than in the 500 m case, because both the intercell and the intracell interferences at the cell edge are smaller when the cell radius is high, provided that the transmitted power is the same for the two radii.

4.4.5.1 Multiple Cell Coupling Characterisation

Similarly to Section 4.4.3.1 for the uplink direction, here the coupling existing in the downlink among the different cells in a multicell scenario is shown explicitly by means of a expression that relates the downlink transmit power of the different cells.

Let us consider a scenario with $K + 1$ cells, $j = 0, \ldots, K$, each of them having n_j simultaneous users. The reference cell is the central cell numbered 0, whose transmitted power is given, according to Equation (4.45), by:

$$
P_{T0} = \frac{P_{p0} + \sum_{i_0=1}^{n_0} \frac{(P_N + \chi_{i_0})}{W} \frac{W}{\left(\frac{E_b}{N_o}\right)_{i_0} R_{b,i_0}} L_{i_0,0} + \rho}{1 - \sum_{i_0=1}^{n_0} \frac{\rho}{\frac{W}{\left(\frac{E_b}{N_o}\right)_{i_0} R_{b,i_0}} + \rho}}
\tag{4.49}
$$

where $P_{p,0}$ is the power devoted to common control channels, $L_{i_0,0}$ is the path loss between the 0th cell site and the user i_0 (i.e. the ith user of the 0th cell). χ_{i_0} is the intercell interference observed by the user i_0, given as a function of the transmitted power by the neighbouring cells P_{Tj} $j = 1, \ldots, K$ and the path loss from this user to the jth cell site $L_{i_0,j}$. In particular:

$$
\chi_{i_0} = \sum_{j=1}^{K} \frac{P_{Tj}}{L_{i_0,j}}
\tag{4.50}
$$

Substituting Equation (4.50) into Equation (4.49):

$$
P_{T0} = \frac{P_{p0} + P_N \sum_{i_0=1}^{n_0} \frac{L_{i_0,0}}{\frac{W}{\left(\frac{E_b}{N_o}\right)_{i_0} R_{b,i_0}} + \rho} + \sum_{i_0=1}^{n_0} \sum_{j=1}^{K} \frac{L_{i_0,0}}{L_{i_0,j}} \frac{P_{Tj}}{\frac{W}{\left(\frac{E_b}{N_o}\right)_{i_0} R_{b,i_0}} + \rho}}{1 - \sum_{i_0=1}^{n_0} \frac{\rho}{\frac{W}{\left(\frac{E_b}{N_o}\right)_{i_0} R_{b,i_0}} + \rho}}
\tag{4.51}
$$

It is useful to introduce the following definitions:

$$S_{0,0}^{DL} = \sum_{i_0=1}^{n_0} \frac{1}{\dfrac{W}{\left(\dfrac{E_b}{N_o}\right)_{i_o} R_{b,io}} + \rho} \tag{4.52}$$

$$S_{0,0}^{DL*} = \sum_{i_0=1}^{n_0} \frac{L_{i_0,0}}{\dfrac{W}{\left(\dfrac{E_b}{N_o}\right)_{i_o} R_{b,io}} + \rho} \tag{4.53}$$

$$S_{0,j}^{DL} = \sum_{i_0=1}^{n_0} \frac{L_{i_0,0}}{L_{i_0,j}} \frac{1}{\dfrac{W}{\left(\dfrac{E_b}{N_o}\right)_{i_o} R_{b,io}} + \rho} \tag{4.54}$$

Note that Equations (4.52) and (4.53) are two terms that depend on the own-cell transmissions without and with propagation conditions, respectively. In turn, Equation (4.54) represents the influence of the jth cell over the reference 0th cell, similar to Equation (4.37) for the uplink case. Nevertheless, it is worth observing that the term $S_{0,j}^{DL}$, and thus the influence of the jth cell over the 0th cell, depends on how the users are distributed in the 0th cell, while in the uplink the term $S_{j,0}^{UL}$, and thus the influence of the jth cell to the 0th cell, depends on how the users are distributed in the jth cell.

Therefore, Equation (4.51) is eventually expressed as:

$$P_{T0} = \frac{P_{p0} + P_N S_{0,0}^{DL*} + \sum_{j=1}^{K} P_{Tj} S_{0,j}^{DL}}{1 - \rho S_{0,0}^{DL}} \tag{4.55}$$

This form explicitly reflects the existing coupling in a WCDMA cellular system, where the resulting transmitted power level in a given cell depends on all the other cells' respective transmitted power levels.

4.5 RRM FUNCTIONS

Taking into account the constraints imposed by the radio interface, RRM functions are responsible for taking decisions regarding the setting of the different parameters influencing the air interface behaviour. From the previous section, the following elements have been identified:

1. The number of active users, N.
2. The number of simultaneous users transmitting, n.
3. The corresponding transmission rates $R_{b,i}$ for each user.
4. The transmitted power levels corresponding to every simultaneous user.

In a cellular layout, the above elements will be referred to on a cell-by-cell basis, the impact of neighbouring cells being captured in the form of a certain intercell interference level.

Clearly, the number of parameters to be controlled, as well as their different nature, require a set of several RRM functions whose joint behaviour should lead to overall radio access network optimisation. RRM functions need to be consistent for both uplink and downlink, although the different nature of these links introduces some differences in the suitable approaches.

Furthermore, the radio access network is dynamic in nature. There are many sources of dynamism in the network at very different time scales. For example:

1. The number of camping users M changes not only as a result of service penetration increase, which is noticeable at a rather long-term scale (e.g. months), but also as a result of spatial traffic variations (e.g. increase of M in business areas during working hours).
2. The number of active users N changes as M changes, but also as a result of the call/session generation characteristics for the different services that users are accessing.
3. The number of simultaneous users n changes as long as N and/or M change. Furthermore, changes on n will follow depending on the traffic generation characteristics for the different services that users are accessing (e.g. video streaming presents much a higher activity factor than web browsing). Note also that in CBR (Constant Bit Rate) services, the traffic source generates a sustained flow while in VBR (Variable Bit Rate) services the traffic source offers different flow rates at different times.
4. All these n, N and M users are moving around the network, so that propagation conditions vary through time. Within propagation conditions, different variation time scales occur: path loss component related to mobile to cell site distance; slow fading (shadowing) due to obstacles in the scenario; and fast fading due to close environment. Clearly, the speed of the mobile terminal influences the radio network dynamism.

Note also that, even if all the above issues were constant in a reference cell, as long as changes occur in neighbouring cells, the reference cell would be affected by dynamic variations caused by the changes in the intercell interference level.

Since the different RRM functions will track different radio interface elements and effects, RRM functions can be classified according to the time scales they use to be activated and executed. Since short/ long term time scales variations are relative concepts, the approach preferred here is to associate typical time scale activation periods with the different RRM functions. Therefore, Table 4.1 captures the list of RRM functions with corresponding typical time scales between consecutive activations of the algorithm (i.e. the time between when an action is carried out by a specific RRM algorithm and the next time that the same algorithm needs to operate). In the following sections, the different RRM functions are elaborated, starting with the less dynamic ones, except for the outer loop power control, which is covered together with inner loop power control within a generic section on power control (Section 4.5.6).

4.5.1 ADMISSION CONTROL

Admission control decides the admission or rejection of requests for set-up and reconfiguration of radio bearers. The request should be admitted provided that the QoS requirements can be met and that the QoS

Table 4.1 Time scale of the different RRM functions

Typical time scale of the time between algorithm activations	RRM function
1 slot	Inner loop Power control
1 frame	UE-MAC
	Packet scheduling
Tenths to thousands of frames	Admission control
	Handover
	Code management
	Congestion control
	Outer loop power control

requirements of the already accepted connections are not affected by the new request acceptance. Admission control is particularly relevant in WCDMA because there is no hard limit on the maximum capacity. Consequently, the design of a proper admission control in UMTS is much more challenging than, for example, GSM, where there is a fixed number of available channels and users may be accepted as long as there are channels available.

Since the maximum cell capacity is intrinsically connected to the amount of interference or, equivalently, the cell load level, the use of admission control algorithms is based on measurements and/or estimates of the current network load situation as well as on the estimation of the load increase that the acceptance of the request would cause.

It is worth noting that admission control decisions are taken at the specific moment a new request is performed, so that the decision may be based on the radio network situation at that time as well as on the recent past history. Nevertheless, the admission decision can in no way anticipate exactly the future network load, so that additional radio resource management functions are necessary to cope with the dynamic network evolution and to keep the QoS requirements under control.

The randomness associated with a cellular radio environment (e.g. propagation, mobility, traffic, etc.) allows admission control to play the role of thick tuning in the management of the radio resources. If decisions are too soft, and too many users are being accepted, an overload situation may follow and further RRM mechanisms will need to be activated. If decisions are too strict, and too few users are being accepted, the operator will be losing revenue and a tuning of the admission control algorithm will be necessary.

Admission control algorithms are executed separately for uplink and downlink because of the different issues impacting on both communication directions. However, a connection request can be admitted only after gaining permission from the corresponding uplink and downlink algorithms. In the case of the downlink, a hard limiting factor – the OVSF codes availability – has to be considered in the admission control.

In contrast to 2G, where the network is accessed mostly by real time voice users with equal quality requirements, in 3G WCDMA multimedia services (e.g. video-telephony, streaming video, web browsing, etc.) with diverse QoS requirements (e.g. business segment, consumer segment, etc.) are expected. Therefore, admission control algorithms must take into consideration that the amount of radio resources needed for each connection request will vary. Similarly, the QoS requirements in terms of real time or non real time transmission should also be considered in an efficient admission control algorithm. Clearly, admission conditions for non real time traffic can be more relaxed on the assumption that the additional radio resource management mechanisms complementing admission control will be able to limit non real time transmissions when the air interface load is excessive.

In addition to the connection set-up request, admission control may also be triggered by handover procedures, transport channel type switching, etc., so other radio resource management functions are closely related to admission control. The particular admission control algorithm may depend on the specific triggering function, so that, for instance, a special consideration may be required with handover requests when compared to set-up requests.

The cell to which the UE will request acceptance is derived from the initial cell selection (reselection) procedures, which are well standardised, and will be further described in Chapter 5.

In terms of signalling, the procedures related with admission control are shown schematically in Figure 4.21. A transaction set-up request in UMTS is always triggered from the UE side, either because it is the UE itself that is initiating an interaction with the network or because the UE is answering a paging message. Prior to the transaction set-up procedure, a signalling path from the UE towards the CN needs to be established, which in the case of the UTRAN is accomplished by means of an RRC connection and the RANAP (Radio Access Network Application Part) protocol, which takes care of the UTRAN-CN interactions.

With the help of the RRC and RANAP protocol, a transaction set-up request message reaches the CN. Therefore, it is always the CN that triggers a RAB establishment by means of the RAB assignment message sent from the CN to the SRNC (see Figure 4.21). With the arrival of such a message, the

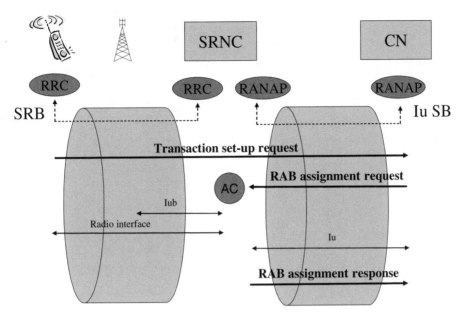

Figure 4.21   Messages and protocols involved in the admission control

Admission Control (AC) algorithm is executed. If the connection can be admitted, the SRNC establishes the resources in the radio and Iub interfaces by means of the RRC radio bearer establishment procedure (see Section 3.5.3.2). Similarly, the required connections are established in the Iu interface. If the establishment procedure succeeds, a positive response is given in the message *RAB assignment response* and the RAB is eventually set-up.

4.5.2 CONGESTION CONTROL

Congestion control, also denoted as load control, faces situations in which the QoS guarantees are at risk due to the evolution of system dynamics (mobility aspects, increase in interference, traffic variability, etc.). For example, if several users in a cell suddenly move far from the Node B, there may not be enough power to satisfy all the links' qualities simultaneously and some actions are required to cope with this situation. Note that, although a strict admission control could be carried out, as long as the radio network behaviour has strong random components, there is always some probability that these overload situations occur and, consequently, congestion control mechanisms must be included in the set of radio resource management techniques.

Congestion situations in the radio interface are caused by excessive interference. Thus, congestion control algorithms need continuously to monitor the network status in order to correct overload situations when they are present. The monitoring will be based on network measurements, such as downlink transmitted power, uplink cell load factor, etc., which need to be suitably averaged to avoid both false congestion detections (i.e. triggering congestion resolution mechanisms when the air interface is not really overloaded) and congestion non-detection (i.e. do not trigger congestion resolution mechanisms when the air interface is really overloaded). Additionally, the congestion control algorithm needs to exhibit a fast reactivity under overload conditions in order to prevent degradation of the quality of the connections.

Similarly to the admission control algorithm, congestion control is closely related to other RRM functions. In particular, congestion resolution actions will be supported, for example, by admission

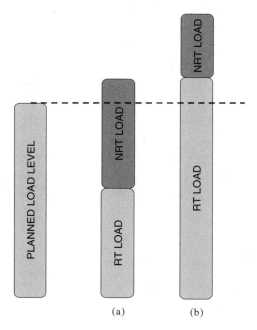

Figure 4.22 Congestion control in the presence of difference service types

control (e.g. by refusing new connections while the network is congested) or handover (e.g. by transferring connections from a congested cell to neighbouring cells). More precisely, the actions to be carried out may depend on the situation, the origin of the congestion and the services mix present at the congestion time. As reflected in Figure 4.22, a maximum load level is planned in the network. In general, the current load will be the result of both RT (Real Time) plus NRT (Non Real Time) services transmissions. If the RT traffic load is below the planned value, it is possible to solve the congestion situation by acting on NRT users, as shown by Figure 4.22a. Congestion may be alleviated by reducing the transmission rates of a number of NRT users. An extreme case would be the inhibiting of all NRT transmissions in the cell. If this is not enough, the help of neighbouring cells could be enlisted, by reducing NRT intercell interference contribution. If this is still not enough, as shown in Figure 4.22b, it would be necessary to reduce the RT load. In the case of conversational services, this should be accomplished by dropping some calls.

The congestion or load control (LC) algorithm will reside in the network side (RNC) and will be based on measurements acting as algorithm inputs (e.g. uplink cell load factor, downlink transmitted power, etc.). When a congestion situation is triggered, congestion resolution actions are implemented with the aid of the RRC protocol, as shown in Figure 4.23. For example, when a reduction on the transmission rate or equivalently of the Transport Format Set for a specific interactive user is decided, an RRC protocol message indicating the fact (i.e. a Transport Format Combination Control or a Transport Channel Reconfiguration message) will be transferred to that UE. Upon arrival of the RRC message at the UE side, proper actions will be taken by reconfiguring the lower layers.

4.5.3 CODE MANAGEMENT

Code management is devoted to managing the downlink OVSF (Orthogonal Variable Spreading Factor) code tree used to allocate physical channel orthogonality among different users. Clearly, the advantage of

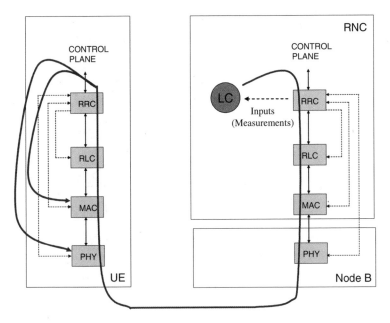

Figure 4.23   Protocols involved in the congestion control algorithm

the OVSF codes used in the UTRAN downlink is perfect orthogonality. However, the drawback is the limited number of available codes. Therefore, it is important to be able to allocate/reallocate the channelisation codes in the downlink with an efficient method, in order to prevent 'code blocking'. 'Code blocking' indicates the situation where a new call could be accepted on the basis of interference analysis and also on the basis of the 'spare capacity' of the code tree but, due to an inefficient code assignment, this spare capacity is not available for the new call that must, therefore, be blocked. This situation is depicted in Figure 4.24, where two transmissions with $SF = 4$ and two transmissions with $SF = 8$ are

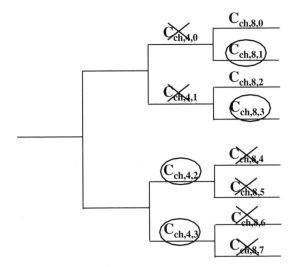

Figure 4.24   Example of code blocking

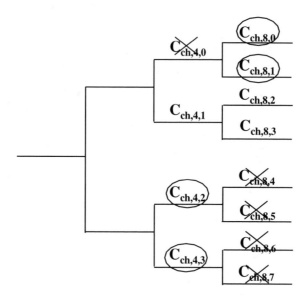

Figure 4.25   Example of code allocation preventing code blocking

assumed to have been assigned the corresponding code sequences $C_{ch,4,2}$, $C_{ch,4,3}$, $C_{ch,8,1}$ and $C_{ch,8,3}$, respectively, which prevent the use of the codes marked with a cross in Figure 4.24. It is worth noting that, with such OVSF code tree occupancy, the arrival of a new call requesting for $SF = 4$ would experience code blocking, since no code at that layer is available. On the contrary, if the code allocation shown in Figure 4.25 was used, it would allow the support of the two $SF = 4$ users, the two $SF = 8$ users and still would provide room to support a new $SF = 4$ request with code $C_{ch,4,1}$.

In general terms, a code allocation strategy would aim at minimising code tree fragmentation, preserving the maximum number of high rate codes and eliminating code blocking. Nevertheless, since the purpose of the code allocation/reallocation strategies is to prevent code blocking, this may require 'code handover', that is, a call using a given code is forced to use a different code belonging to the same layer.

4.5.4 HANDOVER

The purpose of this strategy is to optimise the cell or set of cells (i.e. the Active Set) to which the mobile is connected. Although handover is an inherent functionality with cellular systems, in WCDMA more possibilities are open as long as the mobile terminal can be connected to more than one cell simultaneously, provided that these cells operate at the same frequency. Thus, a distinction between hard and soft handover can be made.

Regardless of the handover type, the handover mechanism in WCDMA is controlled by the network with the assistance of measurements reported from the terminal side. Handover involves three different steps: measurements, decision and execution.

Measurements carried out by the mobile terminal can be transferred to the network either periodically or they can be event-triggered. The former option may consume radio resources unnecessarily if no changes in the radio interface conditions are observed between consecutive reporting periods. Measurements, which are very precisely specified, may be of different categories: intra-frequency (on the same UTRAN carrier), inter-frequency (on a different UTRAN carrier) or inter-RAT (on a radio access technology other than UTRAN), as described in Section 3.5.3.4.

Intra-frequency measurements are used to measure neighbouring cells on the same hierarchical layer since usually cells belonging to different layers (microcells, macrocells, etc.) operate at a different frequency. Inter-frequency measurements may be carried out either for measurements on the same hierarchical layer when multiple carriers are allocated to a given cell to increase capacity or, for example, when hierarchical cell structures are deployed by the network operator. Inter-RAT measurements are carried out when more than one RAT is deployed in the considered service area.

The handover decision is carried out by means of a handover algorithm, which is not standardised. Nevertheless, some algorithm examples are collected in 3GPP specifications [4], where decisions are taken as a result of relative comparisons on CPICH measurements. Thus, the pilot channel plays an important role in handover decisions and, by increasing or decreasing its transmitted power level, more users can be attracted to or refrained from joining the cell.

Although the handover concept is simple, its practical realisation is complex because it is a multidimensional problem with a large number of parameters. Furthermore, the suitability of the handover decisions and the time when decisions are taken has a strong impact on the overall radio access network performance, so that handover constitutes a critical procedure.

4.5.5 UE-MAC AND PACKET SCHEDULING

These algorithms are devoted to deciding the suitable radio transmission parameters for each connection in a reduced time scale and in a very dynamic way. They operate on a frame by frame (or TTI) basis to take advantage of the short term variations in the interference level. Taking into account its operation in short periods of time, in this book these strategies are referred to as *short term RRM strategies*. Their operation is illustrated in Figure 4.26. Specifically, Figure 4.26a reflects a situation in which the current load, including both RT (real time) and NRT (non real time) users, is below the planned load level, thus a certain spare capacity exists in the cell. Therefore, the purpose of short term RRM strategies will be to bring the cell to the situation reflected in Figure 4.26b, in which the spare capacity has been filled with a certain amount of NRT load. This can be achieved by allowing the transmission of other users not included in Figure 4.26a and/or the increase in the bit rate of other users already included in Figure 4.26a.

In the case of the uplink, this functionality is decentralised, so that the MAC layer of every UE executes the so-called UE-MAC algorithm to select the instantaneous bit rate. In the downlink, the operation is naturally centralised and carried out by the packet scheduling algorithm.

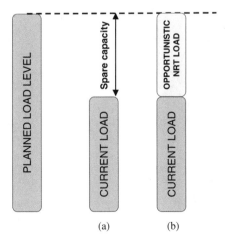

(a)   (b)

Figure 4.26 Operation of short term RRM strategies with NRT load

4.5.5.1 UE-MAC algorithms in the uplink

To ensure specific QoS figures, a strict coordination of access in the uplink direction could only be achieved by means of centralised strategies. This requires intensive signalling from the network side, so that the network would indicate in each frame who is allowed to transmit and at what transmission rate. However, if access in the uplink direction was not coordinated at all and only a pure decentralised random access was used with few signalling requirements, it would not be possible to guarantee QoS figures as long as there was no control over the air interface interference conditions. Thus, the performance/complexity trade-off seems to indicate that a suitable solution should combine both centralised and decentralised components.

In particular, the UE-MAC algorithm is allocated at the UE side and autonomously decides the instantaneous Transport Format or, equivalently, the instantaneous bit rate to be applied in each TTI for a given RAB, and therefore it implements the decentralised component. However, the eligible transmission rates are only those defined within the TFS or TFCS if a combination of RABs exists (see Section 3.2.2.3), which is controlled by the network and therefore implements the centralised component. The TFCS is defined at connection set-up, with the allocation of the RAB at admission control, and can be dynamically adjusted as a result of, for example, congestion control actions. Therefore, such a solution allows operation in the short term without signalling load in order to take full advantage of the time varying interference conditions.

4.5.5.2 Downlink Packet Scheduling

Downlink packet scheduling is responsible for scheduling non real time transmissions over shared channels in the downlink. In UTRAN FDD, this functionality manages the occupation over the FACH, DSCH and HS-DSCH channels.

As explained in Chapter 3, the separation among users is achieved in the downlink by means of the OVSF codes. The OVSF code tree is a limited resource and, consequently, code management strategies as described in Section 4.5.3 must be considered. Then, services with non real time constraints and with non constant bit rate generation should be suitably managed by means of scheduling mechanisms and a rational occupancy of the OVSF code tree would follow.

The typical assignment of OVSF codes in the downlink is shown in Figure 4.27. It can be observed that some OVSF codes belonging to the layer $SF = 256$ are devoted to common channels (i.e. pilot channel CPICH and P-CCPCH containing the broadcast). In addition, another part of the code tree is used by Dedicated Channels (DCH) and is allocated depending on the bit rate that is required by each service. Similarly, a branch starting from a certain spreading factor SF_{root} may also be reserved to the PDSCH carrying the DSCH.

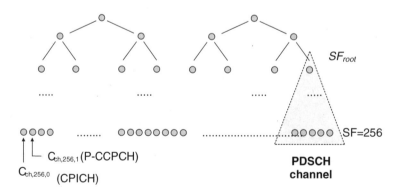

Figure 4.27 OVSF tree split among DSCH, common channels and dedicated channels

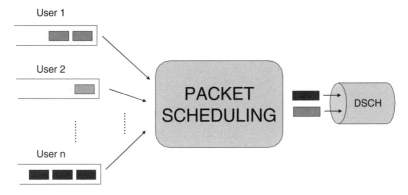

Figure 4.28 Packet scheduling algorithms

The packet scheduling strategy may follow a time-scheduling approach (i.e. multiplex a low number of users simultaneously with relatively high bit rates), a code-scheduling approach (i.e. multiplex a high number of users simultaneously with relatively low bit rates) and combinations of both. Furthermore, prioritisation mechanisms can be considered in the scheduling algorithm. Priorities can be established, for example, at service class level (e.g. interactive traffic is assigned higher priority than background traffic) or at user type level (e.g. business users are assigned higher priority than consumer users).

The number of users mapped to a shared channel managed by scheduling is not limited in the sense that, the higher the number of users sharing the resources the higher the experienced delays. Nevertheless, if some reference soft-QoS figures should be targeted even for non real time users, a proper dimensioning of the resources should be carried out, thus balancing the offered traffic to the packet scheduler with the available channel capacity, which is represented in Figure 4.28 by the width of the DSCH channel pipe.

It is worth noting that, with this procedural approach, the interference level associated with non real time traffic is bounded, as long as the channel capacity is limited by the code branch allocated to the physical shared channel.

4.5.6 POWER CONTROL

The purpose of this strategy is to optimise the mobile transmitted power (uplink) and node B transmitted power (downlink), as already described in Chapter 2.

The inner loop power control is responsible for adjusting, on a fast time basis (i.e. each slot within a UTRAN FDD 10 ms frame), the transmitted power in order to reach the receiver with the required E_b/N_o target. In turn, the outer loop power control is responsible for selecting a suitable E_b/N_o target depending on the BLER (BLock Error Rate) or BER (Bit Error Rate) requirement, and operates on a much slower time basis than the inner loop power control, adapting power control to changing environments.

4.5.7 INTERACTIONS AMONG RRM FUNCTIONS

The overall behaviour of the air interface at a given time will be the result of the decisions taken by the different RRM functions. Clearly, the different RRM functions are highly interrelated and coupled as long as they are all influencing the air interface. Let us consider, for example, the case reflected in Figure 4.29. Let us assume that user (1) is transmitting voice through a dedicated channel. Inner loop power control is operating to keep the link quality. User (2) is requesting the set-up of a videoconference, so that the admission control algorithm is running to decide whether to accept or reject this request. User (3) is involved in a soft-handover procedure, adding the neighbouring cell to its Active Set. User (4) is

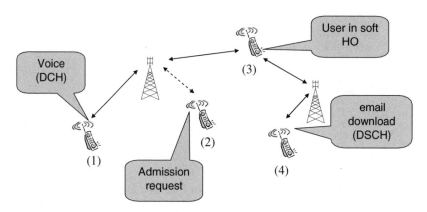

Figure 4.29   Illustration of interactions between RRM functions

receiving an email through downlink packet scheduling. Assume that, once user (2) has been admitted, it causes an interference increase beyond that expected by the admission control algorithm. In this case, congestion control mechanisms are activated and the action is to download the email of user (4) at a lower bit rate.

The example above reveals the dynamism in the cellular system and the need for a controlled interaction among the different RRM functions, which follow action/reaction principles. Therefore, consistency needs to be ensured among the different actions that will be undertaken by the different functions and mechanisms need to be included to solve conflicts deriving from contradictory actions/reactions. A proper design of the RRM functions will also consider that some functionalities rely on the actions/reactions of other functionalities to achieve a global performance according to the operator's targets.

The expected effects of applying RRM strategies in a coordinated and consistent way can be better explained by comparison with the situation where there is not tight control over the use of radio resources, for example in a WCDMA packet network in the uplink direction such as the ones considered in Reference 5, based on a S-ALOHA/CDMA access scheme. The typical uplink behaviour of such a network expressed in terms of throughput and delay is shown in Figure 4.30. Two regions can be distinguished: in region A the offered load is low and the interference is also low, so that packets are correctly transmitted, whereas in region B the offered load is high and the interference is also high, so that packets are incorrectly transmitted and the throughput decreases at the time that delay increases due to retransmissions. This behaviour is due to the lack of coordination among mobile terminals. Although,

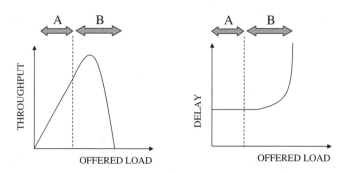

Figure 4.30   Operation with no RRM (e.g. S-ALOHA WCDMA network)

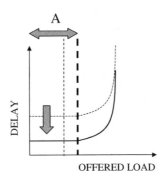

Figure 4.31   Operation when RRM strategies are applied to a WCDMA network.

in strict sense, the WCDMA networks considered in Reference 5 are inherently unstable due to the random access mechanism, in practice the system operation point may provide a controlled performance.

When applying RRM, the purpose of admission and congestion control would be to keep the system operation point in region A, otherwise the system becomes unstable and no QoS can be guaranteed. Smart admission and congestion control strategies will shift region A to some extent to the right side, so the system capacity is increased. Additionally, the performance achieved under region A is dependent on the access mechanism and in some cases it could happen that the system operation is access-limited instead of the more efficient case, which is interference-limited. A suitable UE-MAC strategy should try to take full advantage of the load conditions by pushing the system into an interference-limited situation in which users transmit at the highest possible bit rate in order to improve delay performance (see Figure 4.31). The challenge is to achieve a good balance between improving the performance (for example, in terms of decreasing the delay under low load situations by increasing the transmission rate) and maintaining the interference level manageable by the congestion and admission control algorithms. Moreover, RRM can control and exchange the gain levels between capacity and delay: if desired, the admission region can be extended at the expense of some reduction in the delay gain or the reverse, the delay gain can be increased at the expense of some reduction in the admission region.

In order to close this section, Figure 4.32 summarises the set of RRM functions, the parameters for which they are responsible, their interactions and the time scales at which they operate. Admission control is activated every time a new connection request or a handover request reaches the system. The serving cell is decided, a code sequence is assigned and the maximum bit rate for the allocated RAB is fixed at this stage. Handover procedure is associated to a change in the serving cell and a code change may also be needed, so that handover is related to both admission control and code management.

Congestion should not occur very often if the RRM parameters are correctly adjusted. Nevertheless, when congestion does occur, the reduction of the maximum bit rate is the first action that can be taken, thus modifying the transmission capabilities granted at the admission phase. Furthermore, for severe congestion situations, blocking new connection requests may help to prevent further overload in the network, so that interactions between congestion and admission control are also feasible. Similarly, the admission control algorithm may accept a high priority request relying on the fact that congestion mechanisms will be able to manage the possible resulting overload by taking actions on less prioritised users. In turn, another possibility during congestion is the execution of a handover, for example, to cells using another carrier or to other RATs, which involves the relationship between congestion control and handover management, as shown in Figure 4.32.

The TFCS (i.e. the maximum allowed bit rate) is set by long term RRM functions, while the instantaneous bit rate is managed by short term RRM mechanisms in order to take advantage of the interference dynamics. Therefore, packet scheduling in downlink and UE-MAC in uplink are hierarchically

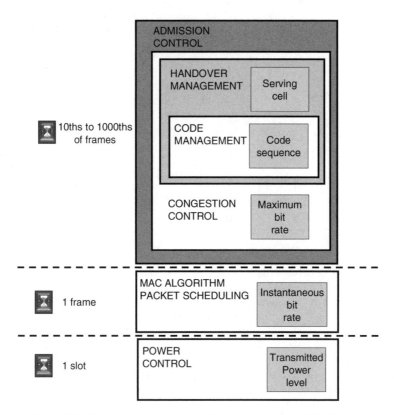

Figure 4.32   Summary of parameters and interactions between RRM functions

conditioned to admission and congestion control decisions fixing the maximum allowed bit rate and have the freedom to choose the bit rate on a frame by frame basis keeping the TFCS constraint.

Finally, power control operates at the shortest time scale, which is the time slot period.

4.6 SYSTEM CHARACTERISTICS RELEVANT AT RRM LEVEL

The inherent characteristics of wireless communications make the maintenance of a consistent radio link supporting an end user service, a challenge that has been coped with for many years. Radio engineering techniques such as synchronisation, channel estimation and equalisation, channel coding and interleaving, space or polarisation diversity, etc. are just some examples of how the human being is able to overcome difficulties and constraints.

Assuming the availability of the physical layer techniques that make a radio link feasible, the next step is to provide large-scale availability of services over time and space. Again, radio engineering envisages the necessary functionalities to make the objective possible. Therefore, certain network architectures and protocols can be developed, and concepts such as frequency reuse pattern, sectoring, seamless handover, etc. are applied.

At this stage, when all the necessary technologies are available, radio network planning is the engineering exercise to come up with a network deployment meeting the requirements in terms of coverage, capacity and quality, as previously stated in this chapter. Furthermore, once the network is in operation, RRM will be in charge of managing the provisioned resources to set all the targets at a satisfactory level. The feasibility of achieving the contradictory objectives simultaneously with the conjunction of radio network planning and RRM algorithms will definitiely depend on the proper design and configuration of the deployed network as well as the implemented RRM algorithms.

Network planning and RRM algorithms require detailed studies and analysis before implementation. Therefore, given that the studies will be conducted with a model simplifying the real world, it is clearly critical that the working assumptions and hypothesis in the study phase are capturing the essential elements affecting the performance in practice. The purpose of this section is to stress the importance of some system characteristics that are certainly present in a 3G framework and must be taken into account for a successful network planning and RRM algorithms design. In particular, the following sections develop further some selected characteristics:

1. In 3G WCDMA, there will be a multiplicity of services with different QoS requirements, which impact on the network and RRM algorithms design that must cope with this heterogeneity.
2. Traffic is not homogeneously distributed over time and space, with hotspots being a particular example. Although this characteristic is inherent to any cellular system, in WCDMA particular implications will follow, conditioning the network design and operation principles.
3. Indoor traffic is also an inherent characteristic in mobile communications. Here, WCDMA poses some constraints that must be considered in advance to come up with suitable solutions.

4.6.1 SERVICE AND USER HETEROGENEITY

The broad range of services expected to be supported through 3G networks will exhibit diverse requirements in terms of QoS. The provision of such mobile multimedia services requires a diverse amount of radio resources as well as diverse principles for the management of such radio resources.

Different classes of users can be considered for each service (e.g. consumer or business) with different service usage. For example, a service mix ratio for the business user class is depicted in Figure 4.33 according to the expectations considered in the EVEREST project corresponding to the 6th framework of the IST (European Commission) [6]. Note that each mobile user could use more than one service simultaneously depending of the RAT and terminal characteristics considered. Predictions of services, traffic and users have been made in a variety of fora, for example, ITU-R and the UMTS forum. One

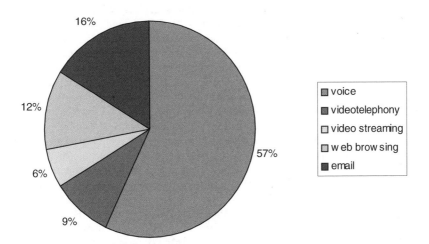

Figure 4.33   Service mix example for the business user class

important example of expected services and corresponding traffic loads in the year 2010 is given in the UMTS report '3G Offered Traffic Characteristics' [7].

In order to stress the importance of the service type, some comments are given below with respect to some of the initial 3G services [6]:

- Video-telephony is a conversational service, always performed between peers of human end users. Video-telephony consists of a continuous sequence of data blocks that will be presented to the user in the right sequence at pre-determined instants. Video-telephony implies a full-duplex system, carrying both video and audio and is intended for use in a conversational environment. The maximum transfer delay is given by the human perception of video and audio conversation. Therefore, the limit for acceptable transfer delay is very strict, as failure to provide low enough transfer delay will result in an unacceptable level of quality. In principle, the same delay requirements as for conversational voice will apply, i.e. no echo and minimal effect on conversational dynamics, with the added requirement that the audio and video must be synchronised within certain limits to provide 'lip-synch' (i.e. synchronisation of the speaker's lips with the words being heard by the end user). The human eye will not notice a certain amount of lost information, so some degree of packet loss is acceptable depending on the specific video coder and the amount of error protection used. It is expected that the latest video codecs will provide acceptable video quality with frame erasure rates upto about 1%.
- Video streaming is a real time data flow which always aims at a human destination. It is mainly a unidirectional stream with high continuous utilisation, i.e. having few idle or silent periods. It is also characterised by the fact that time variation between information entities (i.e. samples, packets) within a flow must be preserved, so that an uninterrupted flow is delivered to the user. It does not have any requirements on low transfer delay and, as a matter of fact, it is possible to have large transfer delays provided that the delay variation is preserved, because in this case the user will simply perceive a large initial set-up time. As the stream is normally time aligned at the receiving end in the user equipment, the highest acceptable delay variation over the transmission media is given by the capability of the time alignment function of the application. Acceptable delay variation is thus much greater than the delay variation given by the limits of human perception. Packet video sources may be classified into Constant Bit Rate (CBR) and Variable Bit Rate (VBR) sources. CBR video coding involves generating a fixed bit rate over time, while VBR video coding allows for generating a variable bit rate over time. VBR coding is more desirable for video applications as it can guarantee constant video quality over time.
- WWW browsing is an interactive service characterised by the fact that the end user requests online data from remote equipment that should be delivered in a relatively short period of time in order to enable some degree of interaction between the end user and the remote server. Round trip delay time is therefore one of the key attributes. Another characteristic is that the content of the packets must be transparently transferred, i.e. with low bit error rate. Therefore, the fundamental characteristics for QoS in interactive services are the request/response pattern and the preservation of the payload content. From the user point of view, the main performance factor is how fast a page appears after it has been requested. It should be mentioned that this category includes retrieving and viewing the HTML component of a WWW page, while other components such as, for example, images and audio/video clips, are dealt with under separate categories, like audio/video streaming.
- Email is a service in which the end user sends and receives data-files in the background. The scheme is thus more or less delivery time insensitive. Another characteristic is that the content of the packets must be transparently transferred and essentially error free. It is important to differentiate the communications between the user and the local email server and the server-to-server transfer. When the user communicates with the local server, there is an expectation that the email will be transferred quite rapidly, although not necessarily instantaneously, while if the communication is between servers, there is no such expectation. For most Internet based email services, the incoming messages of a user are stored at a dedicated email server which keeps all messages until the user logs onto the network and initiates the email application.

Table 4.2 Business user QoS values

Parameter	Voice	Video telephony	Video streaming	Web Browsing	email
Typical bit rate	12 kb/s	64 kb/s	128 kb/s	384 kb/s	384 kb/s
Guaranteed bit rate	12 kb/s	64 kb/s	64 kb/s	N/A	N/A
Maximum bit rate	12 kb/s	64 kb/s	256 kb/s	1500 kb/s	1500 kb/s
PER	< 1%	< 1%	< 1%	Zero	Zero
Delay bound	< 100 ms	< 150 ms	< 1s	< 2s	< 4s

The main QoS parameters at the service level are summarised and reference values are specified for each service [6]. Some important QoS parameters to be taken into account are:

- Typical bit rate, which is the average rate that a user or application generates during an activity period.
- Guaranteed bit rate, which is the bit rate that the network ensures to the application, measured as a number of bits delivered within a specific period of time.
- Maximum bit rate, measured as the maximum number of bits that can be transmitted in the shortest period of time (e.g. the maximum number of bits transmitted in one TTI).
- Packet error rate (PER), which defines the number of erroneous packets with respect to the total number of transmitted packets. It is worth mentioning that a packet may have different meanings depending on the layer of the protocol stack in which it is considered. In that sense, from the point of view of the QoS in the radio interface, it is usual to consider a packet as a SDU received from the NAS (see Section 3.1).
- Delay bound, or transfer delay, which indicates the 95th percentile of the delay distribution for all packets, for example, SDUs, delivered during the lifetime of a bearer service. The delay of a packet is defined as the time from a request to transfer it at one network side to its delivery at the other network side.

The values for the selected QoS parameters differ between classes. As an example, Tables 4.2 and 4.3 show some values that are considered in the framework of the IST EVEREST project for the business and consumer user classes [6].

4.6.2 SPATIAL TRAFFIC DISTRIBUTION HETEROGENEITY

One of the main traffic characteristics in cellular networks is the non-homogenous spatial distribution. The fact that 3G multimedia-intensive traffic demand profiles are expected to be different from 2G ones, makes spatial traffic distributions even more important than in 2G networks. Although network planning can consider this fact, the high dynamics associated with traffic clearly need additional mechanisms to cope with the potential problems on the network performance for real traffic profile distributions

Table 4.3 Consumer user QoS values

Parameter	Voice	Video telephony	Video streaming	Web Browsing	email
Typical bit rate	12 kb/s	64 kb/s	64 kb/s	128 kb/s	128 kb/s
Guaranteed bit rate	12 kb/s	64 kb/s	32 kb/s	N/A	N/A
Maximum bit rate	12 kb/s	64 kb/s	128 kb/s	500 kb/s	500 kb/s
PER	< 1%	< 1%	< 1%	Zero	Zero
Delay bound	< 100 ms	< 150 ms	< 2s	< 3s	< 6s

significantly different from those expected in the network planning phase. Traffic variations over time and space, thus resulting in different load levels in different cells and times, can be smoothed, for example, with the aid of the handover algorithm, trying to balance load among cells. Multicellular admission control and/or congestion control mechanisms are other examples. In this framework, mechanisms supporting smart load control actions could be of great interest and could be applied at different levels. A smart load control for a reference cell would be based on the ability to detect those neighbouring cells mostly affecting the reference cell performance.

In order to devise the sensitivity of a reference cell with respect to the different neighbouring cells, the gradient of the reference cell load factor for the uplink and the transmitted power level for the downlink can be derived, taking advantage of the analytical model presented in Sections 4.4.3.1 and 4.4.5.1 for uplink and downlink, respectively.

Assuming a scenario with $K + 1$ cells, $k = 0,1,\ldots,K$, where the central cell 0 is the reference cell, the gradient of the uplink load factor of cell 0 is given by:

$$\vec{\nabla}\eta_0 = \left(\frac{\partial \eta_0}{\partial \eta_1}, \frac{\partial \eta_0}{\partial \eta_2}, \ldots, \frac{\partial \eta_0}{\partial \eta_K} \right) \tag{4.56}$$

Making use of Equation (4.38), it can be obtained that:

$$\frac{\partial \eta_0}{\partial \eta_k} = \frac{1 - S_{0,0}^{UL}}{\left(1 + \sum\limits_{j=1}^{K} \dfrac{S_{j,0}^{UL}}{1 - \eta_j} \right)^2} \frac{\partial}{\partial \eta_k} \left(\sum\limits_{j=1}^{K} \frac{S_{j,0}^{UL}}{1 - \eta_j} \right) = \frac{1 - S_{0,0}^{UL}}{\left(1 + \sum\limits_{j=1}^{K} \dfrac{S_{j,0}^{UL}}{1 - \eta_j} \right)^2} \sum\limits_{j=1}^{K} \frac{S_{j,0}^{UL}}{\left(1 - \eta_j \right)^2} \frac{\partial \eta_j}{\partial \eta_k} \tag{4.57}$$

Similarly, for the downlink direction, and in terms of the transmitted power, it can be defined that:

$$\vec{\nabla}P_{T,0} = \left(\frac{\partial P_{T,0}}{\partial P_{T,1}}, \frac{\partial P_{T,0}}{\partial P_{T,2}}, \ldots, \frac{\partial P_{T,0}}{\partial P_{T,K}} \right) \tag{4.58}$$

and making use of Equation (4.55) yields:

$$\frac{\partial P_{T0}}{\partial P_{T,k}} = \frac{1}{1 - \rho S_{0,0}^{DL}} \frac{\partial}{\partial P_{T,k}} \left(\sum\limits_{j=1}^{K} S_{0,j}^{DL} P_{Tj} \right) = \frac{1}{1 - \rho S_{0,0}^{DL}} \sum\limits_{j=1}^{K} S_{0,j}^{DL} \frac{\partial P_{Tj}}{\partial P_{T,k}} \tag{4.59}$$

Although an accurate analysis based on solving a linear equations system can be carried out from Equations (4.57) and (4.59) to determine the derivatives of the load factor and transmitted power in both uplink and downlink, it is also possible to make an approximate analysis that simplifies the computations while keeping good accuracy, and therefore allows devising practical RRM algorithms for radio network engineering purposes. The approximations are given by:

$$\frac{\partial \eta_0}{\partial \eta_k} \approx \frac{S_{k,0}^{UL} \left(1 - S_{0,0}^{UL} \right)}{(1 - \eta_k)^2 \left(1 + \sum\limits_{j=1}^{K} \dfrac{S_{j,0}^{UL}}{1 - \eta_j} \right)^2} \tag{4.60}$$

$$\frac{\partial P_{T0}}{\partial P_{Tk}} \approx \frac{S_{0,k}^{DL}}{1 - \rho S_{0,0}^{DL}} \tag{4.61}$$

In order to gain insight into the potentials of the presented framework, a set of illustrative results are shown. Similarly, the role of the different elements involved in the formulation is stressed, relating the analytical formulation with the WCDMA air interface physical behaviour. The scenario considered is an area of $4.5 \times 4.5 \, \text{km}^2$, with 23 omnidirectional cells separated by 1 km according to an hexagonal pattern.

4.6.2.1 Homogeneous Scenario

In this case, a uniform traffic distribution is considered in all cells and within each cell. Figure 4.34 plots, for each snapshot, and a total set of 1000 samples, the location of the mobile with the highest contribution to the term $S_{j,0}^{UL}$ given by Equation (4.37) for $j = 1 \ldots K$. These users are those affecting most the reference cell load factor. Note that, for the uplink direction, these users tend to be those closer to the reference cell site.

In turn, Figure 4.35 presents the gradient of the reference cell load factor with respect to the neighbouring cells load factor. The radial plots represent the derivative value for the six neighbouring cells corresponding to the first ring of interfering cells. It should be noted that, since the traffic is homogeneously distributed, the derivatives are similar with respect to all six cells.

For the downlink direction, the user in the reference cell with the highest contribution to the overall interference is plotted in Figure 4.36, according to Equation (4.54). Note that, according to this expression, and as also seen in the derivative given by Equation (4.61), the influence of the different cells with respect to the central cell depends on the user distribution in the central cell itself, while in the uplink, according to Equation (4.37) and the derivative given by Equation (4.60), the influence between cells depends on the user distribution in the neighbouring cells. The downlink derivatives are presented in Figure 4.37. Similar comments to the uplink apply in this homogeneous case, and the derivatives are equal in all the neighbouring cells.

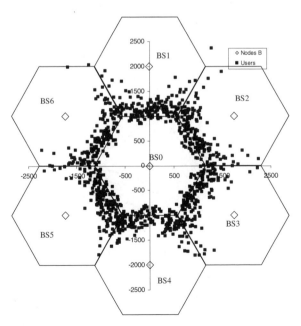

Figure 4.34 Spatial representation of the highest contribution to $S_{j,0}^{UL}$ for $j = 1 \ldots K$

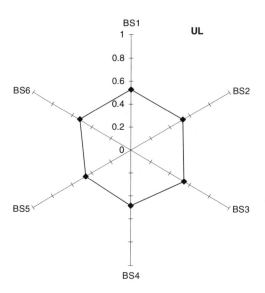

Figure 4.35 Radial plot of $\partial\eta_0/\partial\eta_k$ for the first ring of interfering cells in the uplink

4.6.2.2 Non-homogeneous Scenario

At this stage, two different types of non-homogeneous spatial traffic distribution are considered, referred as inter-cell and intra-cell non-homogeneity. In the former, different numbers of users in the different cells are assumed, although users are uniformly distributed within a cell. In the latter, the same number of users is assumed in each cell, although users are non-uniformly distributed within a given cell.

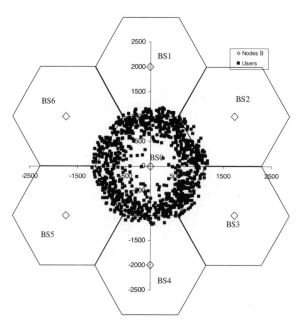

Figure 4.36 Spatial representation of the highest contribution to downlink interference

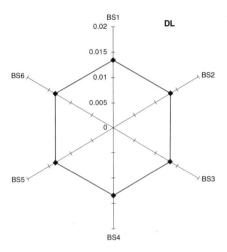

Figure 4.37 Radial plot of $\partial P_{T,0}/\partial P_{T,k}$ for the first ring of interfering cells in the downlink

Inter-cell Non-homogeneity In the case of inter-cell non-homogeneity, some neighbouring cells have been more heavily loaded than the reference cell. In particular, cell BS5 has three times more traffic and cell BS4 has two times more traffic than the rest of cells. Within a cell, traffic is homogeneously distributed.

For the uplink case, the users contributing most to the interference are shown in Figure 4.38 and, consequently, the derivatives shown in Figure 4.39 reflect that BS5 (and also BS4 to a lower extent) are the ones to which BS0 is more sensitive.

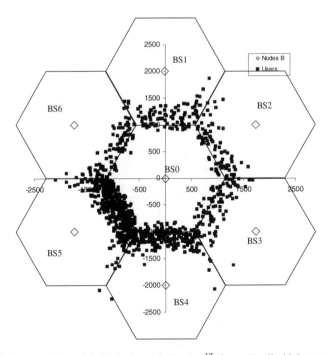

Figure 4.38 Spatial representation of the highest contribution to $S_{j,0}^{UL}$ for $j = 1 \ldots K$ with inter-cell non-homogeneity

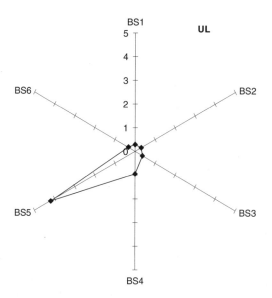

Figure 4.39 Radial plot of $\partial\eta_0/\partial\eta_k$ for the first ring of interfering cells in the uplink with inter-cell non-homogeneity

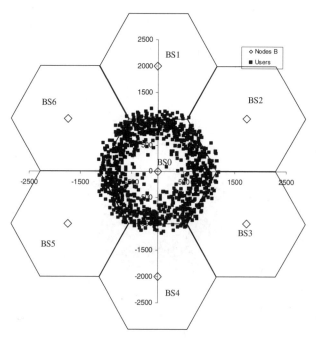

Figure 4.40 Spatial representation of the highest contribution to downlink interference with inter-cell non-homogeneity

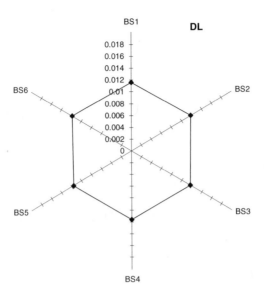

Figure 4.41 Radial plot of $\partial P_{T,0}/\partial P_{T,k}$ for the first ring of interfering cells in the downlink with inter-cell non-homogeneity

However, Figures 4.40 and 4.41 indicate that in the downlink this unequal spatial traffic distribution at the inter-cell level does not influence the distribution of most relevant users as well as the sensitivity from neighbouring cells with respect to the reference cell. Note that the influence of the kth neighbouring cell on the reference cell not only depends on the kth cell transmitted power level (P_{Tk}) but also on the user's distribution within the reference cell. This is because if users in the reference cell are closer to the kth cell, the influence of this cell will be higher. This is reflected in the product term $P_{Tj}S_{0,j}^{DL}$ in Equation (4.55). The relevant users tend to be those far from the cell site, which require higher transmitted power levels. In turn, the sensitivity of the reference cell is similar with respect to the six neighbouring cells as long as users in the reference cell are uniformly distributed.

Intra-cell Non-homogeneity In the case of intra-cell non-homogeneity all cells have the same load. However, users in BS0 are distributed non-homogeneously, with 80% of users concentrated in the lower right area.

For the uplink, Figures 4.42 and 4.43 show similar results to the homogeneous case (shown in Figures 4.34 and 4.35), thus indicating that this intra-cell spatial traffic non-homogeneity is not affecting the sensitivity in the uplink of the reference cell. The reason is that the most relevant users continue to be those users in neighbouring cells closer to the reference cell site.

Nevertheless, different conclusions can be extracted from the downlink analysis. As can be seen in Figure 4.44, the most relevant users from the reference cell perspective are those far from the cell site, which require higher transmitted power levels. The higher user density in the lower right area of the reference cell is clearly reflected in Figure 4.44. This, in turn, is the origin of the higher downlink sensitivity from the reference cell with respect to BS3. Note that the BS3 transmitted power level will have a stronger effect on the reference cell because it will affect more of the reference cell users, as they are closer to this cell. Consequently, the derivative of the reference cell transmitted power level will be most sensitive to changes in BS3 transmitted power level (i.e. $\partial P_{T0}/\partial P_{Tk}$ is maximum for BS3), as shown in Figure 4.45.

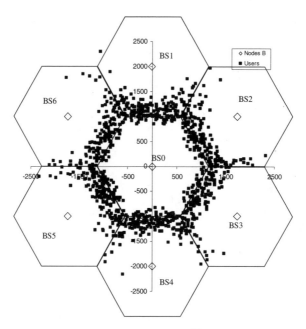

Figure 4.42 Spatial representation of the highest contribution to $S_{j,0}^{UL}$ for $j = 1 \ldots K$ with intra-cell non-homogeneity

4.6.2.3 Traffic Hotspots

A traffic hotspot is a specific area in a scenario where user density and, consequently, traffic load, are much higher than regular conditions. Therefore, hotspot is a particular case of non-homogeneous traffic distribution. Usually, as long as hotspots are clearly identified and concentrated in space, specific infrastructure is deployed in the hotspot area to cope with the traffic excess.

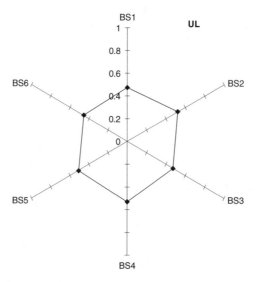

Figure 4.43 Radial plot of $\partial\eta_0/\partial\eta_k$ for the first ring of interfering cells in the uplink with intra-cell non-homogeneity

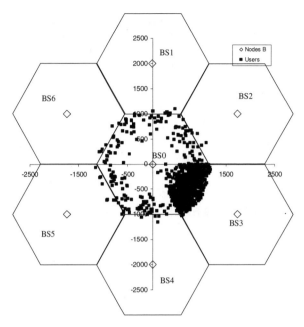

Figure 4.44 Spatial representation of the highest contribution to downlink interference with intra-cell non-homogeneity

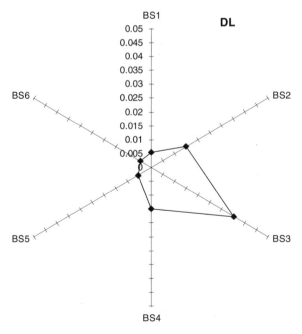

Figure 4.45 Radial plot of $\partial P_{T,0}/\partial P_{T,k}$ for the first ring of interfering cells in the downlink with intra-cell non-homogeneity

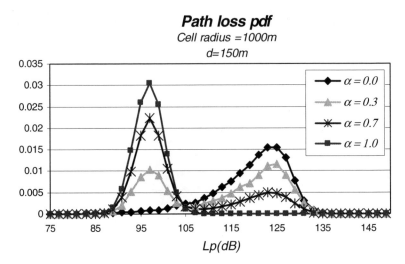

Figure 4.46   Pdf of users' path loss with a hotspot $d = 150$ m far from base station

In order to highlight the effects of a hotspot, a scenario composed of an isolated cell, with a hotspot placed d metres from the base station is considered. There is a total traffic T and a fraction αT is concentrated within the hotspot, while the rest of the traffic is homogeneously distributed around the cell. Different traffic load distributions are analysed by varying α and d. It seems to be clear that different traffic distributions lead to different path loss patterns, or in other words, path loss statistics are tightly dependent on hotspot positions as well as on traffic density [8].

Figures 4.46 and 4.47 show the probability density function (pdf) of the path loss for different α values as well as for different hotspot distances d. According to what has been obtained in Section 4.4, the path loss distribution is one of the key statistics that impact on performance in terms of outage for both uplink and downlink. It is observed that path loss statistics change if α, d or both vary. Path loss pdf with $\alpha = 1.0$ is equal to the path loss of users within the hotspot, which depends on its location and dimension.

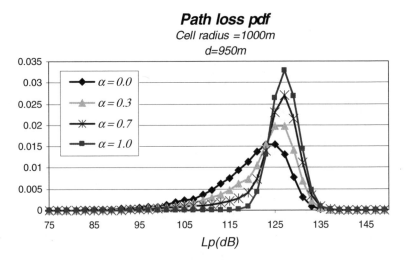

Figure 4.47   Pdf of users' path loss with a hotspot $d = 950$ m far from base station

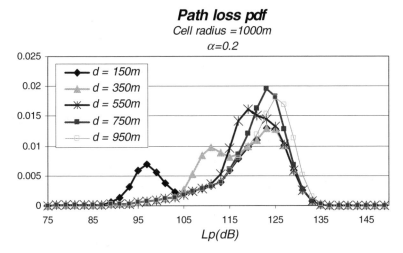

Figure 4.48 Pdf of users' path loss with $\alpha = 0.2$ and different hotspot distances

It is depicted that, for high d, as α increases, the path loss distribution concentrates and shifts to higher values (see Figure 4.47). In turn, for low d, path loss distribution presents a two peak shape (see Figure 4.46). In this case, if α increases, low path loss values become more probable and high values less probable, while if α decreases, low values are less probable and high values are more probable. In the case of high d, high path loss values correspond to hotspot users, while in the case of low d, they are the low path loss values corresponding to the hotspot users.

Figure 4.48 shows the path loss pdf for different values of d and a fixed α. Note that, as d increases, path loss distribution changes from a two peak shape to a shape concentrated around higher path loss values. So, depending on hotspot location and traffic concentration, high path loss situations may appear.

4.6.3 INDOOR TRAFFIC

Indoor traffic is very important in 2G networks, as a remarkably large traffic load originates inside buildings. Typically, in GSM, in-building coverage can be provided as a first step by tuning transmitted power levels. As a second step and also to respond to capacity demands, indoor coverage can be provided by deploying hierarchical cell structures with micro and picocells.

Although 3G traffic demand profiles are expected to be different from 2G ones, because of the different nature of the services, clearly indoor traffic will be even more key in 3G since the envisaged high bit rate services will mainly be of a static nature, in many cases indoors.

Nevertheless, the implications of indoor traffic in 3G WCDMA based systems may significantly differ from 2G TDMA-based solutions because transmitted power levels are the key radio resources in WCDMA. The higher power levels needed for indoor service will reduce cell capacity. Consequently, it can be important for a network operator to quantify the impact that indoor traffic may have on the overall system efficiency in order to devise suitable deployment guidelines (i.e. how fast the transition from outdoor macrocell sites to indoor micro and picocells distributions should be carried out).

Let us consider the two different situations shown in Figure 4.49: case (a) stands for a given mobile in an outdoor location and case (b) stands for the mobile at the same location but indoors.

For the uplink, let $P_{in,UL}$ denote the mobile transmitted power for the indoor case and $P_{out,UL}$ denote the outdoor case. In order to achieve a certain signal to noise plus interference ratio Γ_{UL}, the relations to

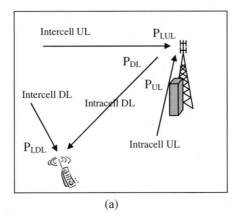

(a)

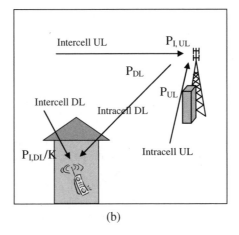

(b)

Figure 4.49 Interference in the outdoor (a) and indoor (b) situations

be ensured in the indoor and outdoor cases are:

$$\frac{\dfrac{P_{in,UL}}{L \cdot K}}{P_{N,UL} + P_{I,UL} + P_{UL}} = \Gamma_{UL} \tag{4.62}$$

$$\frac{\dfrac{P_{out,UL}}{L}}{P_{N,UL} + P_{I,UL} + P_{UL}} = \Gamma_{UL} \tag{4.63}$$

where $P_{N,UL}$ is the noise level, $P_{I,UL}$ the intercell interference, P_{UL} the intracell interference, L the path loss and K the in-building penetration loss. Note that by setting the same $P_{I,UL}$ and P_{UL} in Equations (4.62) and (4.63), it is assumed for simplicity that neither the intercell nor the intracell interference change due to the fact that the mobile under analysis is indoor or outdoor. As a result, the corresponding required transmitted powers are:

$$P_{in,UL} = L \cdot K \cdot \Gamma_{UL}(P_{N,UL} + P_{I,UL} + P_{UL}) \tag{4.64}$$
$$P_{out,UL} = L \cdot \Gamma_{UL}(P_{N,UL} + P_{I,UL} + P_{UL}) \tag{4.65}$$

Thus, the power increase for a user moving from outdoor to indoor in the uplink is given by:

$$\Delta P_{UL} = (P_{in,UL} - P_{out,UL}) = (K - 1) \cdot L \cdot \Gamma_{UL}(P_{N,UL} + P_{I,UL} + P_{UL}) \tag{4.66}$$

For the downlink, let $P_{in,DL}$ denote the base station transmitted power devoted to the user in the indoor case and $P_{out,DL}$ denote the outdoor case. Given a certain noise level $P_{N,DL}$, an intercell interference outdoors $P_{I,DL}$, P_{DL} as the rest of the base station transmitted power (i.e. the DL intracell interference devoted to other users and to common channels), a path loss L, an in-building penetration loss K, an orthogonality factor ρ and a target signal to interference ratio Γ_{DL}, the corresponding relationships are given by:

$$\frac{\dfrac{P_{in,DL}}{L \cdot K}}{P_{N,DL} + \dfrac{P_{I,DL}}{K} + \rho \dfrac{P_{DL}}{L \cdot K}} = \Gamma_{DL} \tag{4.67}$$

$$\frac{\dfrac{P_{out,DL}}{L}}{P_{N,DL} + P_{I,DL} + \rho \dfrac{P_{DL}}{L}} = \Gamma_{DL} \tag{4.68}$$

Again, it has been assumed that neither the intercell nor the intracell interference powers $P_{I,DL}$ and P_{DL} change due to the fact that the mobile under analysis is indoor or outdoor. Therefore, the corresponding required power levels will be:

$$P_{in,DL} = L \cdot K \cdot \Gamma_{DL}\left(P_{N,DL} + \frac{P_{I,DL}}{K} + \rho \frac{P_{DL}}{L \cdot K}\right) \tag{4.69}$$

$$P_{out,DL} = L \cdot \Gamma_{DL}\left(P_{N,DL} + P_{I,DL} + \rho \frac{P_{DL}}{L}\right) \tag{4.70}$$

Note that, in the downlink direction, both the intracell and the intercell interference are reduced by the in-building penetration losses K. As a result, the power increase for a user moving from outdoors to indoors in the downlink is given by:

$$\Delta P_{DL} = (P_{in,DL} - P_{out,DL}) = (K - 1) \cdot L \cdot \Gamma_{DL} \cdot P_{N,DL} \tag{4.71}$$

Comparing the power increase in uplink and downlink given by Equations (4.66) and (4.71), respectively, leads to:

$$\frac{\Delta P_{UL}}{\Delta P_{DL}} = \frac{\Gamma_{UL}(P_{N,UL} + P_{I,UL} + P_{UL})}{\Gamma_{DL} \cdot P_{N,DL}} \tag{4.72}$$

For equal target qualities and noise level in uplink and downlink, Equation (4.72) is simplified to:

$$\frac{\Delta P_{UL}}{\Delta P_{DL}} = 1 + \frac{P_{I,UL} + P_{UL}}{P_{N,DL}} \tag{4.73}$$

Thus, it can be observed that in the downlink the power increase depends mainly on noise power while in the uplink it depends mainly on noise plus system interference. As a result, the power increase will be higher in the uplink direction. Besides, the higher the load in the system the higher the difference with respect to the downlink. Consequently, a lower degradation caused by indoor traffic is expected in the downlink when compared to the uplink. The rationale behind this effect is the higher protection provided by the in-building penetration loss against interference for downlink indoor users [9].

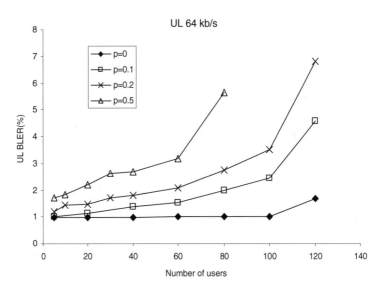

Figure 4.50   BLER degradation in the uplink depending on the fraction p of indoor users for the 64 kb/s service

The higher degradation due to indoor traffic in the uplink is illustrated by comparing Figures 4.50 and 4.51, which depict the performance in terms of BLER for a 64 kb/s service in the uplink and downlink, respectively, for different fractions p of indoor users in the scenario. The scenario consists of seven omnidirectional cells with site separation of 1 km. The in-building penetration loss is $K = 20$ dB, and the BLER target is 1%. It can be observed that the BLER increase is much more significant in the uplink direction, which even for relatively low loads leads to BLER values much higher than the target of 1%. On the contrary, in the downlink the degradation is softer and appears only for high loads.

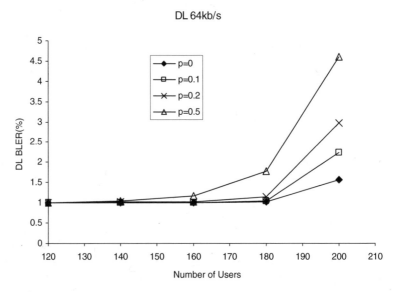

Figure 4.51   BLER degradation in the downlink depending on the fraction p of indoor users for the 64 kb/s service

Table 4.4 Capacity loss (%) relative to the case with no indoor traffic
($p = 0$) for 64/64 kb/s and 64/384 kb/s radio bearer

	UL 64 kb/s (%)	DL 64 kb/s (%)	DL 384 kb/s (%)
$p = 0.1$	19.2	9.3	12.5
$p = 0.2$	37.7	11.6	18.8
$p = 0.5$	88.4	15.3	25.0

Table 4.4 completes the analysis by presenting the capacity loss relative to the case with only outdoor traffic for different percentages of indoor users. In this case, the uplink at 64 kb/s is compared to the downlink at 64 kb/s and 384 kb/s. The capacity is defined by the maximum number of users sharing the air interface while keeping the BLER below 2%. It can be observed that, although the capacity reduction for the DL is higher with the 384 kb/s service than with the 64 kb/s service, it is still much lower than the reduction in the uplink, which, for example, for a 50% of indoor traffic suffers a reduction of 88%.

REFERENCES

[1] J. Zander, 'Affordable multiservice wireless networks – research challenges for the next decade', *The 13th IEEE International Symposium on Personal, Indoor and Mobile Radio Communications (PIMRC-2002)*, September 2002, **1**, pp. 1–4

[2] L. Kleinrock, *Queueing Systems*, John Wiley & Sons Inc., Vols I and II, 1975

[3] H. Holma, A. Toskala, *WCDMA for UMTS*, John Wiley & Sons Ltd, 2nd edition, 2000

[4] 3GPP 25.922 v6.0.1, 'Radio resource management strategies (release 6)'

[5] O. Sallent, R. Agustí, 'A Proposal for an Adaptive S-ALOHA Access System for a Mobile CDMA Environment', *IEEE Transactions on Vehicular Technology*, **47**(3), August 1998

[6] P. Karlsson (editor) *et al.* 'Target Scenarios Specification: vision at project stage 1' Deliverable D05 of the EVEREST IST-2002-001858 project, April, 2004. Available at http://www.everest-ist.upc.es/

[7] '3G Offered Traffic Characteristics', Final Report, UMTS Forum, November 2003

[8] F. Adelantado, O. Sallent, J. Pérez-Romero, R. Agustí, 'Impact of Traffic Hotspots in 3G W-CDMA Networks', *IEEE 58th Semiannual Vehicular Technology Conference (VTC 2004-Spring)*, Milan, Italy, May, 2004

[9] J. Pérez-Romero, O. Sallent, R. Agustí, 'On The Capacity Degradation in W-CDMA Uplink/Downlink due to Indoor traffic', *IEEE 59th Semiannual Vehicular Technology Conference (VTC 2004-Fall)*, Los Angeles, USA, 2004

APPENDIX - PATH LOSS DISTRIBUTION

In this appendix an analytical model is developed to obtain the path loss distribution for a multiple cell scenario with initially only outdoor users and, eventually, when indoor traffic is also considered.

For an outdoor user, the total path loss at a distance r from the serving cell is given by:

$$L(dB) = L_o + \gamma \log r + X \tag{4.74}$$

where X is a Gaussian random variable with standard deviation σ (dB), representing the slow shadowing fading. The path loss L is also a random variable that depends on user location and on shadowing fading.

The cell layout considered in the model is depicted in Figure 4.52, assuming a cell radius R. Although only the first ring of six interfering cells is depicted, the model considers also the twelve cells in the second ring. In such a multicellular scenario, assuming that all the pilot powers have the same level, and that an ideal handover is executed, the users will be connected to the cell with minimum path loss.

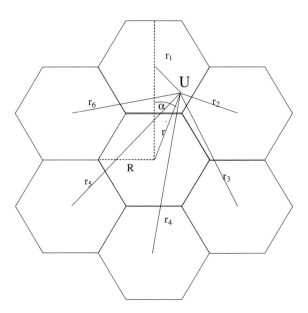

Figure 4.52   Multicellular layout considered in the model

Consequently, the path loss distribution with respect to the central cell should be calculated by taking into account only those locations whose minimum path loss cell is the central one.

For a user located at polar coordinates $U(r,\alpha)$, the path loss with respect to the ith cell ($i = 1...18$) is given by:

$$L_i(dB) = L_o + \gamma \log r_i + X_i \tag{4.75}$$

where X_i ($i = 1...18$) are Gaussian random variables with standard deviation σ (dB).

In order for the user located at U to be connected to the central cell, the condition is:

$$L < L_i \qquad \forall i \in [1, 18] \tag{4.76}$$

or, equivalently:

$$X < \gamma \log\left(\frac{r_i}{r}\right) + X_i \equiv a_i(r, \alpha) + X_i \qquad \forall i \in [1, 18] \tag{4.77}$$

where, for the six cells $i = [1,...,6]$ in the first ring, located at distance $\sqrt{3}R$ from the central cell:

$$a_i(r, \alpha) = \frac{\gamma}{2} \log\left(1 + \frac{3R^2}{r^2} - \frac{2\sqrt{3}R}{r} \cos\left(\alpha - (i - 1)\frac{\pi}{3}\right)\right) \tag{4.78}$$

In turn, for the six cells $i = [7,...,12]$ of the second ring located at distance $3R$ from the central cell:

$$a_i(r, \alpha) = \frac{\gamma}{2} \log\left(1 + \frac{9R^2}{r^2} - \frac{6R}{r} \cos\left(\alpha - (i - 7)\frac{\pi}{3} - \frac{\pi}{6}\right)\right) \tag{4.79}$$

and, for the six cells $i = [13, \ldots, 18]$ of the second ring located at distance $2\sqrt{3}R$ from the central cell:

$$a_i(r, \alpha) = \frac{\gamma}{2} \log\left(1 + \frac{12R^2}{r^2} - \frac{4\sqrt{3}R}{r} \cos\left(\alpha - (i - 13)\frac{\pi}{3}\right)\right) \tag{4.80}$$

Then, the CDF of the path loss L to users connected to the central cell is given by:

$$\Pr(L \le l \mid X < a_i(r, \alpha) + X_i \; \forall i \in [1, 18]) = \frac{F_1(l)}{F_2} \tag{4.81}$$

where, taking into account the conditional probability:

$$F_1(l) = \Pr(L \le l, X < a_i(r, \alpha) + X_i \; \forall i \in [1, 18]) \tag{4.82}$$
$$F_2 = \Pr(X < a_i(r, \alpha) + X_i \; \forall i \in [1, 18]) \tag{4.83}$$

Then, assuming that X, X_i are independent Gaussian variables, that users are uniformly distributed within a maximum distance D to the central base station, and taking into account the symmetry of the problem with respect to the angle α:

$$F_1(l) = \int_0^{\pi/6} \int_0^D \frac{1}{\sqrt{2\pi}\sigma} \int_{-\infty}^{l - L_o - \gamma \log r} e^{-\frac{x^2}{2\sigma^2}} \prod_{i=1}^{18} Q\left(\frac{x - a_i(r, \alpha)}{\sigma}\right) \frac{2r}{D^2} \frac{6}{\pi} dx dr d\alpha \tag{4.84}$$

where

$$Q(x) = \frac{1}{\sqrt{2\pi}} \int_x^{\infty} e^{-\frac{t^2}{2}} dt \tag{4.85}$$

Similarly:

$$F_2 = \int_0^{\pi/6} \int_0^D \frac{1}{\sqrt{2\pi}\sigma} \int_{-\infty}^{\infty} e^{-\frac{x^2}{2\sigma^2}} \prod_{i=1}^{18} Q\left(\frac{x - a_i(r, \alpha)}{\sigma}\right) \frac{2r}{D^2} \frac{6}{\pi} dx dr d\alpha \tag{4.86}$$

Finally, the probability density function (pdf) of the path loss L for outdoor users is given by:

$$f_L(l) = \frac{1}{F_2} \int_0^{\pi/6} \int_0^D \frac{1}{\sqrt{2\pi}\sigma} e^{-\frac{(l - L_o - \gamma \log r)^2}{2\sigma^2}} \prod_{i=1}^{18} Q\left(\frac{l - L_o - \gamma \log r - a_i(r, \alpha)}{\sigma}\right) \frac{2r}{D^2} \frac{6}{\pi} dr d\alpha \tag{4.87}$$

Note that D is set to $D = 3R$ since the probability is negligible that users at higher distances from the central cell are connected to it.

In order to extend the analysis to the presence of indoor users, let us assume that these users are characterised by K dB of additional path loss due to in-building penetration losses. If a fraction p of users are indoor and the remaining are outdoor, the pdf of the total path loss is given by:

$$f_L^{total}(l) = (1 - p) \cdot f_L(l) + p \cdot f_L(l - K) \tag{4.88}$$

where l is the path loss in dB, and $f_L(l)$ is the path loss distribution in dB without considering the in-building penetration losses, i.e. the distribution given by Equation (4.87).

The analytical model has been validated through simulations and the results in terms of the path loss pdf according to Equation (4.88) are shown in Figure 4.53 for $p = 0.5$, $K = 20$ dB, $\sigma = 10$ dB and the

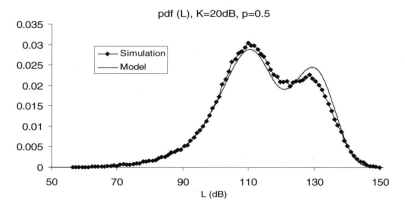

Figure 4.53 Model validation

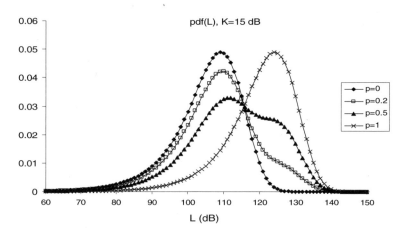

Figure 4.54 Path loss distribution for $R = 0.577$ km, $\sigma = 10$ dB

distance between cell sites is 1 km, which corresponds to a cell radius $R = 0.577$ km. The path loss parameters in Equation (4.74) are $L_o = 128.1$ and $\gamma = 37.6$, with r measured in km. It can be observed that simulations fit well with the analytical model. On the other hand, and in order to see the impact of the fraction of indoor users p over the path loss distribution, Figure 4.54 shows the pdf of the path loss for different situations ranging from $p = 0$ up to $p = 1$ with an in-building penetration loss of 15 dB. As can be observed, there are very significant changes in the distribution, shifted to higher path loss values when p increases.

5

RRM Algorithms

This chapter presents a set of specific algorithms and solutions for the different RRM strategies that have been identified in Chapter 4 in a general WCDMA framework. These solutions are aligned with the 3GPP specifications according to the UMTS radio interface description given in Chapter 3. It should be mentioned that the 3GPP specifications only provide the framework for RRM development, and the specific solutions are implementation dependent so they are not subject to standardisation. In this sense, the purpose of this chapter is the presentation of possible examples and solutions to the RRM problem together with the identification of the most relevant parameters that play a key role in each RRM strategy. They are devised from a practical point of view, based on the theoretical aspects identified in Chapter 4, and try to provide an engineering vision aligned with the 3GPP UTRAN FDD framework.

The chapter starts with a description of the methodologies (Section 5.1) that evaluate RRM algorithms in UMTS networks. Since RRM algorithms operate on a previously planned radio access network, the fundamentals of the UMTS network planning are given as the basis for a subsequent evaluation of the RRM algorithms by means of simulation tools. The different solutions for RRM algorithms will be presented in the next sections. Specifically, Section 5.2 deals with admission control algorithms, Section 5.3 with handover management and Section 5.4 with congestion or load control algorithms, while Section 5.5 presents different short term RRM mechanisms and, finally, Section 5.6 deals with power control.

5.1 RRM ALGORITHM EVALUATION METHODOLOGY

As mentioned in Chapter 4, the RRM problem is of an inherently dynamic nature and should consider system variations in terms of, for example, user mobility and traffic generation. On the other hand, the comparative efficiency of different RRM strategies can normally be only identified in scenarios with heavy load levels, close to the maximum system capacity, because it is in such a situation that the efficiency in the resource allocation process becomes relevant and the capacity gain can be appreciated. Therefore, RRM algorithm evaluation is usually done by means of simulations in order to anticipate the capacity of one or other strategy before it can be implemented in a real system. In that sense, and in order to ensure that the obtained simulation results are close enough to those expected in the real system, it is necessary for the simulation models to capture the relevant aspects of the real network and user behaviour.

RRM algorithm evaluation represents a second step in the UMTS network design and optimisation procedure, since the network planning exercise must be carried out in advance. Network planning is

Radio Resource Management Strategies in UMTS   J. Pérez-Romero, O. Sallent, R. Agustí and M. A. Díaz-Guerra
© 2005 John Wiley & Sons, Ltd

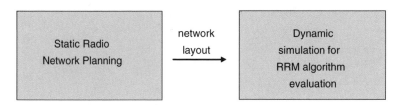

Figure 5.1 Relationship between radio network planning and Radio Resource Management

devised from a static perspective and aims to provide an adequate network dimensioning in terms of, for example, node B's position or initial parameter settings (e.g. maximum transmitter powers, pilot settings, handover thresholds, etc.) according to the expected traffic patterns in a given geographical area. Consequently, network planning provides the radio resource provisioning in this area and RRM strategies will rely on this initial provisioning to ensure that the capacity can be maximised while keeping the QoS and coverage constraints, by dynamically adapting the network operation to the specific traffic conditions at each time instant. Figure 5.1 illustrates the relationship between radio network planning and RRM evaluation. After this two stage procedure, the network optimisation procedure can be carried out to execute the adequate settings in either the RRM parameters or in the radio network planning process, for example, by deciding on the inclusion of new nodes B in the system.

Although the joint consideration of radio network planning and RRM is necessary in any type of mobile network, it is especially critical in WCDMA based networks like UMTS, where users transmit at the same time and on the same carrier. This is due to the lack of a constant value for the maximum available capacity, since it is tightly coupled to the amount of interference in the air interface, which depends on users' dynamics and on the radio access network deployment. Consequently, and compared to 2G systems, much more simulation work regarding 3G networks is necessary because of the multiple issues impacting on the network performance and the much higher degree of coupling among them resulting from the nature of WCDMA.

5.1.1 UMTS RADIO NETWORK PLANNING PROCEDURE

Given the importance of proper network planning prior to an adequate RRM algorithm evaluation, this section presents the fundamentals of the planning procedure for UMTS networks. However, since the object of this book is RRM rather than radio network planning, only the main aspects of the procedure together with some examples including the description of real planning tools will be given. For further details on UMTS planning the reader is referred to References 1 and 2.

The objective of the radio network planning procedure is to decide on the number and location of the sites and carriers that are necessary to provide UMTS services in a given geographical area. This decision determines the amount of RRUs that are provisioned and that RRM strategies will have to manage efficiently according to the network dynamics. The network planning process in 3G networks should take into account the following main characteristics:

- In WCDMA, network capacity and coverage must be considered simultaneously due to the tight relationship that exists between both concepts resulting from the interference-limited nature of the multiple access technique. As a result, analytical studies are rather difficult and, therefore, adequate simulation tools are necessary for a proper estimation of all the involved parameters. With respect to the coverage, the decision to provide in-car and indoor services introduces additional limitations and constraints into the planning procedure, due to the additional propagation losses that have a direct impact in terms of power consumption and interference limitations. On the other hand, in terms of network capacity, each new accepted user increases the interference and represents additional power consumption in the Node B. This resource consumption depends on the geographical proximity of the user to the node B to which it is connected. As a result, the network planning procedure requires knowledge of realistic traffic spatial distributions.

- The multi-service nature of 3G networks also requires the knowledge and characterisation of the services that the operator expects to provide in the short and longer-term, together with the specific QoS requirements of each service. Due to different resource consumption associated with different services, the one with the most stringent requirements will be the one that limits the minimum site density in a given geographical area.

Network planning is carried out in two stages: initial planning, or dimensioning, and detailed planning, which will be described in the following sections.

5.1.1.1 Initial Planning

The initial planning procedure makes use of simple analytical equations to give a rough estimate of the number of required nodes B taking into account the propagation conditions, the expected coverage and the capacity estimation.

The initial planning procedure is executed in two steps. The first one determines the cell radius based on link budget computations, and the second one determines the cell load based on the expected traffic distributions.

Cell Radius Computation   Cell radius computation is done by means of a link budget, which aims to compute the maximum path loss for each cell. Both uplink and downlink are considered in the computation in order to determine which is the limiting link. Once the maximum path loss has been computed, the cell radius is derived based on a specific propagation model.

Table 5.1 presents an example of link budget computation for a 64 kb/s service in the uplink direction. The procedure starts with the computation of the Equivalent Isotropic Radiated Power (EIRP) at the mobile terminal, taking into account the transmit power together with the mobile antenna gain and the losses introduced by the human body. In the case of data or video services, it is usual for the mobile terminal not to be located very close to the human user, and therefore 0 dB are usually considered for the body loss.

Second, the uplink receiver sensitivity S is computed. This corresponds to the minimum power required at the antenna input in order to satisfy a certain E_b/N_0 requirement at the receiver output. In this example, an $E_b/N_0 = 2$ dB has been considered for the uplink direction. The E_b/N_0 requirement is related to service constraints in terms of bit or block error rate that depend on physical layer parameters like the channel coding and interleaving, the modulation scheme and the specific channel impulse response. As was shown in Section 4.4.3, the received power P_i for a certain E_b/N_0 requirement must fulfil the following relationship:

$$\frac{P_i\left(\dfrac{W}{R_b}\right)}{P_N + \chi + P_R - P_i} \geq \left(\frac{E_b}{N_0}\right) \tag{5.1}$$

where χ is the intercell interference and P_R is the total received power from transmissions in the same cell to which the user is connected. The power requirement can then be expressed in terms of the uplink load factor as:

$$P_i \geq \frac{P_N + P_R + \chi}{\dfrac{W}{\left(\dfrac{E_b}{N_0}\right)R_b} + 1} = \frac{P_N}{\dfrac{W}{\left(\dfrac{E_b}{N_0}\right)R_b} + 1} \frac{1}{1 - \eta_{UL}} \approx P_N\left(\frac{E_b}{N_0}\right)\frac{R_b}{W}\frac{1}{1 - \eta_{UL}} \tag{5.2}$$

Table 5.1 Example of uplink link budget

Element	Unit	Value
UL TRANSMISSION		
Mobile transmit poewr (P_T)	dBm	21
Mobile antenna gain (G_m)	dBi	0
Body Loss (L_b)	dB	0
Equivalent Isotropic Radiated Power	dBm	21
$\quad EIRP = P_T + G_m - L_b$		
UL RECEPTION		
Thermal Noise density (KT_0)	dBm/Hz	−174
Noise Figure (F)	dB	5
Bandwidth (B)	MHz	3.84
Noise Power $(P_N = KT_0FB)$	dBm	−103.2
Interference margin (ΔI)	dB	3
Service bit rate (R_b)	kb/s	64
Chip rate (W)	Mchip/s	3.84
Processing gain (W/R_b)	dB	17.8
Required E_b/N_0	dB	2
Receiver sensitivity	dBm	−116
$\quad S = P_N + \Delta I - (W/R_b) + E_b/N_0$		
Node B antenna gain (G_b)	dBi	18
Cable loss (L_c)	dB	4
Fast fading margin (FFM)	dB	4
Slow fading margin (SFM)	dB	9
Soft-HO gain (SHO)	dB	4
In-building penetration loss (L_i)	dB	20
Maximum outdoor planned path loss UL	dB	122
$\quad PLmax = EIRP - S + G_b - L_c - FFM - SFM + SHO - L_i$		

The last term in the previous expression is often referred to as the *interference margin*, since it reflects the power increase that is required due to the interference generated by the rest of transmissions. The interference margin is defined as:

$$\Delta I = \frac{P_N + P_R + \chi}{P_N} = \frac{1}{1 - \eta_{UL}} \tag{5.3}$$

The link budget is then computed for a value of the interference margin corresponding to a maximum planned load factor. In the example, an interference margin of 3 dB has been considered, which corresponds to a maximum uplink load factor $\eta_{max} = 0.5$.

The uplink sensitivity S is then computed by forcing equality into Equation (5.2), so that:

$$S(dBm) = P_N(dBm) + \left(\frac{E_b}{N_0}\right)(dB) - \left(\frac{W}{R_b}\right)(dB) + \Delta I(dB) \tag{5.4}$$

In the previous example, the resulting sensitivity is −116 dBm. Once the sensitivity has been computed, the link budget determines the maximum path loss in the uplink direction taking into account the EIRP, the sensitivity, the node B antenna gain and the cable loss between the antenna and the receiver. Furthermore, the following margins are added to consider different propagation effects:

(a) Slow fading (shadowing) margin. Network planning procedures compute average received signal levels. However, the real received signal exhibits slow fluctuations around the average value due to the existence of obstacles in the propagation path (e.g. the attenuation seen by an indoor mobile terminal may depend on the proximity of the user to the closest window). Usually signal variations are modelled by means of a lognormal distribution. The slow fading margin depends on the following factors:

- Coverage probability. This is the probability that the received signal is above the sensitivity level due to slow fading. Usually values between 90 and 95% are considered.
- Propagation coefficient (γ). This factor reflects the dependence between the path loss and the distance, in the form $1/d^{\gamma}$. Usually values between 2.5 and 4 are considered.
- Standard deviation of the lognormal distribution for the slow fading. This depends on the considered environments. Values between 8 and 9 dB are considered for indoor environments, whereas in outdoor environments the values range between 6 and 10 dB.

In the considered example, a slow fading margin of 9 dB has been assumed.

(b) Fast fading margin. This margin is added for slow moving users in order to compensate for the fast fading fluctuations and the fast power control inaccuracies. In practice, this margin is between 2 and 5 dB, and in the example a value of 4 dB has been considered.
(c) Soft handover gain. When a mobile is in soft handover, i.e. with more than one cell in the active set, the slow fading margin is reduced because of the partial decorrelation existing between the shadowing of the different cells. Furthermore, there also exists a diversity gain that reduces the required E_b/N_0. As a result of these two effects, an additional soft handover gain is used in the link budget. In the example in Table 5.1, a gain of 4 dB has been assumed.
(d) In-building penetration loss. In the considered example, it is assumed that the indoor coverage is provided by means of external macrocells. Consequently, the link budget must assume an additional loss due to the penetration of the electromagnetic signals through the building walls. This attenuation depends on the environment, but a typical value for dense urban zones is 20 dB. Even when indoor coverage is not considered in the planning process, it is usual to add also a certain margin for in-car users, with a typical value of 8 dB for the in-car loss.

Taking into account all the previous effects, the maximum planned path loss for the uplink direction of an outdoor user is 122 dB. For the downlink direction, the link budget computations are presented in Table 5.2. The computations are quite similar to those in the uplink direction. In this example, although the maximum transmitted power available at node B is 43 dBm, a limit in the maximum power available for a connection is usually considered to avoid certain users consuming excessive power. In the example, a maximum power of 35 dBm per connection has been assumed.

On the other hand, note also that in the downlink direction the noise power is usually higher than in the uplink direction. This is because a higher noise figure usually exists in mobile terminals for implementation reasons. This allows a reduction in the terminal cost.

Similar to the uplink direction, the interference margin in the downlink is given by:

$$\Delta I = \frac{1}{1 - \eta_{DL}} \tag{5.5}$$

where the downlink load factor η_{DL} is defined in Equation (4.47) as a function of the number of simultaneous transmissions and their characteristics. It is also worth mentioning that the downlink interference margin is difficult to estimate. This is because the intercell interference is different depending on the specific mobile position with respect to the rest of the cells. On the other hand, with respect to intracell interference, the use of orthogonal codes in the downlink also allows a reduction in the interference. As a result, in the link budget, the downlink load factor is not considered in the initial computations but is extracted after having compared the uplink and downlink path loss.

Table 5.2   Example of downlink link budget

Element	Unit	Value
DL TRANSMISSION		
Node B total transmit power (P_T)	dBm	43
Node B maximum power per connection (P_C)	dBm	35
Node B antenna gain (G_b)	dB	18
Cable loss (L_c)	dB	4
Equivalent Isotropic Radiated Power	dBm	49
$\quad EIRP = P_C + G_b - L_c$		
DL RECEPTION		
Thermal Noise density (KT_0)	dBm/Hz	−174
Noise Figure (F)	dB	7
Bandwidth (B)	MHz	3.84
Noise Power ($P_N = KT_0FB$)	dBm	−101.2
Service bit rate (R_b)	kb/s	64
Chip rate (W)	Mchip/s	3.84
Processing gain (W/R_b)	dB	17.8
Required E_b/N_0	dB	6
Receiver sensitivity	dBm	−113
$\quad S = P_N - (W/R_b) + E_b/N_0$		
Mobile antenna gain (G_b)	dBi	0
Body loss (L_b)	dB	0
Fast fading margin (FFM)	dB	4
Slow fading margin (SFM)	dB	9
Soft-HO gain (SHO)	dB	4
In-building penetration loss (L_i)	dB	20
Maximum outdoor planned path loss DL	dB	133
$\quad PLmax = EIRP - S + G_b - L_b - FFM - SFM + SHO - L_i$		

As a result of the downlink link budget, the maximum acceptable path loss is 133 dB. The comparison between uplink and downlink is shown in Table 5.3. As can be observed, the uplink is the limiting direction in this example, with a difference of 11 dB between both links, so the cell radius will be computed by taking into account the maximum path loss in the uplink direction. Note that, since the downlink interference margin has not been considered in the computations, the link budget reveals that a margin up to the previous value of 11 dB is possible while the uplink remains the limiting direction. In general, in WCDMA systems, the uplink is used as the limiting link for low and medium load conditions due to the higher amount of power available in node B when compared to the maximum transmit power of the mobile terminal. However, for high load conditions, the downlink may be the limiting link because the available power is shared among all the transmissions in a cell.

Table 5.3   Uplink and downlink comparison and downlink interference margin computation

Element	Unit	Value
Maximum outdoor path loss UL	dB	122
Maximum outdoor planned path loss DL	dB	133
Limiting link		UPLINK
Downlink interference margin	dB	11
Maximum planned path loss	dB	122

After the link budget computation, the maximum planned path loss must be translated into a cell radius that allows the position of the different nodes B in the considered geographical region to be decided. As a general rule when deciding the site locations, nodes B should be located close to the expected user positions in order to minimise resource consumption. Furthermore, a certain overlap between cells must exist in order to guarantee the coverage continuity.

Cell Load Computation The next step in the initial planning–cell load computation–consists of taking into account the spatial traffic distribution in order to check whether, for the determined cell radius and site distribution, the resulting cell load is below the planned capacity values or not. This process must then estimate the corresponding cell capacity so that neither the considered uplink and downlink interference margins nor the downlink transmission power exceed the maximum values.

The cell load in the uplink and downlink direction is computed by approximating the cell load factor, taking into consideration the behaviour of the mobile terminals depending on the specific service. Note that, since the network planning is done under static conditions in the scenario, an estimation of the cell load factor is required that properly captures the dynamic behaviour. This leads to a statistical estimation of the uplink cell load factor given by [1]:

$$\eta_{UL} = (1+f) \sum_{i=1}^{N} \frac{1}{\dfrac{W}{\left(\dfrac{E_b}{N_0}\right)_i \alpha_i R_{b,i}} + 1} \tag{5.6}$$

where N users with an ongoing session are assumed in the cell and $R_{b,i}$ and $(E_b/N_0)_i$ are the bit rate and quality requirement of the ith user, respectively. In turn, α_i represents the activity factor of the user, reflecting the fact that, depending on the service, it is possible to have discontinuous transmissions, and the ith user only contributes to the system load when transmitting. The influence of the neighbouring cells is taken into account by means of the factor f, which measures the ratio between intercell and intracell interference. A typical value of approximately $f = 0.6$ is usually considered.

In the downlink direction, similar approximations are required to estimate the cell load factor in the planning process. It is given, from Equation (4.47), by:

$$\eta_{DL} = \sum_{i=1}^{N} \frac{\rho + f_i}{\dfrac{W}{\left(\dfrac{E_b}{N_0}\right)_i \alpha_i R_{b,i}} + \rho} \approx \sum_{i=1}^{N} \left(\frac{E_b}{N_0}\right)_i \frac{\alpha_i R_{b,i}}{W} (\rho + f_i) \tag{5.7}$$

where ρ is the orthogonality factor, which depends on the multipath. Note that the intercell-to-intracell interference factor f_i is user-specific, since it depends on the user position (i.e. users located at the cell edge will perceive a higher intercell interference than users located close to the serving cell). As a result, in the planning process, it is necessary to consider a typical or average value taking into account the different positions.

From the downlink cell load factor, it is also possible to estimate the total downlink transmission power at node B, assuming a representative path loss $\overline{L}_p$ in the cell (a typical value is to consider 6 dB below the maximum path loss obtained in the link budget). The estimation is then given by:

$$P_T = \frac{\overline{L}_p \sum_{i=1}^{N} \left(\dfrac{E_b}{N_0}\right)_i \dfrac{\alpha_i R_{b,i}}{W} P_N}{1 - \eta_{DL}} . \tag{5.8}$$

where P_N is the downlink background noise power.

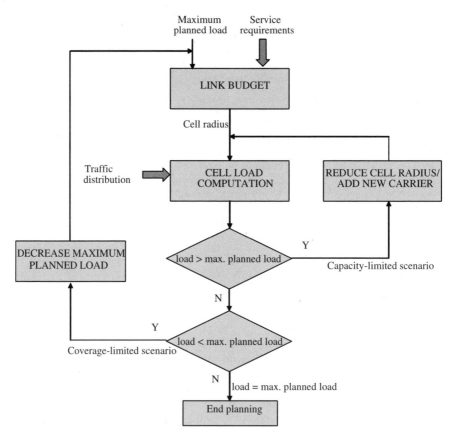

Figure 5.2 Iterative initial planning procedure

From the above expressions and, depending on the spatial traffic distribution per service, it is possible to check if the estimated load factors and transmitted powers are below the maximum admissible planned values for the considered cell range and site distribution. When the maximum values are exceeded, this means that the system is capacity-limited and therefore it is necessary either to reduce the cell radius, thus including more cell sites, or to introduce additional carriers. However, if the load is below the maximum planned value this means that the scenario is coverage-limited because the users could tolerate a higher path loss, so that the cell radius can be increased. This is achieved by reducing the maximum planned value (or equivalently reducing the interference margin in the link budget). The planning procedure is completed when the resulting load equals the planned value. The iterative procedure is illustrated in Figure 5.2.

5.1.1.2 Detailed Planning

The initial network planning provides a rough idea about the cell range and the number of required sites under certain assumptions of maximum planned loads, service characterisation and traffic distribution. Nevertheless, it has been pointed out that some parameters are only considered as average or typical values, for example, the intercell interference or the path loss required to compute the downlink transmitted power. In practice, aspects such as the specific terminal locations and the service characterisations, which have been simplified in the previous analysis, have an important impact on the actual coverage and capacity. Consequently, it is necessary to complete the initial planning by detailed planning procedures that make use of simulation tools to provide more accurate values to the cell range and the cell capacity as well as to check the system sensitivity to possible changes. We now present two examples of planning tools. The first one – the URANO planning tool – is based on propagation models while the second is based on measurements available from the GSM network.

Example 1: URANO Planning Tool   The URANO (UMTS Radio Access Network Optimisation) planning tool is a simulator being used by Telefónica Móviles, the largest mobile operator in Spain, for the operation of third generation systems. URANO incorporates the capacity to simulate scenarios with different network configurations, different mobile terminal distributions and different traffic requirements. In each case, this will impose power demands for both the uplink and downlink, depending on the positions of the other mobiles in the system. The objective of these simulations is to determine whether the simulated network configuration is capable of providing the desired service quality and availability for the traffic distribution associated with the scenario. Analysis is carried out on a sufficient number of simulations, where each is composed of a set of snapshots or non-correlated samples of the mobile positions that correspond to a similar statistical distribution. Naturally, since the mobile positions are random, it is essential to repeat the process a sufficient number of times in order to ensure a proper operation even in the less favourable cases. This type of static analysis, usually known as Monte Carlo simulation, is appropriate for estimating average values and standard deviations, but it does not allow temporal analysis of the mobiles' evolution.

A detailed network planning tool must consider different types of services associated with different types of terminals and with different geographical distributions. In particular, a service is defined in the planning tool by means of a traffic map that is used to define users' positions and a list of terminals that can be used for the considered service. In turn, the concept of terminal is closely related to the type of radio connection that the user establishes with the network. Note that a terminal may be used for various services and a service may also be provided with different terminals. The service bit rate as well as the power constraints and E_b/N_0 requirements are defined at the terminal level. This means that, for example, a given service may be provided with different E_b/N_0 constraints depending on the terminal being used.

The processes of the detailed planning procedure executed by the URANO tool are illustrated graphically in Figure 5.3 and explained below.

The inputs to the simulator are the maps with the spatial distributions and requirements of the different services to be provided. Then, the following steps are executed:

1. Site definition. In this step, the cell radius is estimated as a function of the traffic density foreseen for the different services and the desired degree of coverage for the area being studied. The process is based on link budget computations, under generic propagation conditions, which take into account the service mix. Different cell radius values can be obtained by repeating this procedure, taking into account the non-homogeneity of traffic demand.

   After cell radii computation, the sites are distributed in the zone under study in such a way that the separation between them coincides, approximately, with the cell radii calculated during the previous step. Obviously, possible site locations should respect the restrictions that are naturally imposed by the environment, such as buildings, streets, roads, etc. The latter is of special importance and may be carried out automatically by using geometrical algorithms, for example, to avoid any station falling within a street.

2. Path Loss computation. Basic propagation loss is calculated for all the sectors involved in the area being studied so that the radio propagation conditions are fully characterised. The planning tool must have a wide range of models that are appropriate for the different situations, basically rural and urban

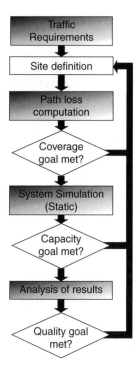

Figure 5.3 URANO detailed planning procedure

macrocells and exterior urban microcells. When not all the area is properly covered, the process goes to step 1 in order to modify the site location.

3. Static system simulation. This process executes a Monte Carlo simulation by evaluating the cell loads and transmitted power levels for different snapshots generated according to the traffic distribution maps. The different mobiles are connected to the sector that requires the lowest transmitted power, so that interference is minimised. When the cell loads are higher than the planned capacity, the process goes to step 1 in order to modify the site location.

4. Analysis of results. Finally, analysis functions enable the user to determine whether or not the proposed network complies with the specified quality and capacity objectives. To do so, different parameters and performance indicators are calculated (e.g. coverage probability, power margins, E_b/N_0, etc.) for each of the services. If any of the requirements is not met, the deployment must be reconsidered.

The results obtained include first and second order statistics and raster maps acquired by moving a test mobile in the zone under study. Different parameters such as the best server, the required power, the soft handover regions and the pilot powers are obtained for each point in the map. A couple of examples are shown in Figures 5.4 and 5.5 including the uplink transmission powers for a voice service and a 384 kb/s service, respectively, corresponding to simulations carried out in a major European city using 3D cartography and ray-tracing models for path loss computation. The legend key with the corresponding power levels is shown in Figure 5.6. Note that, compared to the voice service, the 384 kbit/s service has more out-of-service zones (i.e. black zones) due to it exceeding the transmitted power limit (set at 21 dBm).

Figure 5.4   Uplink transmission power for a voice service

Figure 5.5   Uplink transmission power for a 384 kb/s service

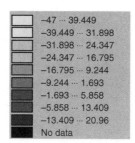

	−47 ⋯ 39.449
	−39.449 ⋯ 31.898
	−31.898 ⋯ 24.347
	−24.347 ⋯ 16.795
	−16.795 ⋯ 9.244
	−9.244 ⋯ 1.693
	−1.693 ⋯ 5.858
	−5.858 ⋯ 13.409
	−13.409 ⋯ 20.96
	No data

Figure 5.6   Legend key with power levels in dBm

Example 2: Planning Tool Using GSM Measurements   The second example of planning tool uses GSM measurements. The introduction and rollout of 3G networks is costly and happens within a very competitive and mature 2G environment. Therefore, operators use their existing GSM network to the fullest possible extent, co-siting 3G sites with existing 2G sites to reduce cost and overheads during site acquisition and maintenance. Obviously, the more accurate the information provided about the network and the characteristics of the users, the more representative the obtained network performance. Therefore, a simulation tool developed for 3G planning based on data extracted from a real GSM network can be very useful in early 3G network development.

Specifically, such a simulation tool can be based on the collection of GSM measurement reports provided by mobile terminals, where the measured received level (Rx_Lev) from the serving cell as well as up to six neighbouring cells is reported [3]. Note that, assuming GSM/UMTS co-siting, this information provides measured propagation data that are useful for UMTS studies with the corresponding propagation corrections when applicable. Also, this approach implicitly captures indoor traffic, traffic from upper floors inside buildings, etc. that otherwise can be rarely represented by statistical modelling. These reports are further processed in order to generate a propagation database that feeds the UMTS simulator, as depicted in Figure 5.7. The resulting simulator provides more reliable information for analysing specific areas of interest for mobile operators. It is worth noting here that, when UMTS networks become fully and commercially operational, the simulator can be easily updated with UMTS measurements. In the meantime, and for initial designs and capacity estimations, traffic distributions inherited from GSM networks may suffice, especially for voice traffic. For data traffic, traces from GSM mobile originated SMS could be considered as a starting point for data traffic spatial distribution, since the locations where users stop to write down an SMS (for example, bus stops, etc.) are also likely locations for UMTS data transfers.

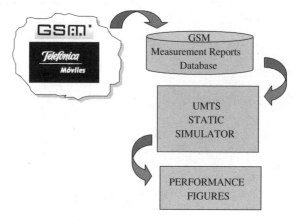

Figure 5.7   Example of a planning tool using GSM measurements

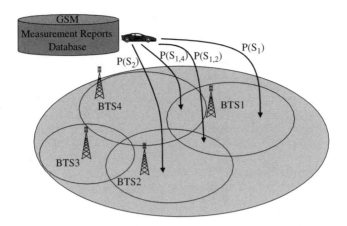

Figure 5.8 Users scattering procedure with a static simulator

In a GSM network, mobile terminals in dedicated mode (i.e. during a call) provide periodical measurements reports through SACCH (Slow Associated Control Channel) with a periodicity of 480 ms. They include the measured *Rx_Lev* from the serving cell and up to six neighbouring cells [4]. If the respective BTS (Base Transceiver Station) transmitted powers are known, it is straightforward to derive the following vector: $[L_1(t), L_2(t), \ldots, L_n(t)]$, $L_i(t)$ being the measured path loss from the ith BTS to the mobile at a certain time. Network monitoring tools may record this information for all the calls in a group of cells simultaneously while keeping the identity of the successive reports belonging to a certain call. Thus, for a certain interest area and once all BTS involved are identified, the first step is to record all measurement reports for a long enough period of time.

To generate a database, a footprint of the traffic distribution in the interest area is obtained by recording the information for all the BTS simultaneously. Thus, defining $P(S_i)$ as the probability that a mobile is in the area of the ith BTS (in the sense that this BTS provides the lowest path loss), this value can be estimated simply by counting how many of the total number of measurement reports have $L_i(t)$ in the first position of the vector. Similarly, $P(S_{ij})$ can be defined as the radioelectrical region where the ith BTS is the best server and the jth BTS is the second best.

Then, the scattering of users in the interest area can be done according to real traffic distributions. That is, users are thrown to the scenario according to the calculated probabilities, with the desired degree of precision $(S_i, S_{ij}, S_{ijk}$, etc.). For example, if there is a hot spot close to BTS1, many of the collected measurement reports will include BTS1 as the best server and so it may be reasonable to consider this spatial traffic distribution for UMTS evaluation purposes. Of course, these probabilities may be modified at will if another traffic distribution is to be studied. Figure 5.8 presents this procedure.

As an example for a real scenario Figure 5.9 plots the distribution of $P(S_i)$ for the downtown of a Spanish city consisting of 13 GSM cell sites. For convenience, each cell is represented by an identification number. It can be observed that traffic is not uniformly distributed in that area and is significantly concentrated around cells #0 and #4. So, for UMTS dimensioning, it could also be reasonable to consider the same traffic distribution. All 13 GSM cell sites or only a subset of them could be assumed in the initial UMTS network deployment.

Taking into account the measurement database and the available traffic patterns, a simulation snapshot would carry out the following steps:

1. Decide the number of users present in the scenario, N (in case of a service mix scenario, decide the number of users for each service).

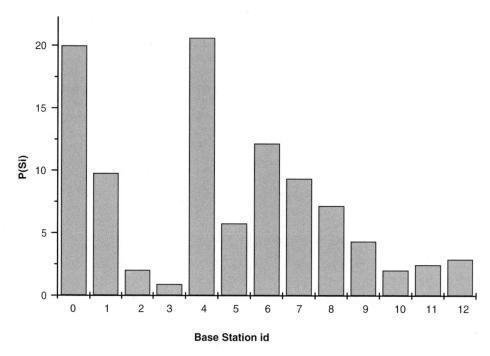

Figure 5.9 Example of $P(S_i)$ for a real case study

2. Decide the subset depth to be considered in the user scattering procedure (i.e., 1 level, S_i, 2 levels, S_{ij}, 3 levels, S_{ijk}, etc.). For illustrative purposes, let us consider two levels in the following.
3. For each user:

  (a) Decide the subset S_i according to $P(S_i)$, that can be either derived from the database or set according to a desired traffic distribution.
  (b) Decide the subset S_{ij} according to $P(S_{ij})$, that can be either derived from the database or set according to a desired traffic distribution.
  (c) Once the subset S_{ij} has been selected, choose randomly a sample from the measurements database belonging to this subset: $[L_i(t), L_j(t), \dots, L_n(t)]$.

4. Once all users are scattered in the scenario, run the power control module to decide the transmitted power levels for all users. Each user aims at achieving a certain target (E_b/N_0), according to the required QoS and service class. Note that this allows an exact analysis of the interference pattern that arose in the snapshot.
5. Collect statistics and performance figures of interest. Statistics for a given scenario are usually only collected in a reference cell in order to avoid border effects from those cells whose neighbouring cells have not been captured by the network monitoring tool. Some system and performance parameters of interest could be:

  (a) Total received power at the reference cell.
  (b) Reference cell load factor.
  (c) Intercell to intracell interference ratio.
  (d) Contribution from each neighbouring cell to the intercell to intracell interference ratio.

(e) Number of users connected to the reference cell whether in soft handover or not.
(f) Percentage of users that achieve the target E_b/N_0.
(g) Downlink transmitted power.
(h) Downlink intercell interference distribution.

5.1.2 RRM ALGORITHM EVALUATION BY MEANS OF SIMULATIONS

Normally, computer simulations constitute the preferred solution for the evaluation and validation of the different Radio Resource Management strategies. Although it is also possible to evaluate them using analytical approaches, this usually requires a high number of simplifying assumptions that reduce the precision of the obtained results. Consequently, the use of analytical approaches is often limited to obtaining rough performances and general trends in RRM strategies. Simulations, however, allow obtaining more precise results provided that the simulation models adequately capture the real system behaviour.

In the RRM algorithm simulation methodology, there exists a clear trade-off between the computational complexity of the simulations and the level of detail considered in them. Therefore, higher numbers of simulated procedures and parameters means longer simulation times but also more precise results. Consequently, a suitable simulation model will be based on extracting the relevant procedures and parameters from the real systems that may have an impact on the evaluated RRM strategies.

Furthermore, to cope with the complexity trade-off, the simulation is usually split into two different types of simulations with different time resolutions. They are denoted as link and system-level simulations. The link-level simulation is responsible for obtaining the physical layer behaviour of the channel observed by a mobile user to communicate with its corresponding base station, either in the uplink or in the downlink. To this end, link-level simulators with a time resolution below the chip time are usually devised (e.g. they typically operate with four or eight samples of the received signals per chip). On the other hand, a system-level simulator evaluates the behaviour of the RRM algorithms in multi-cell, multi-user and multi-service scenarios resulting from the previous network planning procedure. To handle these complex scenarios in moderate simulation times, a system-level simulator makes use of the off-line results obtained by means of the link-level simulator to characterise each link of each user in each cell. Therefore, the system-level simulators typically operate on a slot-by-slot or a frame-by-frame basis rather than on a chip-by-chip basis. The interaction between radio network planning, link and system-level simulators is shown schematically in Figure 5.10.

System-level simulation tools must be able to combine information about the network configuration (e.g. cell sites, transmitted powers, etc.) with information about the position of the mobiles and the traffic that they are likely to generate, in order to build a realistic picture of the network in terms of its coverage and the offered QoS. Users are scattered around the network based on an expected traffic distribution. In the case of static simulators, users do not move and so the tool builds a snapshot of the network for a

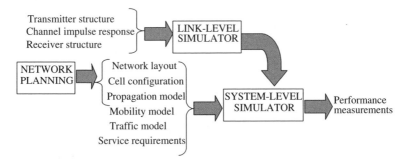

Figure 5.10 Interaction between radio network planning, link-level and system-level simulators

particular user distribution. Many snapshots with different distributions of users are run in order to obtain a composite view of the network performance. However, in the case of dynamic simulations, the users move around and generate traffic, so they behave as much as possible like real users. Consequently, dynamic simulators allow better capturing of the real situations and provide more accurate results than static simulators.

We now present detailed descriptions of two sample link-level and system-level simulators for UTRAN FDD.

5.1.2.1 UTRAN FDD Link-level Simulator

In order to obtain reliable results in RRM algorithm evaluation, it is important to make sure that the simulated higher layers rely on a good model of the link-level. The UTRAN FDD physical layer is a very flexible environment designed to support different QoS requirements. Some of the powerful features that make the WCDMA air interface a key element of UMTS, to provide high capacity, high bit rate and flexibility to adapt to QoS requirements, are: the coexistence of high and low bit rate users with different spreading factors (SF); a cellular reuse factor equal to unity; a fast power control loop updated 1500 times per second; the use of permanently associated control channels to help in channel estimation; and the use of orthogonal variable spreading codes (OVSF) to reduce intra-cell interference. Consequently, the simulation of the UTRAN FDD link layer is a complex task because of the many aspects involved. Also, to obtain results in a reasonable time, some approximations and simplifying assumptions have to be made while maintaining a good degree of accuracy. Therefore, only some selected statistical properties of the UTRAN FDD link can be accurately reproduced.

As was described in Section 3.3, the smallest building block delivered by the MAC layer to the UTRAN FDD physical layer is the Transport Block (TB). It contains a certain amount of information plus RLC and MAC headers. At the physical layer, some CRC bits are added to each TB, so the TB is the minimum block where errors can be detected. Error rate can be measured either at the bit level by means of the BER (Bit Error Rate) or at the TB level by means of the BLER (transport Block Error Rate). Depending on the service under consideration, one or other statistic may be of interest. For instance, for error tolerant services operating in TM or UM RLC modes, it is usual to measure performance by means of the residual BER in the delivered transport blocks. However, in the case of services using the AM RLC mode allowing for retransmissions, the throughput of the retransmission protocol is usually limited by the BLER. On the other hand, for a proper algorithm evaluation, the average values of the BER or BLER alone may not be sufficient because, depending on channel variability, bursts of errors may occur in certain transport blocks and therefore the average BER may not be representative enough. Consequently, in some cases the link-level statistics delivered to the system-level simulators should include not only average BER and/or BLER values but also the statistical distribution of the number of errors per transport block. Such distributions must be given taking into account the main parameters having an impact on them, such as the corresponding channel impulse response, the selected transport format combination for a given radio bearer or the average measured E_b/N_0.

Apart from the performance in terms of BER or BLER, other parameters depending on the physical layer may also be obtained by means of the link-level simulators, such as, for example, the orthogonality factor that measures the loss in orthogonality when orthogonal signals are transmitted through multi-path channels.

As an example, Figure 5.11 shows a block diagram of the link-layer simulator that was developed in the ARROWS project corresponding to the 5th framework of the IST (European Commission) [5]. It is a conceptual block diagram in the sense that some blocks may be implemented in detail while some others may only be taken into account in an indirect way. For example, the CRC bits may not be necessary in the simulation program since the errors can be simply detected by comparison with the transmitted bits. Nevertheless, the effect of the CRC must be considered when calculating the number of bits in each block and the global energy efficiency. The block diagram applies to both uplink and downlink directions

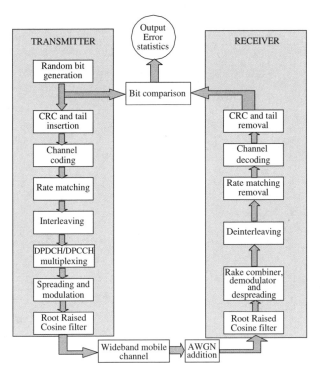

Figure 5.11   General block diagram of a UTRAN FDD link-level simulator

with certain modifications depending on the considered transmit diversity scheme. For details in the simulator, see Reference 6.

As a consequence of its flexibility, the UTRAN FDD physical layer can be configured in many different ways depending on the considered radio bearer for the specific service. Consequently, different link-layer simulations are required to cope with the multiple possibilities of configuration. In particular, there are a sample set of parameters needed to define the link layer performance:

- The radio bearer description including the set of Transport Format Combinations (see Section 3.6 with some examples).
- The value of E_b/N_0 for which the simulation is executed.
- The transmit/receive procedures including the type of diversity, if any.
- The channel impulse response depending on mobile speed and environment (e.g. outdoor/indoor, pedestrian/vehicular).

An example of link-level simulation results is shown in Figure 5.12, where the downlink performance in terms of average BER and BLER as a function of the average Eb/No is presented for different conditions regarding the mobile speed and the use of transmit diversity [6]. The considered RAB corresponds to a 64 kb/s conversational service using DCH with transport block size 640 bits, TTI = 20 ms, 16 bits CRC, and turbo-encoding with rate 1/3. The results are shown for a transport format consisting of the transmission of two transport blocks, corresponding to a spreading factor equal to 32. It can be observed that the use of transmit diversity can reduce the required E_b/N_0 for a given BER or BLER about 2 dB for the 3 km/h case and about 1 dB for the 50 km/h case. Furthermore, the BER and the BLER are higher for the 50 km/h case than for the 3 km/h case.

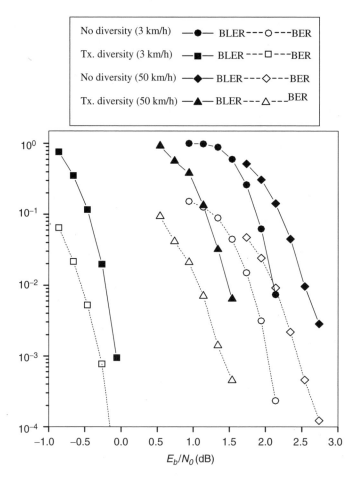

Figure 5.12   Link-layer performance for a downlink conversational RAB at 64 kbit/s

Another example of link-level simulator output is given in Figure 5.13, where the statistical distribution of the number of erroneous bits per transport format is shown. The example considers a downlink transmission through the DSCH channel with a spreading factor of 128 for a transport block size of 336 bits. Note that the probability of having zero errors is approximately 0.55 and, consequently, the average BLER is $1-0.55 = 0.45$. In turn, note also that the number of errors for an erroneous TB is almost uniformly distributed up to a maximum of approximately 70 errors.

5.1.2.2 UTRAN FDD Dynamic System-level Simulator

A system-level simulator is devised to evaluate the performance of a given network in multi-user, multi-cell and multi-service scenarios. The simulator must include an adequate modelling of the relevant aspects that have an impact on the performance of the strategies that are being evaluated. In the context of RRM algorithm evaluation, these aspects include the traffic generation, the user mobility as well as the different network procedures in the radio interface. Taking into account that the RRM problem in WCDMA systems is of an inherent dynamic nature (i.e. the radio resource consumption by a given terminal varies as the terminal moves and depends on the user behaviour during a session), more accurate

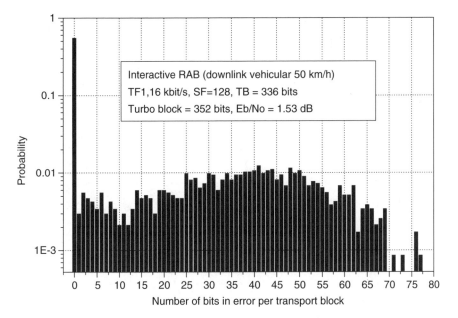

Figure 5.13   Example of error distribution in a transport block

results can be obtained when using dynamic simulators than using static simulators. The former is the preferred option, since the temporal evolution of certain parameters that have an impact over the triggering and operation of RRM strategies (e.g. the E_c/N_0 measurements that lead to a handover, the load measurements that trigger a congestion situation or the session generation dynamics that impact on the admission procedure) can be considered, thus extending the potential to evaluate certain performance indicators.

From a functional point of view, Figure 5.14 presents an example of a UTRAN FDD dynamic system-level simulator architecture developed for the evaluation of RRM strategies. The figure reflects the different procedures to be simulated, together with the interactions with the external link-level simulator and radio network planning tool. The inputs to the simulator are essentially the parameters that characterise the scenario to be evaluated. This includes the number and location of nodes B resulting from the radio network planning tool, the number of users and services to be simulated as well as the service requirements and, finally, the specific parameters of the RRM algorithms under study. On the other hand, at the output, the simulator provides several performance indicators that allow the comparison among the different algorithms.

The simulator operates with a time resolution of one radio frame (i.e. 10 ms), although system-level simulators operating at a time slot level can also be envisaged. Again, the time scale selection responds to the trade-off between computation complexity and accuracy in the results. In particular, radio frame-level operation allows an accurate simulation of all the RRM procedures except for the inner loop power control, which operates at a slot level. Consequently, the received power can only be computed as the average value along a radio frame, which may suffice taking into account the fact that the fast fading channel variations at the time slot level are considered in the link-level simulator.

Figure 5.14 shows how the main core of the simulator contains the RRM procedures to be evaluated. The operation of these procedures is conditioned by the network deployment block and the terminal behaviour in terms of mobility and traffic generation. The network deployment allows the simulator user to specify the network settings and the layout to be studied according to the scenario defined in the planning tool.

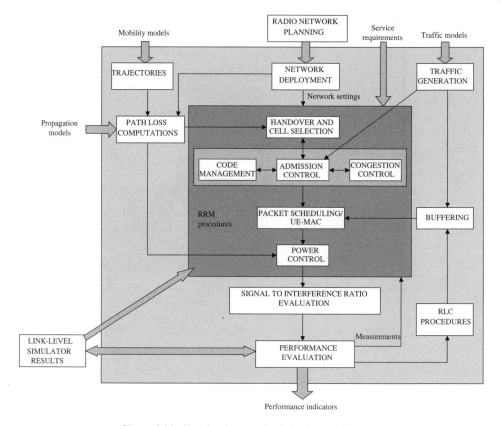

Figure 5.14   Functional system-level simulator architecture

With respect to the mobility aspects, the different terminals are initially scattered in the scenario and follow trajectories that can be either deterministic or result from random mobility models. A summary of some examples of mobility models for different environments is given in Appendix A5.2. The use of deterministic trajectories is valuable when trying to analyse the behaviour of specific mobile terminals under test. Depending on the trajectories and the network layout, the path loss is computed for every mobile and node B taking into account a set of propagation models that depend on the specific scenario. Some examples of propagation models are given in Appendix A5.1. When real measurements for current networks are available, these can be used instead of the propagation models, therefore creating a model closest to the reality for the specific scenario under study. In the RRM module, the path loss measurements are used by the handover and cell selection algorithms to choose the serving cell or cells for each mobile terminal.

With respect to traffic, note that RRM procedures are also conditioned by the traffic generation process. This involves deciding the instants when the users start and finish the sessions as well as the instants when data packets (i.e. RLC SDUs) are generated and buffered at the RLC layer. Depending on the simulated services, different traffic models are used for this purpose. A sample of traffic models is given in Appendix A5.3. Similarly, in this step, it would also be possible to use real traces obtained from specific applications. The start of a new session will trigger an admission control procedure to check if the user can be accepted depending on the system status. Similarly, the number of packets remaining in the buffers will be used by MAC and scheduling algorithms to select the appropriate transport format combinations to be used in each TTI.

In each radio frame, the simulator determines the uplink and downlink transmissions and executes the inner loop power control procedure to obtain the corresponding average transmitted powers for each user. This depends on the path loss computations and the target E_b/N_0 values, which have been selected by the outer-loop power control procedure taking into account the off-line link-level simulation results for a desired error rate level. The inner loop power control procedure is executed by means of an iterative procedure that takes into account all the interferences existing in the system, both intercell and intracell. Once the transmitted powers are devised, it is possible to derive the measured signal-to-noise-and-interference ratios and E_b/N_0 at the receiver output for each transmission.

Depending on the E_b/N_0 value, the performance evaluation block makes use of the off-line link-level simulator statistics, providing the resulting transport block error rate (BLER) for the corresponding transmission. This allows determination of whether or not the different RLC SDUs have been successfully transmitted and, depending on the RLC mode, they can be either retransmitted or delivered erroneously. The performance evaluation block also computes several measurements at both user and system level that are fed back to the RRM procedures. Some examples include the measured uplink cell load factors and downlink transmitted power levels that can be used, for example, to trigger a congestion control procedure. Similarly, the performance evaluation block provides the performance indicators that constitute the simulation outputs.

Output Measurements and Performance Indicators   The high degree of coupling that exists between different RRM strategies in WCDMA scenarios may turn into multiple and, in some cases, unexpected and unpredictable effects resulting from the configuration of a given RRM algorithm, which affect network performance. Consequently, for a proper evaluation of the RRM algorithms, it is necessary to have simulation tools that can provide very different types of measurements and performance indicators, and compile as much information as possible about the network behaviour in order to find out why effects occur so that the influence of the RRM algorithm parameters can be better defined.

A non-exhaustive list of possible output measurements and performance indicators that might be of interest is now presented. We start with *QoS performance indicators*, which are devoted to quantifying the effect of RRM strategies over the service provision. They include:

- Transmission Delay. This corresponds to the time required to transmit a packet between two entities (i.e. the difference between the instants when the packet was generated at one side and when it is completely received at the other side). It is important to note that this definition may include very different types of delay depending on what is considered a 'packet' (e.g. a transport block, a RLC SDU or a packet at the application layer) and on what are the two entities involved in the transmission (e.g. it can be measured between the UE and the RNC if only the radio access network is considered, or between the UE and the terminal equipment with which the end-to-end communication is established). Transmission delay is usually measured not only considering average values but also taking into account the statistic distribution, including second-order statistics and different percentiles (e.g. the 95% delay).
- Throughput. The throughput measures the amount of user data that can be successfully transmitted through a channel per unit of time. Again, different possibilities arise depending on the observation time period to be considered for the measurement. In particular, from the system-level point of view, it is usual to consider an observation period equal to the simulation time. In this case, the throughput depends on the traffic patterns of the different users, so another possibility is to measure the throughput per user taking into account only the activity periods of this user as observation periods. In the case of multi-service environments, it is usual also to have measurements of the individual services together with aggregated throughput measurements (i.e. including all the bits from all the services).
- Packet Error Rate. This statistic measures the ratio between erroneous packets with respect to the total number of transmitted packets. As in the case of the delay, several possibilities exist depending on what is considered a packet (e.g. a transport block, an RLC SDU or a packet at the application layer). When a transport block is considered, this statistic corresponds to the BLER.

- Blocking probability. This statistic is provided at a session level and is directly related with the admission control procedure. It measures the ratio between the number of sessions that cannot be admitted in the system with respect to the total number of sessions that are generated. For an adequate admission control characterisation, it is useful to identify the reasons for rejecting the session (e.g. because of the lack of power in the downlink, the lack of OVSF codes, etc.) and split the blocking probability accordingly (e.g. the blocking due to lack of power, due to the lack of codes, etc.)
- Dropping probability. This statistic measures the ratio between sessions that are dropped with respect to the total number of established sessions. Again it is valuable to identify the reasons for the dropping (e.g. lack of resources during a handover procedure, loss of coverage, etc.).
- Outage probability. In wireless environments, the variability of the link-level and the system-level conditions lead to situations in which the received signal to noise and interference ratio is below some desired value. The probability that this occurs is known as outage probability. Note that this situation may eventually lead to packet errors and even call droppings.

Let us now look at *measurements at the system and user-level*. The following set of output statistics, rather than being of interest from the point of view of service provision characterisation, are necessary to evaluate the behaviour of RRM algorithms from the resource consumption point of view both at the network and terminal level. In some cases, they are the measurements that the RRM strategies use dynamically throughout the simulation to trigger the different events.

- Measurements at system level:

    - Uplink cell load factor in each node B.
    - Total downlink transmitted power in each node B.
    - Number of occupied OVSF codes in each node B.
    - Number of admitted users in each node B.
    - Downlink transmitted power to each transmission in node B.
    - Congestion probability, which corresponds to the ratio between the time a given node B has been declared in congestion by the congestion control algorithm and the total simulation time.
    - Congestion period duration.

- Measurements at user-level:

    - Uplink and downlink transmitted power.
    - Uplink and downlink buffer occupation.
    - E_c/N_0 of the pilot channel.
    - E_b/N_0 for each transmission.

5.2 ADMISSION CONTROL ALGORITHMS

Admission control algorithms in UTRAN FDD aim at deciding the acceptance or rejection of new radio access bearers or the reconfiguration of existing radio access bearers in the access network depending on resource availability and the requested QoS. In a WCDMA scenario, where there is no hard limit on the system capacity, admission control must operate dynamically depending on the amount of interference that each radio access bearer adds to the rest of the existing connections.

From the performance point of view, there are different indicators to evaluate and compare admission control algorithms. Typically, the admission probability (i.e. the probability that a new connection is accepted) or equivalently the blocking probability (i.e. the probability that the new connection is rejected) is used as a measurement of the accessibility to the system provided by a certain algorithm. Nevertheless, the evaluation in terms of accessibility should be complemented with other performance indicators for the on-going sessions, for example, the dropping probability, the packet error rate, the BLER, the delay or the outage probability, in order to see the impact that the interference increase has over the existing connections. Note that a trade-off exists between the blocking probability and the BLER or outage indicators, because by setting very stringent admission control algorithms (i.e. leading to high blocking) the interference and, consequently, the BLER can be better controlled. However, if the admission is softer in order to reduce the blocking, this will be at the expense of a poorer interference control leading to a higher BLER.

In that sense, in the trade-off mentioned above, a good balance is obtained by measuring the probability of occurrence of the following two events, which the admission control should minimise:

- Bad or false rejections, which occur whenever the admission control algorithm rejects a connection request even though there were enough resources in the system to allocate it. In this case, capacity is wasted and operator's revenue is not optimised. Bad rejections can be due to overly stringent admission control algorithms or overly pessimistic interference increase estimations.
- Bad or false admissions, which occur whenever the admission control algorithm accepts a connection request even though there was insufficient capacity in the system to allocate it. In this case, QoS guarantees are not provided and the user's satisfaction is degraded. Bad admissions can be due to overly soft admission control algorithms or overly optimistic interference increase estimations.

The admission control for a given request should be executed taking into account the availability of both uplink and downlink radio resources depending on the QoS requirements in the two links, so that the request can only be accepted if there are available resources in both links. However, due to the different nature of the two radio links and their different limitations, the parameters that determine the algorithms are different in the two cases, and, therefore, they will be presented separately in the following two sections.

5.2.1 UPLINK CASE

In the uplink direction, a new accepted connection represents an increase in the power received at the node B. This increases the interference of the on-going connections so that, depending on this power increase, it is possible that some transmitters located far from the node B may not have enough power to compensate it and, therefore, they may enter in outage. Consequently, admission control strategies in the uplink usually operate taking into account the maximum interference that can be received so that terminals located at the cell edge do not experience power limitations. As shown by Equations (4.29) and (5.3), the interference increase can be measured as a maximum interference margin or, equivalently, a maximum load factor, which can be easily related to the number of connections in the cell. Consequently, it is usual to consider the load factor as the measurement to be used in call admission control [1].

Several schemes have been suggested in the literature for the admission of CDMA systems in the uplink direction [7]. They are based on limiting the interference generated by each connection [8], on measuring the total amount of received power at the node B [9][10] or on the load factor [11][12]. Some of these schemes are considered from a general point of view for WCDMA systems while others are more aligned to UMTS [13][14]. Mathematical models for various call admission schemes have been built and an effective linear programming technique for searching for a better admission control scheme presented [15].

From the implementation point of view, admission control policies can be divided into modelling-based and measurement-based policies [16] depending on whether they take into account estimations of the uplink load factor based on the number of admitted users or they use load factor measurements without considering the number of admitted users. A method to obtain accurate load factor measurements in this way has been presented [17]. The decision whether to consider one type of strategy or another depends on the measurement availability as well as on the desired complexity of the algorithms. In the following, different possibilities are examined in more detail.

5.2.1.1 User Count-based Admission Control

This algorithm constitutes a very simple approach in which the admission decision is based only on the number users already existing in the system, so the interference level in the air interface is neither measured nor estimated. This approach results from a direct translation of the strategies used in 2G systems in which hard capacity limits exist. In particular, the algorithm considers that each user consumes a certain amount of radio resource equivalent units (RREUs), where a RREU corresponds to a user transmitting at a reference bit rate (e.g. the bit rate of a speech service). Depending on the required bit rate, a user then requires more or less RREUs. The maximum number of users that can be accepted will then be given by a certain admission threshold corresponding to the maximum number of acceptable equivalent units R^*. Then, with N users already accepted in the system, the $(N + 1)$th user request is accepted provided that:

$$\sum_{i=1}^{N} R_i + R_{N+1} \leq R^* \tag{5.9}$$

where R_i is the number of RREUs corresponding to the ith user in the cell. The maximum number of units R^* should be related to the maximum load factor that can be accepted. Clearly, since the admission decisions are decoupled from the actual air interference existing in the cell, bad rejections and bad admissions will occur. Nevertheless, the algorithm is very simple to implement and does not require measurement exchange.

In order to assess the performance of the algorithm, some simulation results are presented in terms of the bad admission and bad rejection probability as a function of the number of users N in a reference cell. In particular, a bad admission is assumed to occur when, after the acceptance of the $(N + 1)$th user, the resulting outage probability (i.e. the probability that a user cannot reach the target E_b/N_0) is above 3%. Similarly, a bad rejection occurs when, after rejecting the $(N + 1)$th user, the resulting outage is still below 3%. In the simulations, all the users are assumed to transmit at the reference bit rate and consequently one user equals one RREU.

Figures 5.15 and 5.16 show the bad admission and bad rejection probabilities, respectively, when the admission threshold is set to 40 users. The high variability in the scenario reveals that the user-count mechanism rarely captures these effects and, consequently, false admissions and false rejections occur. Particularly, when the number of users is below 40, users are admitted but in some cases a bad admission

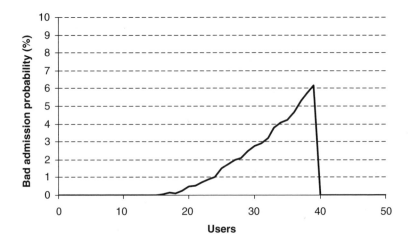

Figure 5.15   Bad admission probability with the user count-based admission control

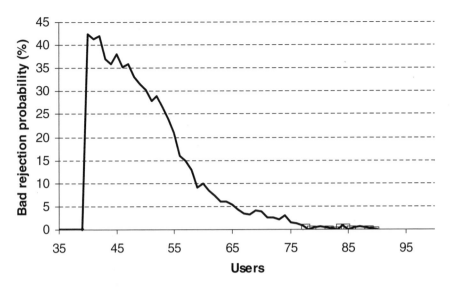

Figure 5.16   Bad rejection probability with the user count-based admission control

exists depending on how users are located in the scenario and on the interference conditions. In turn, when the number of users is higher than 40, users are rejected but a high bad rejection probability exists. This reveals that the threshold of 40 could be increased in many situations while in others it should be decreased as well. Consequently, it does not capture the real WCDMA capacity since it does not take into account the real interference present in the system.

5.2.1.2 Measurement-based Admission Control

An accurate characterisation of the uplink radio interface for admission control purposes can be obtained by measurements of the uplink load factor given by Equation (4.24). These measurements can be carried out from the total power measured at a given node B, P_{TOT}, and the uplink noise power P_N, according to:

$$\eta_{UL} = 1 - \frac{P_N}{P_R + \chi + P_N} = 1 - \frac{P_N}{P_{TOT}} \tag{5.10}$$

Then, assuming that N users are present in the cell, the $(N+1)$th request will be accepted provided that:

$$\eta_{UL} + \Delta\eta \leq \eta_{\max} \tag{5.11}$$

where η_{UL} is the load factor measured and averaged during a certain period T, $\eta_{\max}$ is the admission threshold and $\Delta\eta$ is the load factor increase due to the $(N+1)$th user. $\Delta\eta$ can be estimated depending on the required bit rate $R_{b,N+1}$, target $(E_b/N_0)_{N+1}$ and activity factor α_{N+1} as follows:

$$\Delta\eta = \frac{1+f}{\dfrac{W}{\left(\dfrac{E_b}{N_0}\right)_{N+1} \alpha_{N+1} R_{b,N+1}} + 1} \tag{5.12}$$

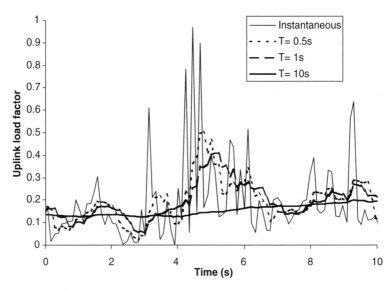

Figure 5.17   Average of load factor measurements in a scenario with a mobile speed 50 km/h and interactive traffic

With respect to the estimation of the new user, a method has been presented [18] to obtain accurate estimations of the interference increase for both the uplink and downlink directions.

For this type of admission control strategies, the measurement-averaging period T becomes a key parameter, as illustrated in Figures 5.17 and 5.18, which present two examples of load factor measurements for different averaging periods T, and for mobile speed 50 km/h and 0 km/h, respectively. The results are obtained for interactive traffic according to the www browsing model presented in Appendix A5.3.3 with an average inactivity period per user of 30 s, an average bit rate per user during activity periods of 24 kb/s and a maximum bit rate of 64 kb/s. It can be observed that the instantaneous value of the load factor exhibits short-term fluctuations depending on the number of simultaneous transmissions that are related to traffic generation.

Furthermore, user mobility also plays an important role since for high speeds, there is a high number of handovers leading to faster interference variations. Note that, for the 50 km/h case, although the average load factor is quite small (i.e. around 0.2), there are sporadic load factor peaks that can achieve very high values. Consequently, if no measurement averaging was done for the admission control, bad rejections could occur in admission attempts executed during the load factor peaks. Therefore, the measurement period must be high enough to smooth the short-term load factor fluctuations. Similarly, depending on the session generation process, the variation in the number of active users will generate variations in the average load factor (see the example in Figure 5.18, where at 4 seconds there is an increase in the load factor). These changes must be appropriately captured by the averaging period, since otherwise bad admissions could occur due to an underestimation of the real load factor. This requires low enough averaging periods to capture these fluctuations (as an example, notice in Figure 5.18 that the highest averaging periods cannot follow the long-term variation). Consequently, the setting of the averaging period results from a trade-off that should take into account the expected traffic patterns and user mobility in a given scenario. Furthermore, the measurement period T impacts on the amount of measurements that should be exchanged between the RNC and node B through the Iub interface.

In order to illustrate the algorithm performance, let us consider a scenario with conversational traffic, which is sensitive to packet losses. Figures 5.19 to 5.21 plot some results in terms of admission probability, dropping probability and BLER for the case $\eta_{max} = 0.75$ with conversational CBR (Constant

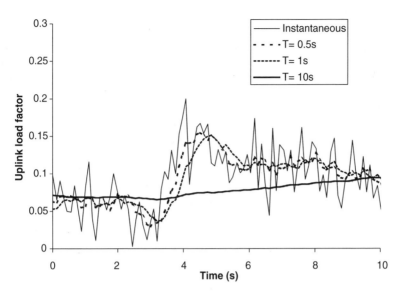

Figure 5.18 Average of load factor measurements in a scenario with a mobile speed 0 km/h and interactive traffic

Bit Rate) 64 kb/s traffic, an average call duration of 180 s and mobile speed 50 km/h. In the case of CBR traffic sources, the cell load fluctuations mainly come from the mobility (i.e. handovers) and the call generation process. However, in contrast to the interactive traffic case, all the users with a call in progress become simultaneous transmissions. It can be observed that, for relatively low loads (20–40 Erlangs), a larger averaging period ($T = 100\,$s) is desirable because it achieves a better admission probability and

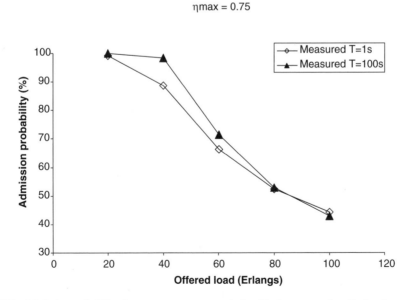

Figure 5.19 Admission probability for two measurement periods with the measured method and conversational traffic

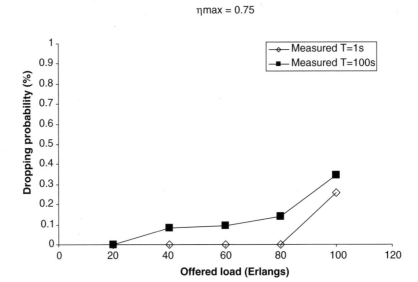

Figure 5.20 Dropping probability for two measurement periods with the measured method and conversational traffic

similar BLER values at the expense of only a slightly higher dropping probability. The reason is that, for shorter averaging periods, the sporadic high interference levels may cause rejection of some calls. However, if the average cell load is relatively low, the call request could have been accepted because this sporadic high load patterns would only have had a very limited impact on the network performance. However, the higher averaging period of $T = 100$ s is able to better capture the real average cell load

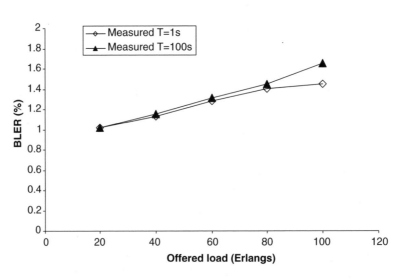

Figure 5.21 BLER for two measurement periods with the measured method and conversational traffic

level compared with the $T = 1$ s case. For a high load situation, there is not much difference between $T = 100$ s and $T = 1$ s and, even $T = 1$ s could provide a somehow better performance with equivalent admission rates.

5.2.1.3 Statistical-based Admission Control

Instead of using load factor measurements, it is possible to make estimations of the load factor depending on the number of admitted users of each service class. This can be done by recalling the relationship between the load factor and the number of instantaneous transmissions given by Equation (4.26), In this case, the admission condition when there are N admitted users is given by:

$$\eta_{UL}^{S} + \Delta\eta^{S} \leq \eta_{\max} \tag{5.13}$$

where load factor estimation with N users η_{UL}^{S} is given by:

$$\eta_{UL}^{S} = (1+f) \sum_{i=1}^{N} \frac{1}{\dfrac{W}{\left(\dfrac{E_b}{N_0}\right)_i \alpha_i R_{b,i}} + 1} \tag{5.14}$$

where f is the ratio between the intercell interference and the total intracell power. In turn, the load factor increase due to the $(N + 1)$th user $\Delta\eta^{S}$ would be given by Equation (5.12) as in the measurement-based method.

It should be emphasized that the resulting estimated cell load, η_{UL}^{S}, which is used for admission purposes, will be different from the real cell load, not only instantaneously but even in average terms, if the admission control parameters are not suitably adjusted. Then, according to Equations (5.13) and (5.14), the statistical-based admission control requires the appropriate setting of the following parameters:

- Transmission bit rate $R_{b,i}$. The estimation of this parameter is required when VBR (Variable Bit Rate) services are considered, which can change the instantaneous transport format from one TTI to another. Consequently, the bit rate will depend on the available TFCS for the considered connections. In this case, by setting $R_{b,i}$ as an estimated average value of the real bit rate that the user will adopt throughout its connection time, the estimated load factor will be closer to the real value. Nevertheless, such a setting would result in higher admission rates that might cause excess interference in certain situations where, due to the randomness in traffic generation, several traffic sources transmit at their highest bit rate. However, by considering $R_{b,i}$ as the highest value associated with the TFCS, the admission control covers the worst interference case at the expense of overestimating the impact of every individual user and, consequently, reducing the admission probability. This effect is illustrated in Figure 5.22, which represents the real instantaneous and averaged load factor evolution and the corresponding estimation with different bit rates (i.e. with the maximum and average bit rates). It corresponds to a scenario with interactive users according to the www browsing model in Appendix A5.3.3 with a maximum bit rate of 64 kb/s. It can be observed how the estimation with the highest bit rate of 64 kb/s provides, in general, an overestimation of the load factor with respect to the instantaneous value, while with the estimation based on average bit rates, the estimated load factor is quite close to the value obtained with the averaged method.
- Activity factor α_i. This factor should be set according to the activity periods of the traffic source. Note that, even for CBR services such as voice, there are periods in which the source has no information to transmit (e.g. because one voice user is simply listening to the user at the other end). As in the case of the transmission bit rate, by setting an appropriate α_i, the admission procedure can be closer to the

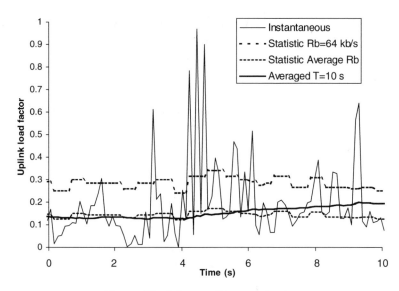

Figure 5.22 Influence of bit rate estimation

real situation of discontinuous activity by relying on the traffic multiplexing. However, an excess of interference could randomly appear within short periods of time whenever many sources are simultaneously transmitting. In turn, $\alpha_i = 1$ covers the worst case at the expense of overestimating the impact of every individual user and, consequently, reducing the capacity.

- Admission threshold η_{max}. By setting the admission threshold, the admission procedure allows for some protection against traffic multiplexing situations above the average value, for example, having more simultaneous transmissions than the expected average number, or having more users making use of high instantaneous bit rate than the expected number. η_{max} should maintain the desired coverage and avoid the well known cell breathing effect. Admission thresholds can be fixed or dynamically adjusted [19].

- Users in soft handover. Another important issue related to admission consists of carrying out a suitable estimation of the contribution of users in soft handover to the load factor in the cells to which they are connected. Note that in the uplink direction, macrodiversity is used, so that the cell in the active set with the highest instantaneous signal-to-noise-and-interference ratio is selected at each time instant. Consequently, during certain periods and in a given cell of its active set, the user will be seen as intracell interference while during other periods it will be seen as intercell interference. Therefore, since intercell interference is already captured in the f factor and so as not to count the user twice, it is adequate to introduce a correction factor in the load factor estimation taking into account the number of cells AS_i that the ith user has in the active set, thus leading to the following estimation [20]:

$$\eta_{UL}^S = (1 + f) \sum_{i=1}^{N} \frac{1}{AS_i} \frac{1}{\dfrac{W}{\left(\dfrac{E_b}{N_0}\right)_i \alpha_i R_{b,i}} + 1} \tag{5.15}$$

The impact of the estimation of the soft handover users' contribution to the load factor is illustrated in Figures 5.23 and 5.24. They present the admission probability and the BLER, respectively, when

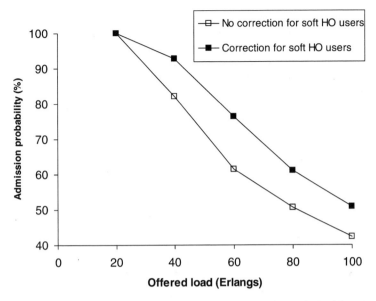

Figure 5.23 Influence of the estimation of soft handover users' contribution to the load factor over the admission probability

the soft handover users are considered in the load factor estimation like the rest of the users according to Equation (5.14) and when the estimation given by Equation (5.15) introduces a correction factor for soft HO users. A conversational CBR service with 64 kb/s is considered to illustrate the effects of this estimation. The admission threshold is set to $\eta_{max} = 0.75$. As can be observed, the admission probability is much lower when no correction factor is introduced for handover users. In turn, the

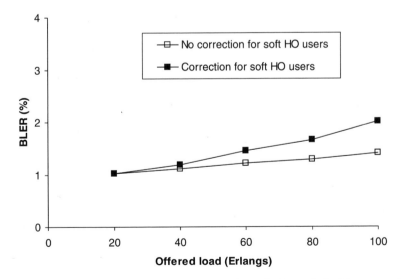

Figure 5.24 Influence of the estimation of soft handover users' contribution to the load factor over the BLER

strategy that considers the correction factor experiences a somewhat higher BLER. Nevertheless, note that, for loads below 40 Erlangs, when there are only slight differences in terms of the achieved BLER for both methods, the admission rate achieved when introducing the correction factor for handover users is significantly higher, thus concluding that, in general, it is advantageous to use this strategy.

- Number of users connected to the cell N. The current number of users connected to the cell site depends, on the one hand, on the call generation process and, on the other hand, on the handover procedures related to user mobility. Consequently, traffic statistical averaging could bring some improvements in the load factor estimation for admission control, since deviations over the average number of users connected to the cell could be smoothed and, therefore, cell blockings would occur less often. In particular, the expressions given in Equations (5.14) and (5.15) can be modified by averaging the estimation during the most recent T frames as follows:

$$\eta_{UL}^S = \frac{1+f}{T} \sum_{j=1}^{T} \sum_{i=1}^{N_j} \frac{1}{AS_i} \frac{1}{\left(\frac{E_b}{N_0}\right)_i \alpha_i R_{b,i}} + 1} \tag{5.16}$$

instead of taking the instantaneous estimation. In the estimation, N_j is the number of users admitted in the cell in the jth frame.

The traffic smoothing effects are illustrated in Figures 5.25 and 5.26, which present the admission probability and the BLER when no traffic averaging is used and when the average over the last 100 s is taken into account. The same situation as in the previous figures is considered with conversational 64 kb/s users and speed 50 km/h. Again, $\eta_{max} = 0.75$ is assumed. It can be noted that admission probabilities improve and negligible variation in the BLER performance is obtained with the traffic averaging. It is remarkable that the main admission improvement is achieved at relatively low loads, where it is better to consider the average number of users in the cell rather than the instantaneous value, which could present disadvantageous peaks leading to unnecessary call rejections.

- f factor. Another important parameter that has a key influence on the statistical admission control is the other-to-own cell interference factor f. If it is not appropriately set, the admission control could

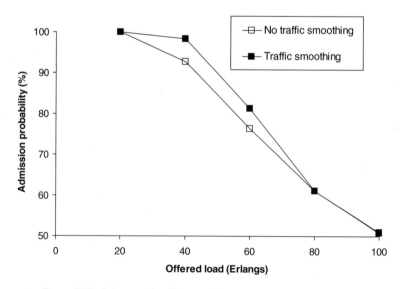

Figure 5.25 Influence of traffic smoothing over the admission probability

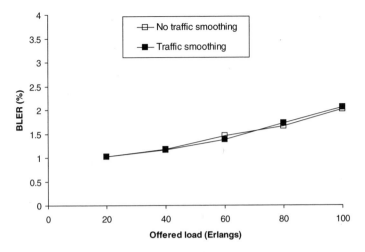

Figure 5.26   Influence of traffic smoothing over the BLER

overestimate or underestimate the existing intercell interference thus causing bad rejections or bad admissions. In order to assess the influence of this parameter and to devise how important an accurate estimation is, let us consider just as an example Figures 5.27 and 5.28, which present the admission and dropping probabilities for the case of three different values of the factor f, namely $f = 0.2$, $f = 0.6$ and $f = 1$. These simulations assume $\eta_{max} = 0.75$ and no traffic averaging. Note that, the lower the f factor estimation, the better the admission probability. However, if the f factor estimation is lower than its real value, so that the cell load is underestimated for admission control purposes, the

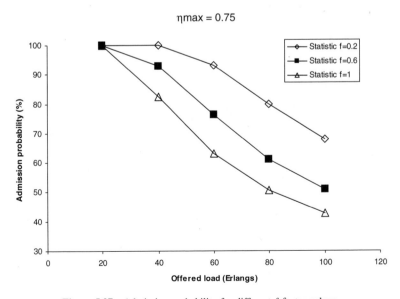

Figure 5.27   Admission probability for different f factor values.

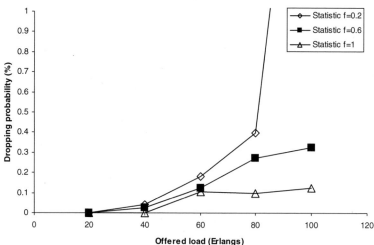

Figure 5.28 Dropping probability for different f factor values.

resulting performance will be worse and a higher dropping probability will be obtained. This is the case, for example, of $f = 0.2$, where the admission procedure has not been conducted properly because too much load has been accepted in the cell and the BLER performance is, consequently, degraded.

The intercell-to-intracell interference factor f should be set for each scenario according to the specific intercell interference patterns that depend on the user speed and the traffic generation process [21]. Some relevant statistics of the f factor distribution are presented in Table 5.4, obtained from the simulations with traffic averaging presented in Figures 5.25 and 5.26 with conversational 64 kb/s traffic and 50 km/h mobile speed. In turn, Figure 5.29 presents the probability density function (pdf) of the f factor for the 20 Erlangs case.

It can be observed that the f factor distribution exhibits a very high dispersion with non-negligible probabilities for values higher than 1, depending on how users are distributed in the scenario at a given time. As a result of this dispersion, high average values are observed for this parameter and the average value is not very representative of the real distribution. For example, the probability that the instantaneous value exceeds the average is only about 30%, as seen in Table 5.4. Therefore, the average value can provide rather pessimistic results in the admission control. However, the most probable value (i.e. the value leading to the maximum of the pdf) or, for example, the 50% percentile (i.e. the value for which the Cumulative Distribution Function CDF is 50%) are significantly smaller than the average value, so they could increase the admission probability.

Table 5.4 Different statistics of the f factor obtained by simulation

Offered load	Value with the highest probability	Average	$CDF_{50\%}$	Prob($f >$ Average)
20 Erlangs	0.30	1.32	0.66	0.26
100 Erlangs	0.58	1.11	0.87	0.34

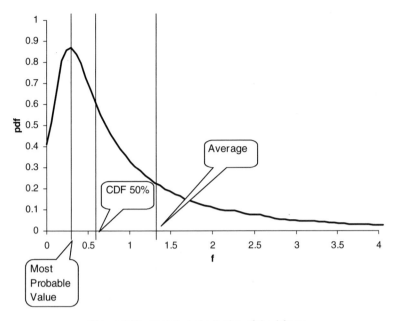

Figure 5.29   Statistical distribution of the f factor

From the above comments, in the load factor estimation, it is more suitable to include, for example, the 50% percentile of the f distribution in the admission condition instead of the average value. Specifically, Figures 5.30 and 5.31 show that, with the 50% percentile, the admission probability for low loads is improved compared to using average values, while the degradation in terms of BLER is

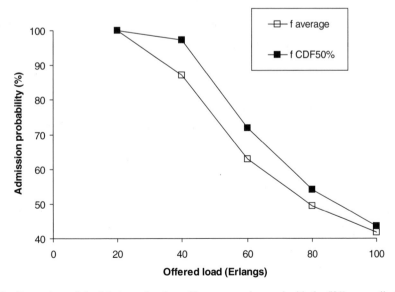

Figure 5.30   Comparison of the f factor estimation with average values and with the 50% percentile in terms of admission probability

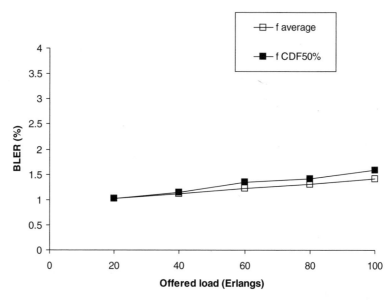

Figure 5.31 Comparison of the f factor estimation with average values and with the 50% percentile in terms of BLER

negligible. Consequently, the setting of the 50% percentile becomes a good tuning of the statistical-based admission control algorithm.

5.2.1.4 Statistical Versus Measured-based Admission Control

The behaviour exhibited by the measured-based and statistical with traffic averaging (denoted as statistic-averaged) uplink admission control approaches presented in the previous sections is compared in Figures 5.32 and 5.33 in terms of admission probability and BLER, respectively. From these two figures, it can be concluded that, if both approaches are suitably adjusted, the resulting performance can be equivalent. Specifically, the guidelines for a good algorithm adjustment are:

- For the measurement-based approach, take a long enough averaging period T in order to smooth traffic and interference fluctuations.
- For the statistical approach, do not consider the average value of f because it leads to rather pessimistic estimations; use, for example, the 50% percentile instead.
- For the statistical approach, average the traffic component over a sufficient period of time in order to avoid undesirable cell load estimation peaks due to traffic fluctuations.

Thus, the final decision on the followed approach will depend more on implementation issues (e.g. the difficulty and accuracy of the cell load measurements, the difficulty of estimating the real f factor distribution, the activity factor, etc.) than on performance considerations.

5.2.1.5 Admission Control During Handover Procedure

The admission control procedure should be executed not only for new connections but also whenever a connection is to be handed over from one cell to another. Furthermore, in a scenario where all the cells

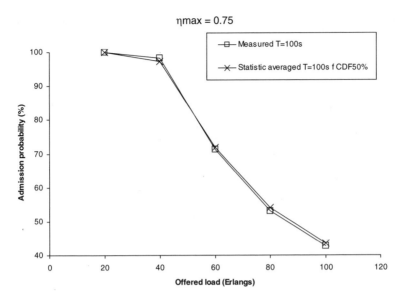

Figure 5.32 Comparison between measured and statistical admission control in terms of admission probability

use the same frequency, handover becomes a critical procedure since it affects the intercell interference patterns. In particular, note that a user connected to a cell that requires a higher transmission power than another, generates an excess of intercell interference to other transmissions.

On the other hand, from the admission point of view it is usual in mobile communication systems to give priority to handover users before new users since existing users perceive a more subjective

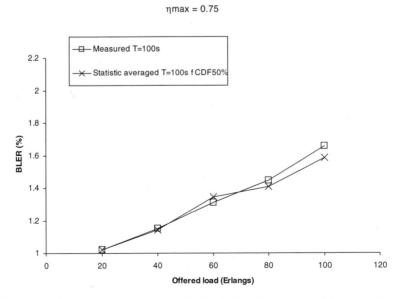

Figure 5.33 Comparison between measured and statistical admission control in terms of BLER

degradation in quality with dropped calls compared to when it is not possible to establish a call [22]. Taking this into account, three different possibilities for the uplink admission control of an incoming handover user are considered [23]:

1. No precedence is given to handover users with respect to new users. In this case, the admission condition to be checked for the handover user is the same that would be considered for a new user, i.e. Equation (5.11) or (5.13), depending on whether measured-based or statistical-based admission control is considered, respectively.
2. Give a certain precedence to handover user. According to this policy, the admission control conditions are modified with a higher admission threshold for handover users. For instance, in the measured-based admission control the condition would be:

$$\eta_{UL} + \Delta\eta \leq \eta_{HO} \tag{5.17}$$

where $\eta_{HO} > \eta_{\max}$ in order that a handover user is more likely to be accepted than a new user.
3. Always accept a handover user. This strategy involves not checking the uplink admission condition in the case of handovers. Note that, in the uplink direction, it is not a problem to always accept a user in handover because each user has its own scrambling code. Therefore, code consumption is not a limit as in the downlink case.

Figures 5.34 to 5.36 show an example of the admission control performance for the three considered cases. A scenario with conversational users at 64 kb/s and mobile speed 50 km/h is examined. The admission threshold for new users is $\eta_{\max} = 0.75$. Therefore, case 1 without precedence to handover corresponds to $\eta_{HO} = 0.75$ while case 2 is executed with $\eta_{HO} = 0.9$. The statistical admission control method is assumed with $f = 0.6$ and without traffic averaging. The figures reveal that, in terms of new users' admission, only a slight reduction in the admission probability is observed when handovers are always accepted in the admission procedure (see Figure 5.34), whereas an important reduction in the dropping probability is achieved (see Figure 5.35). The reduction is higher when handovers are always

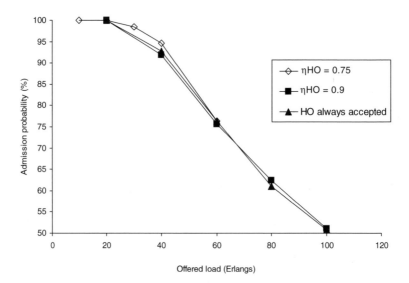

Figure 5.34   Admission probability with three different policies regarding handovers

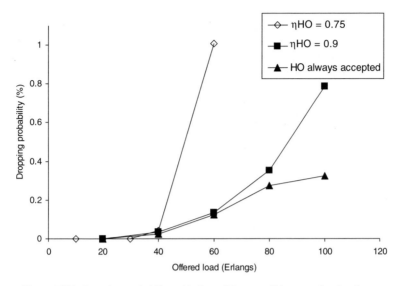

Figure 5.35   Dropping probability with three different policies regarding handovers

accepted than when $\eta_{HO} = 0.9$. It is worth noting that in terms of BLER performance an improvement is also observed when giving precedence to handovers (see Figure 5.36). This is because a reduction in the interference is achieved. Specifically, note that, by facilitating handovers, the situation is avoided where a user moves far from the serving base station and is not allowed to handover the call, thus causing a high interference to the new cell before the call is eventually dropped.

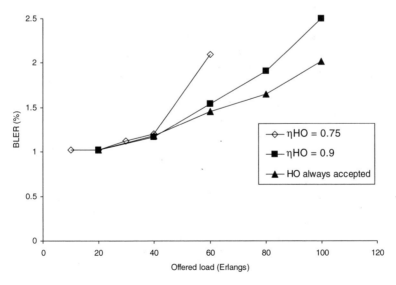

Figure 5.36   BLER for the three different policies regarding handovers

5.2.1.6 Multi-service Case

In a multi-service scenario, it is usual for admission control strategies to make use of service prioritisation [24]. As an example, and without loss of general applicability, let us assume a scenario including conversational and interactive users. In this case, the statistical-based admission control can be used in order to devise a strategy that differentiates between the contributions of the considered services. In particular, the total cell load factor η_{UL} can be statistically split into two contributions: η_C, including the contribution of conversational users, and η_I including the contribution of interactive users, defined as:

$$\eta_C = (1+f) \sum_{i=1}^{N_C} \frac{1}{AS_i} \frac{1}{\dfrac{W}{\left(\dfrac{E_b}{N_0}\right)_i \alpha_i R_{b,i}} + 1} \tag{5.18}$$

$$\eta_I = (1+f) \sum_{i=1}^{N_I} \frac{1}{AS_i} \frac{1}{\dfrac{W}{\left(\dfrac{E_b}{N_0}\right)_i \alpha_i R_{b,i}} + 1} \tag{5.19}$$

where N_C and N_I are the number of conversational and interactive users, respectively, already admitted in the cell. The traffic averaging procedure in Equation (5.16) can also be used in this approach.

By splitting the load factor estimation in this way, the admission control algorithm can be set to reflect different operator policies with respect to the considered services.

- *Policy 1. Admission without prioritisation.* In this case, the admission control algorithm considers that both interactive and conversational services have the same importance for the network operator. To this end, whenever a new connection, either conversational or interactive, demanding $\Delta\eta$ is to be admitted, the condition to check is:

$$\eta_C + \eta_I + \Delta\eta \le \eta_{max} \tag{5.20}$$

It is worth noting that, with this strategy, a conversational call may be blocked by the existing interactive connections. However, this has not been the usual policy followed by operators in 2.5G systems like GPRS, where interactive services use the capacity left by voice users.

- *Policy 2. Admission with prioritisation of conversational users with respect to interactive users.* More priority can be given to conversational calls, taking into account the variability of interactive users, by carrying out a differentiated admission control and relying on congestion control mechanisms to control the actual system load. In this case, whenever a new conversational user demanding $\Delta\eta_c$ requests admission, the condition to check considers only the amount of load coming from conversational users, this is:

$$\eta_C + \Delta\eta_C \le \eta_{max} \tag{5.21}$$

Note that, in this case, a conversational user cannot be blocked by the interactive load. However, also note that this condition does not ensure that the real load after admission was below the expected maximum admission threshold η_{max}, since the interactive load is not taken into account. As a result, in order to ensure that the performance of conversational users is not degraded in case when there is an excessive interactive load, congestion control algorithms will be required to reduce the interference coming from interactive traffic.

In turn, with respect to interactive users, a new request demanding $\Delta\eta_I$ will be admitted according to the condition:

$$\eta_C + \eta_I + \Delta\eta_I \leq \eta_{\max} \tag{5.22}$$

Consequently, the admission for an interactive user is more stringent since both conversational and interactive load are taken into account in the condition. Also different $\eta_{\max}$ thresholds could be used for each type of service.

5.2.1.7 Multi-cell Admission Control

Admission control algorithms are usually applied at single cell level. However, since all the cells operate with the same frequency in WCDMA systems, the acceptance of a new user in one cell generates an increase of the interference in the neighbouring cells. Thus, provided that the interference increase in the neighbouring cells can be adequately estimated, a multi-cell admission control algorithm can be devised in order to guarantee that the set-up of the new connection is affordable at the whole system level therefore achieving better performance, especially in non-homogeneous scenarios [25].

A multi-cell admission control strategy can be developed from the derivatives framework presented in Section 4.6.2. Specifically, whenever a new user is to be accepted in cell 0, the following conditions would be checked in the K neighbouring cells (as well as the condition in Equations (5.11) or (5.13) in cell 0 depending on whether measurement-based or statistical-based approach is considered):

$$\eta_i + \Delta\eta_0 \frac{\partial\eta_i}{\partial\eta_0} \leq \eta_{\max} \quad i = 1,\ldots,K \tag{5.23}$$

where $\Delta\eta_0$ is the estimated load factor increase in the serving cell 0 and η_i is the load factor in the ith cell. The derivative is computed according to the expressions given in Section 4.6.2.

The check of the admission condition in the neighbouring cells allows reduction of the false or bad admissions as reflected in Figure 5.37, which presents the comparison between the admission control that

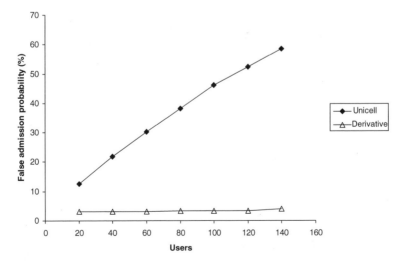

Figure 5.37 Comparison of the false or bad admission probability for the unicell and multicell admission control approaches

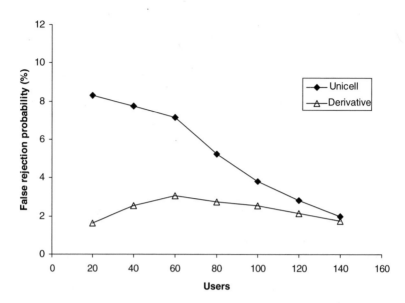

Figure 5.38 Comparison of the false or bad rejection probability for the unicell and multicell admission control approaches

takes only into account the serving cell and the multi-cell based admission control. The effects are illustrated in a situation with only conversational CBR 64 kb/s users. In this case, the false or bad admission probability is measured as the fraction of acceptances in which at least one cell, including the serving and the neighbouring cells, has a load factor higher than the maximum threshold. Similarly, the estimation by means of the derivative approach can reduce the false rejection probability, as depicted in Figure 5.38. It is measured as the ratio of rejections in which the uplink load factor in all the cells would have remained below the admission threshold if the user had been accepted. Consequently, the multi-cell admission control based on the derivative framework allows a more accurate prediction of the influence of a new user on a system-basis.

5.2.2 DOWNLINK CASE

Admission control in the downlink direction is based on checking the availability of code and transmission power consumption, which are the two radio resources that are utilised by each connection. Therefore, the downlink admission control will operate in two steps, and a connection will only be admitted if both conditions are fulfilled.

5.2.2.1 Code Availability

In the downlink direction, transmissions of a given node B are distinguished by different OVSF codes, as explained in Section 3.3.2. This poses a limit in the maximum number of acceptable connections depending on the spreading factor they use. This limit is determined by the Kraft's inequality [26], given in Equation (3.1), which specifies the maximum number of simultaneous transmissions that can be supported by one OVSF code tree depending on their spreading factors.

When checking code availability for a downlink dedicated channel, it should be considered that part of the code tree can be reserved for common and shared channels, thus reducing the number of codes that can be allocated to dedicated channels. This modifies the previous Equation (3.1) as follows:

$$\sum_{i=1}^{N_D} \frac{1}{SF_i} \leq 1 - R_C = 1 - \sum_{j=1}^{C} \frac{1}{SF_j} \tag{5.24}$$

where N_D is the number of allocated dedicated channels, each one with a spreading factor SF_i, and R_C represents the fraction of the code tree that is reserved for a total of C common and shared channels. For instance, when the cell only has one code for the CPICH channel, another for the Primary CCPCH carrying the BCH and a couple of Secondary CCPCH channels carrying the FACH and the PCH, all of them with $SF = 256$, the fraction reserved for the common and shared channels would be $R_C = 4 \cdot (1/256) = 1/64$ and the number of dedicated channels will have to fulfil the relationship:

$$\sum_{i=1}^{N_D} \frac{1}{SF_i} \leq \frac{63}{64} \tag{5.25}$$

This would give room for example, for 63 DCHs with $SF = 64$, or 31 DCHs with $SF = 32$, and so on.

The above inequalities hold for the usual case in that a single scrambling code is available at the node B. When additional secondary scrambling codes are available, the condition to be fulfilled would be:

$$\sum_{i=1}^{N_D} \frac{1}{SF_i} \leq K_S - R_C \tag{5.26}$$

where K_S is the total number of scrambling codes at the node B.

An additional aspect that can be taken into account during the code availability check is the precedence that handover users should have in the admission control in order to guarantee that they are always connected to the cell requiring the lowest transmission power. Section 5.2.1.5 raised this effect in the case of the uplink admission control. For the downlink, this precedence can be given by reserving some part of the OVSF code tree to handover users, so that the code availability condition to admit the new $(N + 1)$th user, provided that there are already N admitted users, would be:

$$\sum_{i=1}^{N} \frac{1}{SF_i} + \frac{1}{SF_{N+1}} \leq K_S - R_C - R_{HO} \tag{5.27}$$

where R_{HO} represents the fraction of the OVSF code tree that is reserved for handover users.

Correspondingly, if the admission is checked for a handover user, the condition becomes:

$$\sum_{i=1}^{N} \frac{1}{SF_i} + \frac{1}{SF_{N+1}} \leq K_S - R_C \tag{5.28}$$

5.2.2.2 Power Availability

In the downlink direction, the available transmitted power is shared among all the transmissions in a node B. As a result, each new accepted connection, apart from increasing the interference to the rest of the connections, reduces the total available power. Due to this power sharing, the instantaneous user locations have a large impact on the performance of the rest of the users in a cell, even for low loads, while in the uplink a particular user location only has impact on its own performance. As a result, the cell load as well as the amount of downlink resources demanded by a specific user varies as this user moves around the cell.

As in the uplink direction, downlink admission control strategies to check the power availability can be based either on measurements or on load statistical estimations. Some downlink admission control strategies that are based on transmission power measurements have been presented [27][28][29]. Below, some of these possibilities are presented, starting with the user count-based admission control algorithm.

User Count-based Admission Control   This algorithm operates according to the same principles as the algorithm presented in Section 5.2.1.1 for the uplink direction. It simply assumes that each user consumes a certain amount of radio resource equivalent units (RREUs) and limits the maximum acceptable number of RREUs in order to avoid power limitations in any of the accepted transmissions. Then, with N users already accepted in the cell, the $(N + 1)$th user request is accepted in the downlink direction if:

$$\sum_{i=1}^{N} R_i + R_{N+1} \leq R^* \tag{5.29}$$

It is worth mentioning that the maximum number of equivalent units R^* in the downlink direction does not necessarily have to be equal to the value in the uplink direction, and it should be related to the total power available at the node B and the power consumption associated with each equivalent unit.

As in the uplink case, the decoupling existing between this algorithm and the actual power consumption leads to a high number of bad rejections and also to some bad admissions, as depicted in Figures 5.39 and 5.40. These results have been obtained in the same scenario as the results of Section 5.2.1.1. The admission threshold R^* has been set to 60 equivalent units. Clearly, when the number of users in the cell is below 60, there is a certain number of bad admissions depending on the user position. Similarly, when the number of users is over 60, new users are rejected although in most cases they could have been accepted because there is actually enough power available for them.

The critical point with this type of algorithm is that, although it allows very simple implementations, suitable at the initial deployment of 3G networks, it is difficult to have an adequate setting of the maximum number of acceptable connections due to the uncertainties existing in the transmission power requirements due to mobility, propagation conditions and traffic generation in the different scenarios.

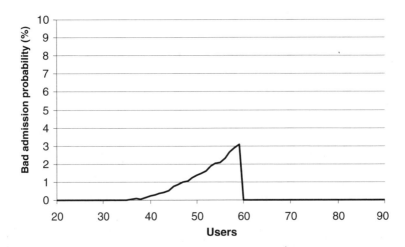

Figure 5.39   Bad admission probability with the user count-based downlink admission control algorithm

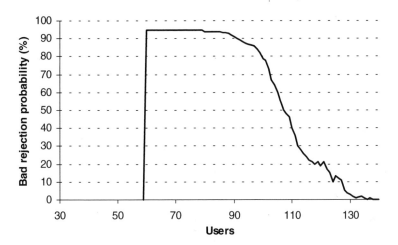

Figure 5.40 Bad rejection probability with the user count-based downlink admission control algorithm

Cell Load-based Admission Control Another strategy is the cell load-based admission control algorithm. This example operates in a similar way to the statistical-based admission control algorithm in the uplink direction, presented in Section 5.2.1.3, in order to estimate the maximum downlink load factor that can be accepted in a given cell. In the downlink direction, the load factor can be related to the total transmitted power at node B by means of the following expression (see Section 4.4.5):

$$P_T = \frac{P_p + P_N X_n}{1 - \eta_{DL}} \tag{5.30}$$

where P_p is the power devoted to pilot and common control channels and X_n depends on the path loss and requirements in terms of bit rate and E_b/N_0 of the n simultaneous users according to:

$$X_n = \sum_{i=1}^{n} \frac{\frac{L_{p,i}}{W}}{\left(\frac{E_b}{N_0}\right)_i R_{b,i}} + \rho \tag{5.31}$$

Furthermore, the downlink cell load factor η_{DL} is given by:

$$\eta_{DL} = \sum_{i=1}^{n} \frac{\rho + f_i}{\frac{W}{\left(\frac{E_b}{N_0}\right)_i R_{b,i}} + \rho} \tag{5.32}$$

where ρ is the orthogonality factor and f_i is the ratio between the intercell interference χ_i and the total intracell power measured at the receiver of user i, given by:

$$f_i = \frac{\chi_i L_{p,i}}{P_T} \tag{5.33}$$

From Equation (5.30), limiting the maximum available power at the node B is equivalent to limiting the maximum downlink load factor according to:

$$\eta_{max} = 1 - \frac{P_p + P_N X_n}{P_{T,max}} \tag{5.34}$$

Consequently, the downlink admission control condition to accept the $(N + 1)$th user could be given, as in the uplink direction, by:

$$\eta_{DL} + \Delta\eta \leq \eta_{\max} \tag{5.35}$$

We now point out some critical issues associated with this downlink admission control strategy:

- Load factor estimation. One important difference between the uplink and the downlink direction is the difficulty of obtaining accurate measurements of the downlink load factor. The reasons are two-fold. On the one hand, the orthogonality factor ρ depends on the multipath characteristics of the channel of each user, so that it can only be estimated depending on the considered environment in which the cell operates. On the other hand, the intercell interference factor f_i is user-specific and depends on the position of each user with respect to the neighbouring cells. Furthermore, the terminals do not report intercell interference measurements. As a result of that, the downlink cell load factor can only be estimated, in contrast to the uplink, where it can be physically measured. The estimation can be carried out as follows:

$$\eta_{DL}^S = \sum_{i=1}^{N} \frac{\rho + f_{DL}}{\dfrac{W}{\left(\dfrac{E_b}{N_0}\right)_i \alpha_i R_{b,i}} + \rho} \tag{5.36}$$

Similarly, the increase associated to the new user would be given by:

$$\Delta\eta = \frac{\rho + f_{DL}}{\dfrac{W}{\left(\dfrac{E_b}{N_0}\right)_{N+1} \alpha_{N+1} R_{b,N+1}} + \rho} \tag{5.37}$$

where α_i is the activity factor and f_{DL} represents an estimate of the intercell to intracell interference factor that should try to capture the real behaviour of this parameter. Just to illustrate the dependence between the f factor in the downlink and the distance, Table 5.5 presents the average and standard deviation values of this factor at different distances of the node B in a scenario with seven omnidirectional cells of radius 500 m and 40 users distributed in it, transmitting at 64 kb/s. The general observation is that the average value is not a good estimation due to the high deviation that this parameter exhibits. As a matter of fact, f_{DL} really depends on several factors apart from distance, such as the offered traffic or the considered environment, which are hardly reflected when only an average value is considered. For instance, when tri-sectorial antennas are used the main intercell interference source becomes the other sectors of the same site, which modifies the f factor distribution and its dependence on the distance. In this case, users close to the cell site may experience a higher intercell interference than users in the cell edge, just the opposite than when omnidirectional antennas are used [30].

Table 5.5 Average and standard deviation of the downlink f factor at different distances

Distance	Average	Standard deviation
150 m	0.31	1.09
300 m	0.78	1.87
500 m	1.31	2.39

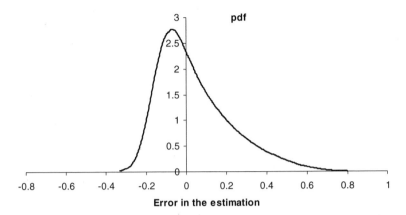

Figure 5.41 Distribution of the error between the measured and the statistical estimation of the downlink load factor.

As a result of the uncertainties in setting the intercell to intracell interference factor, the estimation of the downlink load factor may substantially differ from the real value. This fact is illustrated in Figure 5.41, which plots the pdf of the error between the real load factor and the statistical estimation. The results are obtained under the same conditions as Table 5.5. The estimation has been done with $f_{DL} = 0.72$, since this is the total average value that has been observed in the scenario. It can be observed that the estimation provides an average error around zero, but there may exist significant differences up to 0.6 depending on how terminals are located at a given instant.

- Maximum downlink load factor (η_{max}) setting. The maximum downlink load factor threshold considered in the admission control should be related to the maximum power available at the node B, $P_{T,max}$ according to Equation (5.34). Consequently, η_{max} should be adjusted according to the parameter X_n, which depends on the user location and the traffic generation process.

   Figure 5.42 plots the downlink load factor admission threshold η_{max} as a function of the variable X_n for the case where $P_p = 33$ dBm, $P_N = -100$ dBm and $P_{T,max} = 43$ dBm. Figure 5.43 presents the statistical distribution of parameter X_n for different numbers of simultaneous 64 kb/s users in a cell

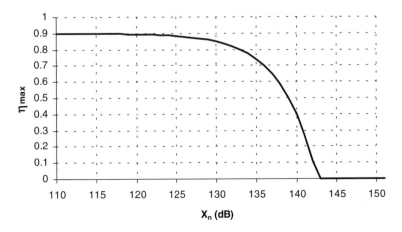

Figure 5.42 Maximum downlink load factor

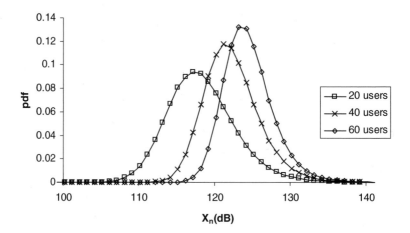

Figure 5.43   Statistical distribution of the factor X_n

with radius 500 m. Note that X_n may take a broad range of values, depending on the specific positions of the users to be served. These variations are directly translated into variations of the admission control threshold. For example, note in Figure 5.42 that on one hand the maximum cell load factor decreases rapidly for X_n higher than 125 dB and, on the other hand, the probability of X_n being higher than 125 dB is not negligible, as can be observed in Figure 5.43. With respect to the estimation of X_n it should be mentioned that the mobile terminals report their path loss measurements, so the computation of X_n is feasible from a practical point of view.

Taking into account the above settings, it is possible to establish a downlink admission control based on load factor estimations in order to control the transmitted power. However, and since the maximum cell load factor and the transmitted power are related by means of a time varying factor (i.e. the term X_n) that depends on user locations, it appears to be more reasonable to control the downlink operation through transmitted power measurements rather than through the cell load factor, as used to be the case in the uplink.

Transmitted Power-based Admission Control   We now consider the transmission power-based admission control algorithm, which checks the following condition to decide the acceptance of a new connection request in the system, arriving at the ith frame:

$$P_{AV}(i) + \Delta P_T(i) \leq P_T^*(i) \tag{5.38}$$

$$P_{AV}(i) = \frac{\sum_{j=1}^{T} P_T(i-j)}{T} \tag{5.39}$$

where $P_{AV}(i)$ is the average total transmitted power during the last T frames, $\Delta P_T^*(i)$ is the power increase estimation due to the new request and $P_T^*(i)$ is the admission threshold. It is assumed that not only the average transmitted power but also the power increase and the admission threshold may vary with time in order to adapt to different load situations.

Despite the simplicity of the algorithm, it can offer an efficient performance provided that the design parameters are suitably set. Therefore, we now analyse the three crucial design criteria that need to be set: the measurement period; the admission threshold; and the power increase estimation.

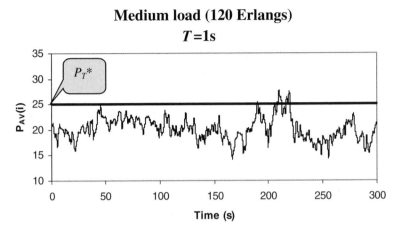

Figure 5.44 Plot of the time averaged transmitted power for a medium load situation with $T = 1$s

In order to overcome the high variability of the mobile radio channel as well as the different interference patterns, transmitted power measurements should be time averaged. This poses a trade-off in the selection of the measurement period, since low values will not be able to smooth sporadic transmission peak powers while high values may not be able to capture the variations due to the load increase/reduction.

The measurement averaging period T must be set in accordance with the mobility patterns, the propagation conditions and the average load in the system. The dependence of the existing load is plotted in Figures 5.44 to 5.47, which analyse the impact of different averaging periods. The simulations assume a scenario with seven omnidirectional cells with radius 500 m, pilot power 10 dBm and maximum available power $P_{T,\max} = 43$ dBm. Conversational CBR traffic at 64 kb/s is considered, and the mobile speed is 3 km/h. The admission threshold is set to $P_T^* = 25$ dBm, represented in the figures by means of a solid line, and the power increase is estimated as the average value among the transmitted power of all the users. In all cases, a representative 5 minute segment is shown. The figures analyse a medium and a

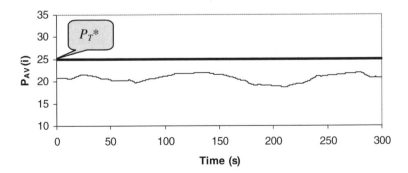

Figure 5.45 Plot of the time averaged transmitted power for a medium load situation with $T = 50$s

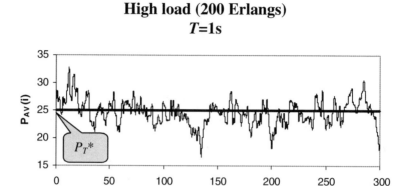

Figure 5.46 Plot of the time averaged transmitted power for a high load situation with $T = 1$ s

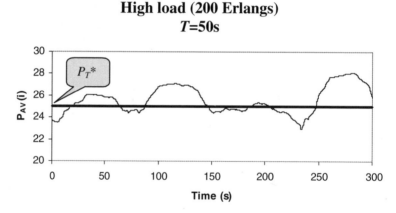

Figure 5.47 Plot of the time averaged transmitted power for a high load situation with $T = 50$ s

high load situation. The corresponding performance results in terms of admission rate and BLER are shown in Table 5.6.

For the medium load, where generally the required power would be low, a relatively high averaging period T is desirable to avoid effects from instantaneous and rare high-transmitted power situations that could lead to the rejection of some calls. This can be observed in Figure 5.44 for the case $T = 1$ s, where

Table 5.6 Admission probability and BLER for different measurement periods.

	Admission probability (%)		BLER (%)	
	$T = 1$ s	$T = 50$ s	$T = 1$ s	$T = 50$ s
Medium load	98	100	1	1
High load	41	35	1.1	1.1

short periods exist in which the average power is higher than the admission threshold and consequently some calls are rejected. For this case, the measured admission probability is 98% and no degradation in the block error rate with respect to the target value of 1% is observed. However, if T is high as in Figure 5.45, the smoothing due to the longer averaging periods avoids unnecessary call rejections, and keeps a 100% admission probability without experiencing BLER degradation (see Table 5.6).

The situation changes for high loads where the required transmitted power is in general close to the admission threshold, as observed in Figures 5.46 and 5.47. In particular, in this case a shorter averaging period T is desirable to avoid long memory effects, which would produce a high rejection rate. Note that for the $T = 50$ s case in Figure 5.47, a wave-like effect arises in the averaged power and the periods in which calls are accepted are reduced. However, $T = 1$ s in Figure 5.46 allows advantage to be taken of the periods where the required node B transmitted power is lower than the threshold in order to increase the acceptance ratio without significant degradations in terms of error rate.

Taking the above comments into account, it can be concluded that in the transmission power-based admission control the measurement-averaging period required for $P_{AV}(i)$ evaluation depends on the load level relative to the admission threshold P_T^*. For low loads, long averaging periods are more suitable (in the order of a minute), while for high load situations shorter averaging periods (in the order of a second) lead to a better performance.

The setting of the admission control power threshold P_T^* allows the execution of a softer or stricter admission control at the expense of the performance perceived by the admitted users. The threshold value should be set below the maximum available power at node B, $P_{T,max}$, giving room for fluctuations of the real power over the threshold and trying to avoid power limitations occuring due to mobility and traffic variations. Note that while high values of the threshold tend to increase the admission rate, they may also cause certain power limitations depending on how users are located, thus degrading the user performance in terms of error rate. However, low threshold values will decrease the admission probability but power limitations will seldom occur.

This effect is shown in Figures 5.48 and 5.49 in terms of admission probability and block error rate for different admission thresholds $P_T^* = 25$, 30 and 35 dBm. Conversational users with 64 kb/s are considered in the scenario with seven cells of radius 500 m and mobile speed 3 km/h. The maximum power at node B is 43 dBm. Note that for a restrictive value such as 25 dBm, many requests are rejected

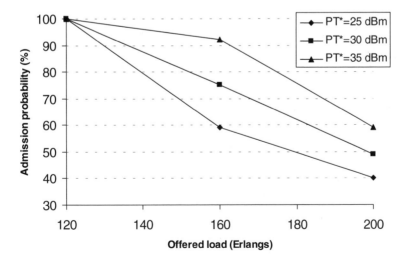

Figure 5.48   Admission acceptance ratio for different power thresholds

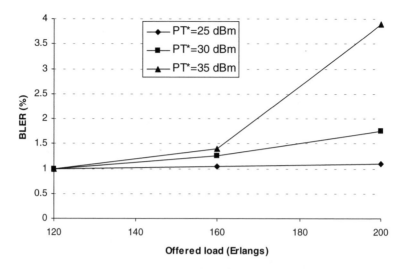

Figure 5.49 Packet Error Rate for different power thresholds

when 160 Erlangs is the offered load in the scenario. On the other hand, softer admission policies (e.g. with P_T^* 35 dBm) provide a much higher power limitation probability, and as a consequence the BLER increases (see Figure 5.49).

There is a trade-off that exists in the admission threshold between admission probability and BLER. This is in the fact that high threshold values are of interest for low load levels in order not to degrade unnecessarily the admission probability, while lower thresholds values would be required for high load levels in order not to degrade the BLER performance. This suggests that an optimum admission threshold can be found for each load level and, therefore, adaptive strategies could be adequate to set the threshold to the most adequate level according to the system load. In such a case, the admission threshold at the ith frame can be set as a function of the average measured power, given by:

$$P_T^*(i) = f(P_{AV'}(i)) \tag{5.40}$$

where $P_{AV'}(i)$ is the average transmitted power during a certain period T' that does not necessarily coincide with the averaging period T used to measure $P_{AV}(i)$ in Equation (5.38). Note that the average $P_{AV'}(i)$ gives an indication of the load existing in the system. Figures 5.50 and 5.51 plot two possible functions for the adaptive admission threshold. The first considers a progressive transition in the threshold between 25 dBm for high loads and 43 dBm, equal to the maximum power at the node B, for low loads. In turn, the second function introduces a hysteresis margin to smooth the transition between different admission thresholds. As depicted in Figure 5.51, when the average power increases the admission threshold follows the right-side function, while when the power decreases the admission follows the left-side function.

The corresponding results are shown in Figures 5.52 and 5.53 for both the admission probability and the BLER. For both adaptive functions the resulting admission probability is quite similar. Note that in both cases, the results represent an intermediate situation with respect to the ones obtained with fixed thresholds of 25 and 30 dBm. However, in terms of BLER performance, the adaptive function with hysteresis presents a better behaviour than the one without hysteresis. In this case, the obtained BLER is closer to the one of the lowest admission threshold of 25 dBm for high loads, and the admission

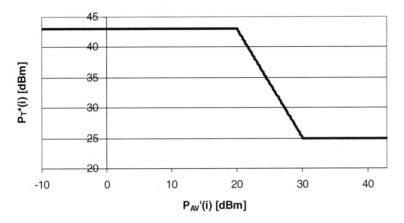

Figure 5.50   Adaptive admission threshold function without hysteresis

probability is closer to the highest admission threshold of 43 dBm for low loads. Therefore, introducing the hysteresis margin allows a somehow better control.

The estimation of the power increase demanded by the user requesting admission should be done by trying to anticipate as much as possible the real power that will be required by the user, which is highly dependent on the mobility patterns of the user and the transmission bit rate of the service. Note that, due to mobility, the power consumption for a given user changes along the connection lifetime and, consequently, the estimation done in the admission instant may not be equal to the real power, which may turn into bad admissions or bad rejections. To illustrate these effects in different conditions, let us consider two power-based admission control algorithms, which differ in the way they estimate the power increase of a new user [31].

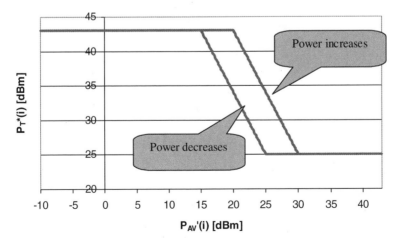

Figure 5.51   Adaptive admission threshold function including hysteresis.

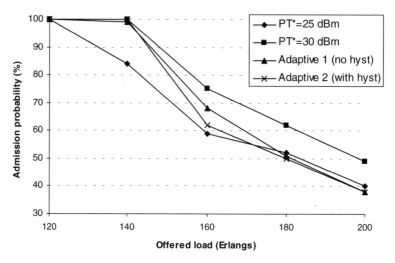

Figure 5.52   Admission probability for fixed and adaptive thresholds

The PABAC (Power Averaged-Based Admission Control) algorithm assumes that the power demand is estimated as a time average along the last T frames of the required transmitted power for the already accepted users belonging to the same service type. The estimation would then be given by:

$$\Delta P_T(i) = \frac{1}{T}\sum_{j=1}^{T}\frac{1}{N_{i-j}}\sum_{n=1}^{N_{i-j}}P_{Tn}(i-j) \qquad (5.41)$$

where N_{i-j} is the number of users admitted and $P_{Tn}(i-j)$ the power required by the nth user in the $(i-j)$th frame.

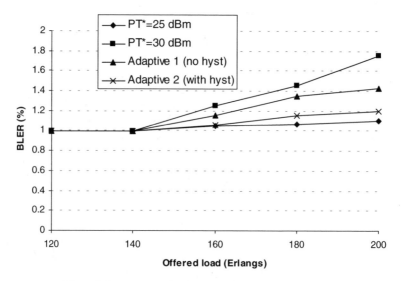

Figure 5.53   BLER for fixed and adaptive admission thresholds

The rationale behind this algorithm lies in the fact that, when mobility is not known a priori, user power consumption may vary along the connection lifetime depending on the user location and the location of the rest of users. As a result, an average estimation may provide a good trade-off between bad rejections and bad admissions. Note also that the above estimation assumes that all the users transmit at the same bit rate. When different bit rates existed, different estimations would be needed for each one.

The PLEBAC (Path Loss Estimation-Based Admission Control) algorithm estimates the power increase of the user taking into account the user measurement reports provided during the call set-up process, which include the total path loss with respect to the serving cell. To this end, the algorithm defines a set of $M + 1$ path loss ranges $\{PL_0, PL_1, \ldots PL_M\}$ with resolution Δ:

$$PL_k(dB) = PL_0(dB) + k\Delta(dB) \tag{5.42}$$

The kth range ($k = 1 \ldots M - 1$) includes all the path loss values higher than or equal to PL_k and lower than PL_{k+1}. For the special cases $k = 0$ and $k = M$, they include the values lower than PL_0 and higher than PL_M, respectively.

A correspondence is established between each path loss range and the power increase estimation. In particular, for the kth range this correspondence is obtained from the average of the transmitted power to the already accepted users whose reported path loss falls within this range. This average is carried out with a slide window of T frames. Therefore, the power demand estimation in the ith frame for the kth range is defined as:

$$\Delta P_T(k, i) = \frac{1}{T} \sum_{j=1}^{T} \frac{1}{N_{k,i-j}} \sum_{n=1}^{N_{k,i-j}} P_{Tn}(i - j) \tag{5.43}$$

where $N_{k,i-j}$ is the number of accepted users at the $(i - j)$th frame whose last path loss report falls within the kth range. In the case of users with different bit rates, the measurements should be done separately for each bit rate.

It is worth mentioning that, when users are static, $\Delta P_T(k, i)$ provides a good estimate of the power that will be transmitted to the user along the connection lifetime.

The essential difference between PLEBAC and PABAC behaviour is presented in Figure 5.54, which shows the ΔP_T value that is used as a function of the path loss in both algorithms for a scenario with

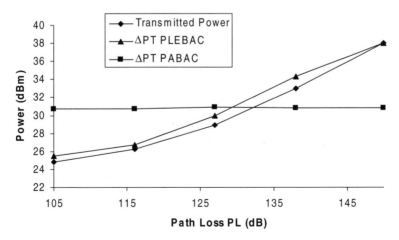

Figure 5.54 Average power increase estimated by PABAC and PLEBAC algorithms as a function of the path loss with a 384 kb/s service in static conditions

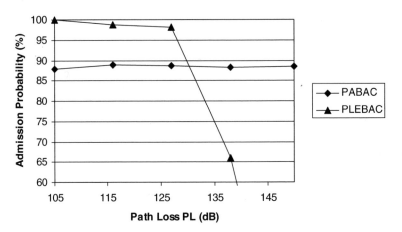

Figure 5.55 Admission probability for the PABAC and PLEBAC estimations as a function of the path loss with a 384 kb/s service in static conditions

seven omnidirectional cells with radius 2 km, maximum available power 43 dBm and users transmitting at 384 kb/s. With PABAC, the estimation of the power increase demanded by a new user does not depend on the path loss. In contrast, with PLEBAC the estimated value of the power increase required by the new user increases with the path loss. For comparison purposes, the figure also plots the real transmitted power if the users were static. As can be observed, PLEBAC fits much more efficiently the power increase to the real required power.

Continuing with the case of static users, Figures 5.55 and 5.56 illustrate the impact of the admission control algorithm on the admission probability and the BLER, represented as a function of the measured path loss. Both cases assume a 384 kb/s service. With PABAC, the admission probability does not depend on the path loss. With PLEBAC, however, admission probability is improved for low path losses (i.e. for users close to the cell) and it is reduced for high path losses (i.e. for users at the cell edge). Nonetheless, note in Figure 5.56 that PLEBAC tends to reject those users whose QoS in terms of BLER cannot be

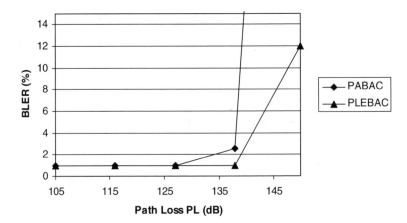

Figure 5.56 BLER for PABAC and PLEBAC as a function of the path loss with a 384 kb/s service in static conditions

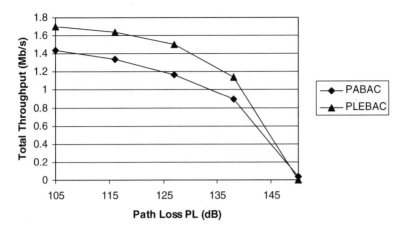

Figure 5.57 Downlink total throughput with PABAC and PLEBAC as a function of the path loss with a 384 kb/s service in static conditions

guaranteed (i.e. those with high path loss values) while it admits those users whose BLER is guaranteed at the target value of 1%.

Therefore, although the overall admission probability remains similar in both cases, PLEBAC reduces both bad admissions (i.e. the users with high path loss whose BLER is not assured and are accepted by PABAC) and bad rejections (i.e. the users with low path loss whose BLER could be assured but are rejected by PABAC), which creates a higher system throughput, as depicted in Figure 5.57.

The power increase demand is highly dependent on the scenario in terms of cell radius and bit rate, in the sense that higher cell radius or higher bit rates require higher transmission power levels and, consequently, accurate power increase estimations at the admission control become even more critical. This is illustrated in Figures 5.58 and 5.59, which present the gain in terms of system throughput of PLEBAC with respect to PABAC for different cell radii and for different service bit rates. Note that, for the lowest radii and bit rate values, the differences between both algorithms are quite small, which

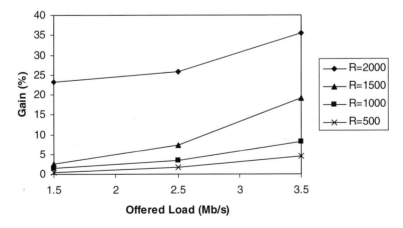

Figure 5.58 Throughput gain of PLEBAC vs PABAC for different cell radii with a 384 kb/s service in static conditions

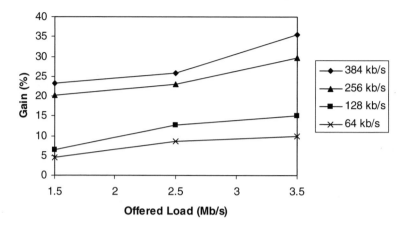

Figure 5.59 Throughput gain of PLEBAC vs PABAC for different bit rates with 2 km cell radius in static conditions

reveals the fact that under these circumstances the power increase estimation is not as important as with higher radii or higher bit rates.

The above results have been presented under static conditions, which will be a quite typical situation when considering high bit rate services, usually received by users using, for example, a laptop and thus having limited mobility. When that mobility is considered, the estimation provided by PLEBAC is not as accurate since, depending on the user speed, the power demanded at the beginning of the session may substantially differ from the one required during the connection lifetime, especially for long sessions. This is illustrated with some results in Table 5.7, where the throughput gain of PLEBAC strategy versus PABAC is presented with a cell radius of 1000 m for two different call durations and three different mobile speeds. Results are presented for both 384 kb/s and 64 kb/s. For high mobile speeds, the estimation used by PLEBAC at admission control is not representative of the real power demanded by the service, and therefore in this case the average estimation given by PABAC provides a better behaviour. This effect is more significant for longer call durations, and it can be observed that at a mobile speed of 50 km/h PLEBAC still performs somewhat better than PABAC if the calls are 30 seconds long, while it performs worse than PABAC in the case of 3 minute calls. In the 64 kb/s case the differences between both algorithms are less significant than in the 384 kb/s case.

Admission Control During Handover Procedure   We now examine the admission control during handover procedure. Whenever a handover is to be executed for a certain user, the admission control conditions must be checked in both the uplink and downlink directions. It has been pointed out in Section 5.2.1.5 that, from the uplink point of view, it is necessary to give priority to handover users in the admission control phase. In turn, in the downlink, the admission control should take into account a

Table 5.7   Throughput gain of PLEBAC vs PABAC for different call durations and speeds

Throughput gain	384 kb/s		64 kb/s	
	30 seconds	3 minutes	30 seconds	3 minutes
Static	9.45%	8.90%	6.12%	5.30%
3 km/h	2.64%	2.49%	2.60%	1.82%
50 km/h	1.06%	−2.75%	0.58%	−2.94%
120 km/h	−5.49%	−8.40%	−3.8%	−4.30%

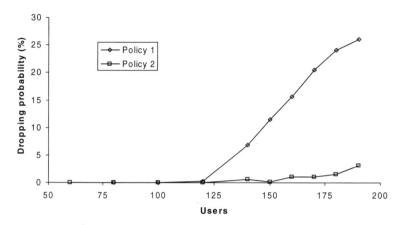

Figure 5.60   Dropping probability for two policies dealing with handover in downlink admission control

hard-limit posed by the code availability (see Section 5.2.2.1) as well as a soft-limit posed by the power availability. In this context, similar policies to those considered in the uplink can be considered for the downlink power availability check.

In particular, let us compare Policy 1, in which soft handover requests check the power condition in the admission control given by Equation (5.38) and Policy 2, in which the power availability condition is not checked and the handover user is accepted provided that there are available codes.

Figure 5.60 presents the performance in terms of dropping probability, assuming a cell radius of 500 m and a power admission threshold of 40 dBm. It is assumed that a call is dropped when the measured Eb/No is 3 dB below the target for 1 s. Note that, for load levels higher than 120 users, the dropping increases significantly with Policy 1. The reason is that, when handover requests are rejected in the admission control algorithm and the mobile terminal maintains the connection with the old serving cell, it will require higher transmitted power levels, thus causing a higher interference to the rest of mobiles in both the old and the new cell. As a result, the overall cell quality level in terms of BLER degrades and some calls are eventually dropped. In contrast, this situation is avoided with Policy 2, thus maintaining lower dropping values even for high loads.

As a result, and as in the uplink direction, Policy 2 also appears to be more suitable in the downlink, which can be understood as a prioritisation mechanism for handover users; not only for their own benefit, to avoid on-going call dropping, but from the overall cell performance resulting from lower interference patterns. In any case, it is worth mentioning that the acceptance of a handover user without checking the power availability is subject to the condition that there are available codes in the new cell. If this is not the case, the high degradation associated with the excess of power and interference involved in keeping the user connected to the old cell suggests that it might be better to drop the call rather than to keep the connection in the old cell.

Influence of the Maximum Power per Connection   A parameter that plays a key role in the downlink direction is the maximum power that can be allocated to a connection. The main purpose of this parameter is to avoid users consuming an excess of power that could degrade the rest of transmissions in the cell. In the admission control phase, this maximum power per connection can be taken into account simply by checking that the power increase due to the new user is below the maximum power that can be allocated to the user.

Some illustrative results of limiting the maximum power per connection are now shown. Figures 5.61 and 5.62 plot the BLER and dropping performance as a function of the maximum power per connection for a 64 kb/s conversational service. It is assumed that a call is dropped when the measured Eb/No is 3 dB

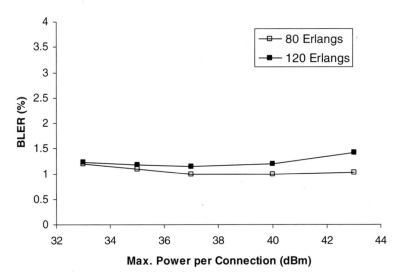

Figure 5.61 Performance in terms of BLER as a function of the maximum power per connection

below the target for 1 s. A cell radius of 2 km has been considered, pointing out that the maximum power per connection is especially relevant for high coverage areas. Note that, if the maximum power is set too low, the cell coverage is not assured and, consequently, certain users far from the base station may not achieve the required signal-to-interference ratio because of power limitations, thus leading to an increase in both the BLER and the dropping probability. However, if the maximum power per connection is set too high, excessive power expense may be required for some users, thus preventing a good service for the rest of accepted users and therefore also causing degradations in terms of both BLER and call droppings.

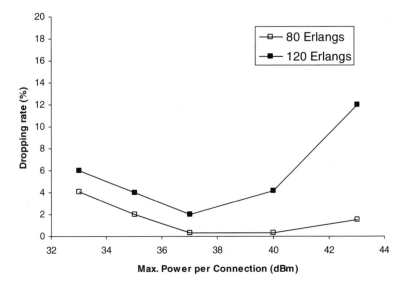

Figure 5.62 Performance in terms of dropping as a function of the maximum power per connection

This trade-off results in an optimum value for the power per connection that will depend on the specific scenario. In particular, for the presented sample case, an optimum of around 37 dBm per connection is found.

Multi-service Case The downlink admission control in multi-service scenarios must take into account the different requirements and different resource consumptions associated with the different services, as well as the operator policies with respect to the priorities established between services, as was stated in Section 5.2.1.6 for the uplink direction.

In this sense, note that the downlink cell load-based admission control strategy allows a direct application of the same policies presented for the uplink case by distinguishing the contribution of each service class to the load factor. In particular, assuming without loss of generality a mix of conversational and interactive users, the total downlink load factor could be split into the following two contributions:

$$\eta_{DL} = \eta_C + \eta_I = \sum_{i=1}^{N_c} \frac{\rho + f_{DL}}{\left(\frac{E_b}{N_0}\right)_i \alpha_i R_{b,i}} W + \rho + \sum_{i=1}^{N_I} \frac{\rho + f_{DL}}{\left(\frac{E_b}{N_0}\right)_i \alpha_i R_{b,i}} W + \rho \tag{5.44}$$

where N_c and N_I are the number of admitted conversational and interactive users, respectively.

Similarly, when the transmitted power-based admission control strategy was used in the downlink direction, the service differentiation could be done by using different power admission thresholds associated with the different services, thus allowing the prioritising of some of them over others. The power increase estimation ΔP_T should be different depending on the required service. In this sense, strategies based on averaging the power devoted to the already admitted users will have to take into account in the averaging process only those connections with the same bit rate as the requesting user.

Multi-cell Admission Control The final control strategy to examine is multi-cell admission control. The admission of one user in a WCDMA scenario affects not only the serving cell but also the neighbouring cells due to the increase in intercell interference. This effect can be included in the admission control by means of the derivatives framework presented in Section 4.6.2 for the downlink direction, which allowed the computation of the derivative of the transmission power of one cell with respect to another.

Therefore, when a new user is to be admitted in cell 0, the following set of conditions would be checked in the K neighbouring cells:

$$P_{T,i} + \Delta P_{T,0} \frac{\partial P_{Ti}}{\partial P_{To}} \leq P_T^* \tag{5.45}$$

where $P_{T,i}$ is the average transmitted power at the ith cell and $\Delta P_{T,0}$ is the power increase estimation in cell 0 due to the new user. Note that the product $\Delta P_{T,0} \cdot (\partial P_{T,i}/\partial P_{T,0})$ provides an estimate of the power increase that the new user will cause in the ith cell.

5.3 HANDOVER AND CELL SELECTION ALGORITHMS

This section describes the algorithms and procedures that are used by the mobile terminal to select a suitable cell to be connected to. When the mobile terminal is switched on, it must first select the PLMN (Public Land Mobile Network) and RAT (Radio Access Technology). Then it searches for the most adequate cell as which to camp and to perform the registration procedure if required. As long as the terminal remains in idle mode, it will continuously execute the cell reselection procedures in order to choose the appropriate cell according to the terminal position in each instant. Depending on the PLMN availability and network configuration, it is possible that cell reselection procedures involving a change to other RATs, mainly to the 2G networks GSM/GPRS, are executed.

The coexistence between the UMTS and GSM networks poses certain restrictions on allowing both proper operation and that dual terminals can be connected to the two networks. These aspects will be presented in Section 5.3.1, while Section 5.3.2 will present the cell selection procedures in idle mode.

In turn, when the mobile is in connected mode, meaning that it has an ongoing RRC connection, it will continue to execute cell reselection procedures if it has not allocated dedicated resources (i.e. in Cell_FACH, Cell_PCH or URA_PCH states). When the mobile is in Cell_DCH state and there is a dedicated channel allocated to it, the handover procedures will be executed in order to choose the successive cells where dedicated resources are being allocated. This allows retention of the continuity in the dedicated connection and the transferring of the service from cell to cell in a transparent way for the final user. Handover algorithms will be covered in Section 5.3.3.

5.3.1 REQUIREMENTS FOR GSM-UMTS INTEROPERATION

The deployment of 3G networks is being carried out in most of the cases by incumbent GSM operators with an extensive network of 2G base stations. As a result, the coexistence between 2G and 3G networks has been taken into account in the specification process of UMTS by establishing different procedures that allow the operation of dual mobile terminals that can be connected either to the GSM or to the UMTS access network.

There are several possibilities for cooperation between GSM and UMTS networks, although some of them might bring difficulties in the terminal design or even changes in the existing GSM standard. As a result, some restrictions have been established regarding the coexistence between both systems:

- A dual terminal can only have connection with one of the two networks at a time.
- The required information about the UMTS network to be transmitted in GSM should be encapsulated in existing GSM messages, thus minimising the changes required in the standard.
- The mobile should always be able to make measurements from both networks, i.e. when it is connected to GSM/GPRS, it must be able to measure the UMTS network and vice versa.

The measurement procedure varies depending on whether the terminal is in idle or in dedicated mode. In particular it is fairly simple in idle mode to make measurements no matter whether the mobile is camping on a GSM or on a UTRAN cell, because it has plenty of free time to execute measurements. Nevertheless, such measurements may reduce the battery lifetime because the mobile must leave the low power consumption stand-by state to execute them. In order to avoid this problem, it is possible to control when a terminal must start measuring the other network.

In turn, when the terminal is connected to GSM in dedicated mode, the TDMA technique used allows free transmission time intervals in which the mobile can perform UTRAN measurements. However, these time intervals are the same that are used to measure the GSM neighbouring cells and, consequently, it reduces the available time to measure GSM and may affect the resolution of the measurements. Therefore, it is also a good solution to control the instants when a mobile connected to GSM must start executing UTRAN measurements.

When the mobile is connected to UTRAN in dedicated mode, the WCDMA access technique being used requires, in principle, continuous transmission. Consequently, in order that a dual terminal connected to UTRAN can make GSM measurements, it is necessary either to have two receivers for the GSM and UMTS bands, or to make use of the compressed mode available in the transmission in dedicated channels (see page 79).

The terminal measurement and multi-mode/multi-RAT capabilities are indicated when the mobile connects to the UMTS network in the RRC Connection Setup Complete message. An example of the fields related to inter-RAT measurements included in this message is given in Table 5.8. The example

Table 5.8 Measurement capabilities included in the RRC Connection Setup Complete message

	Field		Value
UE-MultiMode/Multi-RAT	multiRAT Capability List		
Capability	support of GSM		TRUE
	support of Multicarrier		FALSE
	multiMode Capability		FDD
Measurement Capability	downlink Compressed Mode		
	fdd measurements		FALSE
	gsm measurements	gsm900	FALSE
		dcs1800	TRUE
		gsm1900	FALSE
	uplink Compressed Mode		
	fdd measurements		FALSE
	gsm measurements	gsm900	FALSE
		dcs1800	TRUE
		gsm1900	FALSE

considers a mobile terminal that can only operate with UTRAN, GSM900 and 1800 and has two separate receivers, one for GSM and the other for UTRAN. Therefore, the mobile indicates the support of GSM by setting the corresponding field of the message to TRUE. It also indicates that it supports the FDD mode of UTRAN in the field *multiMode Capability*. Furthermore, the different sub-fields in *uplink/ downlink compressed mode* indicate the requirement of the compressed mode to measure the different RATs. Note that, since the terminal has two separate receivers, it can measure the GSM900 band without requiring compressed mode (i.e. the field gsm900 is set to FALSE in both uplink and downlink). However, measurements in the 1800 MHz band cannot be carried out simultaneously with UTRAN FDD transmission and reception because of the proximity between the uplink and the downlink bands of both systems. Therefore, compressed mode is required for DCS 1800 (i.e. the field *dcs1800* is set to TRUE in both links). With respect to GSM 1900, since it is not supported by the terminal of this example, no compressed mode is required. Also, the terminal indicates by setting the field *fdd measurements* to FALSE that it does not require compressed mode to measure other UTRAN FDD carriers.

When the mobile terminal is connected to GSM, its capabilities are indicated to the network in the Classmark Change message, which is sent during the call establishment procedure. In particular, an example of some of the fields contained in this message is shown in Table 5.9 for a mobile supporting both UMTS FDD and GSM.

5.3.2 PLMN, RAT AND CELL SELECTION ALGORITHMS

This section deals with the procedures carried out by a terminal, after switching on, to select the appropriate PLMN and RAT to operate with as well as the specific cell to camp on. The overall process is illustrated in Figure 5.63 and will be covered in detail in the following sections.

Table 5.9 Capabilities included in the Classmark Change message

Field	Value
RF Power Capability	(0) class 1
Multiband supported	(6) E-GSM/R-GSM + GSM 1800
Multislot class	8
UMTS FDD radio access technology capability	(1) UMTS FDD supported
UMTS TDD radio access technology capability	(0) UMTS TDD not supported
CDMA 2000 radio access technology capability	(0) CDMA 2000 not supported

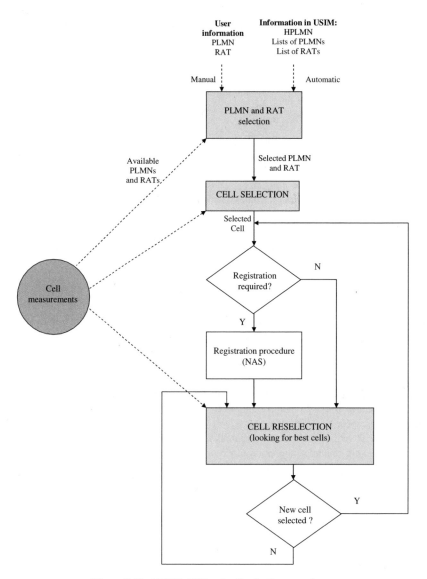

Figure 5.63 PLMN, RAT and cell selection procedures

5.3.2.1 PLMN and RAT Selection

When a mobile terminal is switched on, it must first choose the mobile network to which it will be connected as well as the specific RAT. To this end, it must perform some initial measurements to identify which networks and radio access technologies are available.

In the case of UTRAN, the terminal searches in the UTRAN carrier frequencies for the primary SCH and executes the synchronisation procedure explained in Section 3.3.4.4 that leads to the detection of the primary scrambling codes of the existing cells. This allows, on the one hand, the measurement of the CPICH channel to determine if the received signal quality is acceptable and, on the other hand, the retrieval of the information in the broadcast channel, which indicates whether the mobile can or cannot camp on the cell. Similarly, the broadcast channel informs about different parameters to be used in the cell selection and reselection procedures, as will be explained later in this section.

In the case of GSM, a similar procedure is carried out by scanning the GSM frequencies and synchronising with the available cells, so that the information in the broadcast channel can be retrieved and signal measurements can be carried out.

A network is identified by its network code (PLMN code) and the Radio Access Technology it uses. Typically, a dual operator having both GSM and UMTS networks has a single network code and two radio access technologies. However, the specifications also allow having two different network codes, so that the mobile in fact detects two equivalent networks. The PLMN code is broadcast both in GSM and UTRAN.

As is shown in Figure 5.63, the PLMN and RAT selection procedure is based on the measurements and information retrieved from the network and it can be either manual or automatic, depending on how the terminal has been configured by the user.

In the automatic selection procedure, a dual GSM/UMTS terminal continually tries to be connected to the last combination of PLMN and RAT it camped on before the switch off procedure. This is done by making use of the information about the last frequencies and codes that were used, which is stored in the USIM card. When the last combination of frequency and code is not available, for example, because the terminal has changed its location, it scans all the available networks in order to find one that fulfils the minimum quality criterion, given by UTRAN RSCP $>= -95$ dBm and GSM RxLev > -85 dBm. The network choice is then done according to the following prioritisation criteria [32]:

1. HPLMN (Home PLMN). The first network that the mobile looks for is the HPLMN, which is the network where the terminal normally operates. In this case, the RAT is initially selected by means of the field 'access technologies priority list' in the USIM, which includes a prioritisation of the RATs. When this information is not available, the mobile starts by default with GSM.
2. Preferred PLMNs included in the field 'user controlled PLMN list' of the USIM. If the HPLMN is not available, the terminal selects the network from this list, which is established by the user and stored in the USIM. The priority between networks is given by the ordering specified in this field. As in the previous case, the RAT is selected by means of the 'access technology priority list' field.
3. Preferred PLMNs included in the field 'operator controlled PLMN list' of the USIM. If neither the HPLMN nor the PLMNs in the list established by the user are available, the terminal chooses the network according to the list established by the operator and stored in the USIM. The RAT is again selected by means of the 'access technology priority list' field.
4. PLMNs that fulfil the minimum quality criterion. If neither the HPLMN nor the PLMNs stored in the USIM are available, the terminal chooses among all the PLMNs fulfilling the minimum quality criterion given by UTRAN RSCP $>= -95$ dBm and GSM RxLev > -85 dBm. In this case, the networks are randomly ordered, so the mobile terminal selects the first one that can be found. Furthermore, the mobile searches in all the RATs.
5. Rest of networks. Finally, if none of the previous networks was available, the terminal orders the detected PLMNs in decreasing order according to the received power level (i.e. RSCP for UTRAN and RxLev for GSM) and selects the best one. Again, the mobile searches in all the RATs.

In any case, the forbidden networks indicated in the USIM and the forbidden registration areas, in which the terminal is not allowed to register, are not considered in the previous network search procedure.

When the network selection is done manually, the mobile terminal presents to the user an ordered list of the available PLMNs according to the same prioritisation order of the automatic mode but also including the forbidden networks and registration areas. The selection is then executed by the user. The mobile may also indicate for each network the available RATs.

It is also possible that the user configures the terminal to operate either with UTRAN or GSM by setting the 'system mode'. In this case, the mobile terminal will always select the RAT according to the user preferences and, if the RAT is not available, the user will not get service.

Once the terminal has selected the network and the RAT, it must select the appropriate cell to camp on, according to the cell selection procedure described in Section 5.3.2.2. The mobile may then initiate a NAS registration procedure if necessary (e.g. if the registration area has changed or if the terminal has been switched off for some time) in the registration area of the chosen cell. The mobile will then keep continuously looking for better cells by means of the cell reselection procedure that will be described in Section 5.3.2.3. After a cell reselection, a new registration procedure may be triggered if the new cell belongs to another registration area. The overall procedure is summarised in Figure 5.63.

5.3.2.2 Cell Selection in UTRAN

The cell selection procedure in UTRAN is executed mainly after switching on the mobile and after having selected the network. There are two different search procedures for cell selection:

1. Initial cell selection. This is done by scanning all the UTRAN bands to find a suitable cell of the selected PLMN. The strongest cell in each carrier is selected.
2. Cell selection based on stored information. In this case, the mobile terminal looks only for cells in carrier frequencies stored in the USIM from previous measurements. If no cell is found, the initial cell selection procedure is carried out by scanning all the UTRAN bands.

There are two criteria that must be fulfilled by a cell in order to be considered valid in the selection procedure [33], dealing with signal level and quality level. The first is the *signal level criterion*, for which the following condition must be fulfilled:

$$Srxlev = Qrxlevmeas - Qrxlevmin - Pcompensation > 0 \qquad (5.46)$$

where Qrxlevmeas is the RSCP (Received Signal Code Power) in the CPICH measured in dBm and Qrxlevmin is the minimum RSCP level, configurable by the operator in the range -115 to -25 dBm in steps of 2 dB [34]. Furthermore, Pcompensation is a parameter that takes into consideration the maximum transmitted power allowed for the access in the uplink direction maxTxPowerUL together with the maximum power P available at the UE. It is defined as:

$$Pcompensation = max(maxTxPower\,UL - P, 0) \qquad (5.47)$$

The second criterion is the *quality level criterion*. In this case, the condition to be fulfilled is:

$$Squal = Qqualmeas - Qqualmin > 0 \qquad (5.48)$$

where Qqualmeas is the Ec/No value measured in the CPICH, and Qqualmin is the configurable minimum Ec/No in the CPICH, which takes values from -24 dB to 0 dB [34].

The selected cell is the first one that meets both signal and quality criteria. Furthermore, the cell must neither belong to a forbidden registration area nor be barred. A cell is barred if the terminal is not allowed to camp on it because of cell access restrictions. The cell barring condition is indicated in the broadcast messages.

When the mobile cannot find a cell that fulfils the above requirements in the current UTRAN carrier, it should start an initial cell selection procedure by scanning different frequencies. If no cell is finally found, it may remain connected to another network but with service limitations (e.g. only emergency calls may be allowed). The mobile would then continuously look for another allowed network to which it could connect.

5.3.2.3 Cell Reselection

Having selected the cell to camp on, the mobile terminal continuously makes measurements in order to find a better cell. It ranks the cells from all the supported RATs (i.e. GSM and UTRAN), according to the quality and signal measurements, as explained below.

It is worth mentioning that cell reselection algorithms can also be used to distribute the traffic among different layers (e.g. micro and macro cells) and radio access technologies.

Mobile Camped on a UTRAN Cell   When the mobile is camped on a UTRAN cell, it continuously measures other UTRAN cells. It will also start making GSM measurements if the quality of the UTRAN cell is below a certain level according to the following condition:

$$\text{Squal} = \text{Qqualmeas} - \text{Qqualmin} < \text{sRATSearch} \tag{5.49}$$

For instance, if sRATSearch = 4 and Qqualmin = −18, the UE starts GSM measurements whenever the CPICH Ec/No is below −14 dB. This condition is set in order to avoid that the mobile continuously measuring GSM cells, which would increase the power consumption.

The parameter sRATsearch is broadcast by the UTRAN network. When it is not present, GSM is always measured by default. The GSM cell selection parameters as well as the GSM neighbours are also broadcast in the BCH. There are up to 32 GSM neighbour cells and, if not properly defined, GSM cannot be reselected. Consequently, definition of the GSM neighbour cells becomes a key procedure in the system optimisation. Note that, if the mobile loses the UTRAN coverage and the reselection has not been made because the cell reselection parameters are not properly configured, the mobile would start an initial cell selection procedure, looking for other networks and RATs. Then, it could eventually connect to GSM, but after a long period without service.

As an example, let us assume that a mobile terminal enters a building or turns a corner, thus experiencing a sudden reduction in the received signal strength. In such a case, if sRATsearch is set too low, GSM cells may not have been measured yet and, consequently, the service is lost.

The cells that are considered in the ranking for the cell reselection procedure are those that fulfil the signal and quality requirements in UTRAN, while for GSM cells only the signal level criterion based on the C1 parameter is considered [4], given by:

$$\text{C1} = \text{A} - \max(\text{B}, 0) > 0 \tag{5.50}$$

where A depends on the averaged received signal level RLA in dBm and the minimum required received signal level RxLev_Access_Min:

$$\text{A} = \text{RLA} - \text{RxLev\_Access\_Min} \tag{5.51}$$

and B depends on the maximum power that is allowed by the GSM network in the access procedure MS_Txpwr_Max_CCH and the maximum power available at the terminal P:

$$\text{B} = \text{MS\_Txpwr\_Max\_CCH} - \text{P} \tag{5.52}$$

The RxLev_Access_Min parameter is usually set to the minimum possible value of −110 dBm, which corresponds to a typical mobile sensitivity. This reduces the probability of losing service (i.e. C1 < 0) and avoids, for example, roaming terminals in foreign countries re-selecting another operator because they lost service with the operator they had initially selected.

The cell ranking for GSM cells together with UTRAN cells is done according to the R criterion, which defines two different values for each cell depending on whether they are the server (*s*) or a neighbour (*n*):

$$R(s) = \text{Qmeas}(s) + \text{Qhyst}(s) \tag{5.53}$$

$$R(n) = \text{Qmeas}(n) - \text{Qoffset}(s, n) \tag{5.54}$$

where, for UTRAN cells, Qmeas can be the CPICH RSCP if the ranking is based on signal level or the CPICH Ec/No if the ranking is based on quality level. In turn, for GSM cells, Qmeas is the averaged received signal level RLA. In order to avoid continuous cell changes, some hysteresis parameters are considered: Qhyst for the measurement of the server and Qoffset for the measurement of the neighbour n with respect to the server s.

Taking the above into account, the cell ranking is executed in two steps.

1. Initially, the ranking according to the R criterion is based on signal level (i.e. Qmeas for UTRAN is the CPICH RSCP). This allows comparing UTRAN and GSM cells under the same measurement type, because for GSM cells the R criterion is always based on the signal level. Then, if a GSM cell is the first one in the ranking, the mobile reselects to GSM.
2. If the best cell is not a GSM cell, the mobile stays in UTRAN and the specific UTRAN cell is selected according to the R criterion based on CPICH RSCP or on CPICH Ec/No depending on the parameter qualMeasQuantity, which is broadcast by the network. Therefore, the mobile camps on the selected cell and performs the cell update or URA update procedures in each case, depending on the specific RRC state.

When hierarchical cell structures are used in the network, the H criterion is used initially in order to restrict the list of cells that are ranked afterwards according to the R criterion. This allows some preference to the selection of certain layers over others. For specific details, the reader is referred to Reference 33.

Mobile Camped on a GSM Cell   When the mobile is camped on a GSM cell, a criterion is also introduced to avoid it continuously executing UTRAN measurements. In particular, the parameter Qsearch_I, broadcast in GSM, indicates whether UTRAN measurements should be performed when the received signal level RLA of the GSM serving cell is below or above a given threshold, as shown in Table 5.10 [4][35].

The UTRAN neighbours to be measured must be defined by the carrier frequency and the scrambling code. There can be up to 32 UTRAN neighbours per carrier frequency and up to three carriers. The information about which UTRAN neighbours to measure is given in the message System Information 2ter, which is broadcast by the network. As a matter of fact, this message has been specifically included in GSM specifications to cope with UTRAN measurements [35].

The mobile executes a ranking of the UTRAN cells together with the server and neighbour GSM cells. So that a UTRAN cell is included in the ranking, the following condition has to be fulfilled:

$$CPICH\ Ec/No > FDD\_Qmin \tag{5.55}$$

where FDD_Qmin was defined (up to version 8.17 of GSM specification TS 05.08) as a value between 0 and 7 corresponding to a minimum Ec/No between -20 and $-13\,dB$ in steps of 1 dB. However, these

Table 5.10   Values of Qsearch_I parameter

Qsearch_I	Threshold to measure UTRAN	Qsearch_I	Threshold to measure UTRAN
0	$<-98\,dBm$	8	$>-78\,dBm$
1	$<-94\,dBm$	9	$>-74\,dBm$
2	$<-90\,dBm$	10	$>-70\,dBm$
3	$<-86\,dBm$	11	$>-66\,dBm$
4	$<-82\,dBm$	12	$>-62\,dBm$
5	$<-78\,dBm$	13	$>-58\,dBm$
6	$<-74\,dBm$	14	$>-54\,dBm$
7	Always measure UTRAN	15	Never measure UTRAN

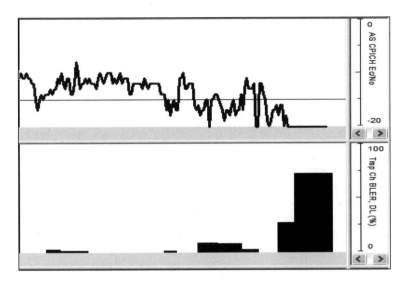

Figure 5.64   Example of Ec/No and DL BLER time evolution for a voice service

values appeared to be somewhat low because the typical quality values for a good UTRAN connection are around -15 dB. This is illustrated in Figure 5.64, which presents the temporal evolution of the Ec/No and downlink BLER for a voice service. It can be observed that the BLER increases very significantly when the measured Ec/No is below -15 dB. Consequently, a request was made to increase the range of FDD_Qmin and, from version 8.18 of GSM TS 05.08, the corresponding minimum Ec/No values were changed (Table 5.11).

After carrying out the ranking, the mobile terminal will camp on the UTRAN cell if this cell has remained in the first position of the ranking for 5 s, which means that the following condition must be fulfilled for this cell:

$$\text{CPICH RSCP} > \text{RLA(s} + \text{n)} + \text{FDD\_Qoffset} \tag{5.56}$$

where $\text{RLA(s} + \text{n)}$ represents the maximum received power of any GSM cell and FDD_Qoffset applies an offset for cell reselection involving a RAT change. This offset takes values between 0 and 15 where $0 = -\infty$ threshold (i.e. always select a UTRAN cell if it is acceptable), $1 = -28$ dB, $2 = -24$ dB, and

Table 5.11   Minimum Ec/No values for each FDD_Qmin value

FDD_Qmin	Ec/No min
0	-20 dB
1	-6 dB
2	-18 dB
3	-8 dB
4	-16 dB
5	-10 dB
6	-14 dB
7	-12 dB

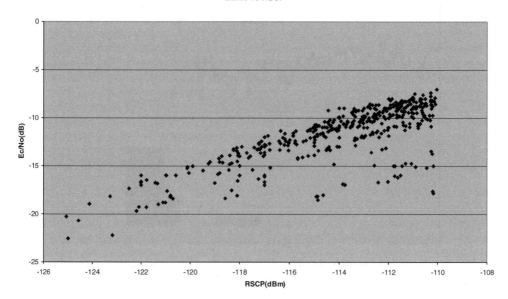

Figure 5.65   Relationship between Ec/No and RSCP measurements

so on in steps of 4 dB until the value 15, corresponding to 28 dB. Furthermore, if during the previous 15 s another cell reselection has occurred, the FDD_Qoffset is increased in 5 dB.

A new access condition based on the RSCP level has been included in GSM specification TS 05.08 v8.20 for cell reselection from GSM to UMTS. The reason is that the Ec/No used in the previous criterion corresponds to a measurement which does not take into account the received signal level. Consequently, even if the Ec/No is good, it is possible to have a low RSCP or equivalently a high path loss that in this case would degrade performance mainly in the uplink direction. An example showing the correspondence between Ec/No and RSCP measurements is plotted in Figure 5.65 for a voice user located at the cell edge. Note that in all cases, the RSCP is very low but in some situations the reported Ec/No may reach values even higher than −10 dB. In such a case, if the mobile moves to UTRAN based only on Ec/No and then the quality is not enough, it may move again to GSM, causing a ping-pong effect between both systems. Therefore, the additional condition to be fulfilled in order to select a UTRAN cell is:

$$\text{CPICH RSCP} \geq \text{FDD\_RSCP\_threshold} \tag{5.57}$$

5.3.3 HANDOVER ALGORITHMS

The handover is a critical procedure in any cellular mobile communications system, since it allows the transferring of on-going calls from one cell to another in a transparent way for the user, thus enabling user mobility in the coverage area. Therefore, handover is executed for those users having dedicated resources in any cell.

The decision to execute a handover procedure is mainly based on measurements reported by the UE regarding the downlink and measurements done by the network regarding the uplink as well as the overall system status. The handover decision is not necessarily based only on radio conditions, and it can be related to other reasons, such as controlling the traffic distribution between the different cells. To this end, there is a set of different handover-related parameters that can be configured by the operator in the different cells to control the handover decision.

There are several possible causes that might trigger the handover procedures: the quality in the uplink or downlink in terms of BER or BLER; the received signal level in either uplink or downlink; the distance to the serving node B; a change of service; or it may even result from a decision of congestion control strategies.

In a scenario where several RATs coexist, as is the case for UMTS and GSM/GPRS systems, the handover may be done either by changing the cell within the same RAT, denoted as Intra-RAT or horizontal handover, or by changing the RAT, which is denoted as Inter-RAT or vertical handover. In the following sections both cases are further developed.

Within UTRAN, two types of Intra-RAT handover can be distinguished, denoted as soft handover and hard handover. The soft handover occurs when the mobile is connected simultaneously to several cells and benefits from diversity by combining the signals received from each one. The set of cells to which the mobile is simultaneously connected is denoted as Active Set, and the different cells are progressively added and removed from this set depending on user mobility. When the user is simultaneously connected to different sectors of the same site, a special case of soft handover occurs, denoted as softer handover.

Soft handover is only possible when the involved cells operate with the same carrier frequency. However, when the mobile must hand over between cells operating with a different carrier frequency, the only possibility is to execute a hard handover, in which the mobile is only connected to one cell at a time. Consequently, the handover procedure involves removing the connection with the old cell and establishing the connection with the new one. In any case, hard handover may also occur between cells operating at the same frequency, for example, because the active set size is limited to one cell.

5.3.3.1 General Steps in a Handover Procedure

A handover procedure involves the following functions [23]:

(a) Realisation of measurements. Measurements are carried out by the mobile terminal in the list of neighbouring cells specified in the Cell_Info_List [34], which is broadcast by the network in the System Information Blocks Type 11 and Type 12. It can also be sent in the RRC Measurement Control message that configures the measurements for a specific terminal. The main fields in the Cell_Info_List are the following:

- Intra-frequency Cell Info. This contains the list of cells to be measured operating at the same carrier frequency of the cell to which the mobile is connected. For each cell, the following information is provided:

  - Primary CPICH scrambling code.
  - Primary CPICH transmitted power.
  - Indication of whether diversity transmission is applied or not.
  - Timing difference between the primary CCPCH of the current cell with respect to the neighbouring cell.
  - Cell individual offset: an offset between -10 and $10\,dB$ in steps of $0.5\,dB$ is added over the measured quantity. This parameter can be used to ensure the terminal is more likely to connect to some specific cells (i.e. the higher the offset the higher the probability that one cell is selected). By default this parameter is $0\,dB$, which means that all the cells are treated equally.
  - Cell selection and reselection info. This contains the different parameters Qrxlevmin, Qqualmin, Pcompensation and Qoffset explained in Section 5.3.2 used in the cell selection and reselection procedures.

- Inter-frequency Cell Info. This contains the list of UTRAN cells to be measured that are operating at a different carrier frequency. The information provided is the same as in the Intra-frequency Cell info in addition to the UARFCN of the carrier frequency.

- Inter-RAT Cell Info. This contains the list of cells belonging to other RATs that should be measured and reported. For the case of GSM cells, the following information is provided:

  - Cell selection and reselection info. This includes the offsets Qoffset to be applied in the R criterion and the RxLev_Access_Min and MS_Txpwr_Max_CCH parameters for measuring C1 parameter, as explained in Section 5.3.2.
  - Frequency of the BCCH, given by the ARFCN (Absolute Radio Frequency Channel Number).
  - BSIC (Base Station Identity Code).

Measurements should be averaged and adequately reported by the UE. In addition to the averaging done at the physical layer during a measurement period, it is possible to execute a L3 filtering at the RRC layer of the UE according to the following expression [34]:

$$F_n = (1 - a)F_{n-1} + aM_n \qquad (5.58)$$

where F_n is the resulting filtered measurement in the current measurement period n, F_{n-1} is the previous filtered measurement and M_n is the current measurement provided by the physical layer. The weight a is defined as:

$$a = \frac{1}{2^{k/2}} \qquad (5.59)$$

where k is the *filter coefficient* configured in the RRC Measurement Control message. When $k = 0$, no L3 filtering must be carried out. L3 filtering is not applied in cell reselection.

The reporting is done by the RRC Measurement Report message [34]. The measurement reporting may be configured either on a periodical or an event-triggered basis. This is controlled by the *Reporting Interval* parameter in the RRC Measurement Control message, which specifies the time between consecutive reports. The number of reports that must be provided is controlled by the *Amount of Reporting* parameter (a value of infinity means that measurements must be reported indefinitely).

When *Reporting Interval* takes the value 0, measurements are configured on an event-triggered basis. In this case, a set of events are defined, each of them with a specific triggering condition. If the condition is fulfilled during a certain period, a report is transmitted indicating the corresponding event. In some cases, it is possible to repeat the report more than once, for example, because the RNC has not been able to process it. This is done by activating the periodical reporting after the triggering condition has been fulfilled for a given event. In this case, the periodical reporting is stopped if the triggering condition is no longer fulfilled.

(b) Handover algorithm. Based on the measurements reported by the mobile terminal, the network carries out the handover algorithm to decide the specific instant at which a soft or hard intra-RAT or an inter-RAT handover must be executed.

(c) Handover execution. Once a handover has been decided, the execution phase involves the required signalling procedures to activate or remove radio links in the different cells or RATs according to the appropriate RRC procedure depending on the type of handover [34].

In the following sections, specific details about the different handover types in UTRAN are presented.

5.3.3.2 Intra-frequency Handover

The intra-frequency handover in UTRAN may be either soft or hard. The different UTRAN cells operating with the same carrier may belong to three following mutually exclusive sets, specific at each terminal:

- Active Set: includes the cells to which the mobile is simultaneously connected.
- Monitored Set: includes the cells that the terminal measures and reports but that are not included in the Active Set. Therefore, they constitute the candidate cells to be added to the Active Set.
- Detected Set: includes the rest of the cells that the terminal is able to detect (i.e. the measured RSCP and the Ec/No are above specific thresholds) but are not included in the Cell_Info_List. Therefore, no measurement reports are provided for these cells.

The intra-frequency handover controls how the different cells are added or removed from the active set based on the measurement reports, which can include the CPICH RSCP, the CPICH Ec/No and/or the path loss for each cell. Note that by establishing a maximum active set size of one cell, the intra-frequency handover turns into hard handover, while if the active set size is higher than one cell, soft handover is used.

In general, it is recommended to use the CPICH RSCP when the network load is low, while the use of CPICH Ec/No is especially indicated when the load is high or when the network is not well optimised.

When measurements are event-triggered, the occurrence of the different events depends on how measurements are configured in terms of the measured quantity to be reported (i.e. CPICH Ec/No, CPICH RSCP or path loss) and on specific thresholds, which can be absolute (i.e. an absolute threshold is broadcast by the network) or relative to the best cell of the active set (i.e. a reporting range is broadcast by the network).

The set of events is:

- Reporting event 1A: a primary CPICH enters the reporting range.
- Reporting event 1B: a primary CPICH leaves the reporting range.
- Reporting event 1C: a non-active primary CPICH becomes better than an active primary CPICH.
- Reporting event 1D: change of best cell.
- Reporting event 1E: a primary CPICH becomes better than an absolute threshold.
- Reporting event 1F: a primary CPICH becomes worse than an absolute threshold.

The measurement reports are triggered when the previous events occur during a certain period of time denoted as time-to-trigger, also configured by the network. Furthermore, different hyteresis margins can be considered for both the absolute and relative thresholds. The use of the time-to-trigger and the hysteresis avoids sporadic fluctuations in the measurements leading to unnecessary decisions regarding the active set update.

Based on the above events, the decisions regarding when a cell should be added or removed from the active set are taken according to a specific soft handover algorithm that is implementation dependent. The Active Set Update RRC procedure is then used to execute the decisions. The procedure is initiated by the *Active Set Update* message sent by the RNC and confirmed with the *Active Set Update Complete* message issued by the mobile.

We now examine two examples of soft handover algorithms based on relative and absolute thresholds.

Soft Handover Algorithm Based on Relative Thresholds   A possible relative threshold soft handover algorithm based on the previous events 1A, 1B and 1C has been defined [23]. It operates according to the following parameters:

- R_{1A} (dB): reporting range to be considered in event 1A (cell addition to the active set).
- R_{1B} (dB): reporting range to be considered in event 1B (cell removal from the active set).
- H_{1A} (dB): hysteresis to be considered in event 1A (cell addition to the active set).
- H_{1B} (dB): hysteresis to be considered in event 1B (cell removal from the active set).
- H_{1C} (dB): hysteresis to be considered in event 1C (cell replacement).
- β: weight factor.
- ΔT: time-to-trigger.
- AS_Max_Size: maximum size of the active set.

Let us assume that the measured quantity configured by the network is the CPICH Ec/No and that Nc is the current number of cells in the active set for a given terminal. Depending on the measurements provided by this terminal when the different events occur, the following actions can be taken by the algorithm:

(a) Add cell i to the Active Set. This action is executed when the active set is not full (i.e. $Nc <$ AS_Max_Size) and event 1A is triggered according to:

$$10 \log \left(\frac{E_c}{N_0}\right)_i > (1 - \beta) 10 \log \left(\frac{E_c}{N_0}\right)_{bestAS} + \beta 10 \log \left(\sum_{j=1}^{Nc} \left(\frac{E_c}{N_0}\right)_j\right) - \left(R_{1A} - \frac{H_{1A}}{2}\right) \tag{5.60}$$

and, during the following period ΔT, it is fulfilled that:

$$10 \log \left(\frac{E_c}{N_0}\right)_i > (1 - \beta) 10 \log \left(\frac{E_c}{N_0}\right)_{bestAS} + \beta 10 \log \left(\sum_{j=1}^{Nc} \left(\frac{E_c}{N_0}\right)_j\right) - \left(R_{1A} + \frac{H_{1A}}{2}\right) \tag{5.61}$$

where $(Ec/No)_{bestAS}$ is the Ec/No of the best cell in the Active Set.

Factor $\beta(0 < \beta < 1)$ is used to weight the contribution of the different cells belonging to the active set in the triggering condition. When $\beta = 0$, only the best cell is considered. Note that when β increases, it is more difficult to add a new cell in the active set.

(b) Remove cell i from the Active Set. This action is executed when event 1B is triggered according to:

$$10 \log \left(\frac{E_c}{N_0}\right)_i < (1 - \beta) 10 \log \left(\frac{E_c}{N_0}\right)_{bestAS} + \beta 10 \log \left(\sum_{j=1}^{Nc} \left(\frac{E_c}{N_0}\right)_j\right) - \left(R_{1B} + \frac{H_{1B}}{2}\right) \tag{5.62}$$

and, during a period ΔT, it is fulfilled that:

$$10 \log \left(\frac{E_c}{N_0}\right)_i < (1 - \beta) 10 \log \left(\frac{E_c}{N_0}\right)_{bestAS} + \beta 10 \log \left(\sum_{j=1}^{Nc} \left(\frac{E_c}{N_0}\right)_j\right) - \left(R_{1B} - \frac{H_{1B}}{2}\right) \tag{5.63}$$

Note that in this case, when β increases, it is easier to remove a cell from the active set.

(c) Replace the worst cell of the active set with the best cell in the monitored set. This action is executed when the active cell is full (i.e. $Nc =$ AS_Max_Size) and event 1C is triggered according to:

$$10 \log \left(\frac{E_c}{N_0}\right)_{bestMS} < 10 \log \left(\frac{E_c}{N_0}\right)_{worstAS} + \frac{H_{1C}}{2} \tag{5.64}$$

and, during a period ΔT, it is fulfilled that:

$$10 \log \left(\frac{E_c}{N_0}\right)_{bestMS} < 10 \log \left(\frac{E_c}{N_0}\right)_{worstAS} - \frac{H_{1C}}{2} \tag{5.65}$$

where $(Ec/No)_{bestMS}$ is the Ec/No of the best cell in the monitored set and $(Ec/No)_{worstAS}$ is the Ec/No of the worst cell in the active set.

An example of the operation of this algorithm is shown in Figure 5.66 assuming an active set with a maximum of two cells (i.e. AS_Max_Size $= 2$) and $\beta = 0$.

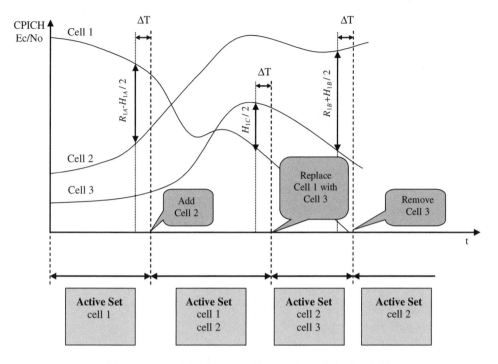

Figure 5.66   Soft handover algorithm based on relative thresholds

With respect to reporting range R_{1A}, used to add cells to the active set, typical values are between 3 and 5 dB. In turn, R_{1B}, used to remove cells from the active set, takes typical values between 5 and 8 dB. In general, when the network has been accurately optimised and the load is high, low values are recommended for these parameters, in order to reduce the number of users in soft handover and therefore the resource consumption. This is particularly important when high bit rate services are considered (e.g. 384 kb/s), since they are high power-demanding services and they consume high fractions of the OVSF code tree. However, when the load is low or when the network is not yet optimised, higher values of R_{1A} and R_{1B} are preferred, in order to benefit from macrodiversity and make combining gains in regions with low received signal levels.

In turn, typical values for hysteresis factors H_{1A}, H_{1B}, and H_{1C} are between 1 and 2 dB, in order to cope with signal fluctuations, and the time-to-trigger parameter is usually set between 200 and 500 ms.

In terms of the maximum number of cells in the active set, a typical value is 3. Higher values turn into excessive resource consumption because of a high number of users in soft handover. As a matter of fact, the planning procedure should try to avoid locations with more than three cells that could be included in the active set, because this means that the overlapping between cells is excessive. This situation is known as *pilot pollution*, and it can be controlled by adjusting the antenna downtilt of the different cells, as illustrated in Figure 5.67.

Soft Handover Algorithm Based on Absolute Thresholds   Another possible soft handover algorithm inspired by the one presented in Reference 7 makes use of events 1E and 1F operating with absolute thresholds. Let us assume again that the measured quantity is the CPICH Ec/No and that the following parameters apply:

- T_ADD (dB): absolute threshold to be considered by event 1E.
- T_DROP (dB): absolute threshold to be considered by event 1F.

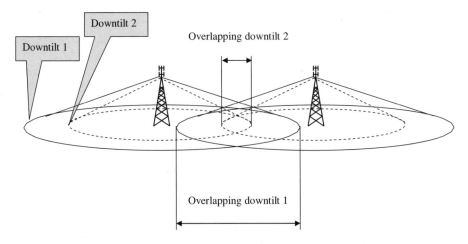

Figure 5.67   Effect of antenna downtilt over the cell overlapping

- H_A (dB): hysteresis margin with event 1E.
- H_D (dB): hysteresis margin with event 1F.
- H_R (dB): replacement hysteresis.
- ΔT: time to trigger.
- AS_Max_Size: maximum size of Active Set.

In this case, the following actions are taken by the algorithm:

(a) Add cell i to the Active Set. This action is executed when there are less than AS_Max_Size cells in the active set and event 1E is triggered according to:

$$10 \log(Ec/No)_i > T\_ADD + H_A/2 \tag{5.66}$$

and, during a period ΔT, it is fulfilled that:

$$10 \log(Ec/No)_i > T\_ADD - H_A/2 \tag{5.67}$$

(b) Remove cell i from the Active Set. This action is executed when event 1F is triggered according to:

$$10 \log(Ec/No)_i < T\_DROP - H_D/2 \tag{5.68}$$

and, during a period ΔT, it is fulfilled that:

$$10 \log(Ec/No)_i < T\_DROP + H_D/2 \tag{5.69}$$

(c) Replace the worst cell of the active set with the best cell in the monitored set. This action is executed when there are AS_Max_Size cells in the active set and event 1E is triggered according to the following two conditions:

$$10 \log(Ec/No)_{bestMS} > T\_ADD + H_A/2 \tag{5.70}$$

$$10 \log(Ec/No)_{bestMS} > 10 \log(Ec/No)_{worstAS} + H_R/2 \tag{5.71}$$

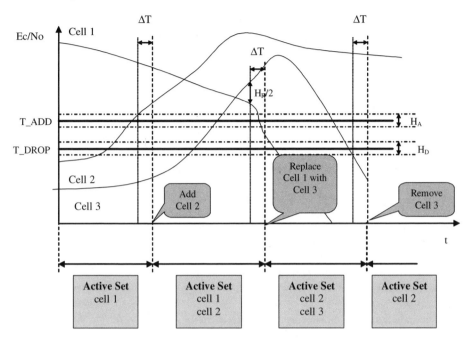

Figure 5.68   Soft handover algorithm based on absolute thresholds

and, during a period ΔT, it is fulfilled that:

$$10 \log(Ec/No)_{bestMS} > T\_ADD - H_A/2 \tag{5.72}$$
$$10 \log(Ec/No)_{bestMS} > 10 \log(Ec/No)_{worstAS} - H_R/2 \tag{5.73}$$

where $(Ec/No)_{bestMS}$ is the Ec/No of the best cell in the monitored set and $(Ec/No)_{worstAS}$ is the Ec/No of the worst cell in the Active Set.

Figure 5.68 shows an example of this algorithm with a maximum active set size of two cells.

5.3.3.3 Inter-frequency Handover

The inter-frequency handover is executed in UTRAN in two different scenarios:

1. Cells with multiple carriers. The most immediate way to increase the cell capacity is by including new carriers in the current nodes B. Then, upon a service request, if the admission condition does not hold in the carrier to which the mobile is connected, it may be possible to establish the service in a different carrier. Similarly, when some carriers are more loaded than others, it is possible to handover some connections from one carrier to another in order to balance the load among carriers.
2. Hierarchical Cell Structures (HCS). Hierarchical cell structures including macro, micro and picocells have been widely deployed in 2G networks in order to increase the capacity by using cells with reduced sizes. However, in the case of WCDMA networks, the high overlapping that exists between layers, which causes an excess of intercell interference, normally requires that each layer operates with a different carrier. In this case, an inter-frequency handover may be required to transfer traffic

from one layer to another, trying to tie up the service requirements with the performance that each layer can offer. In that sense, the lowest range cells (e.g. indoor picocells) would typically provide high bit rate services to users with reduced mobile speed while users with higher speeds, and necessarily lower bit rates, would be allocated to higher layers such as micro or macrocells.

These mechanisms require the support of measurements in the different carriers, provided by both the network and the mobile terminal, in order to decide the appropriate instant at which to execute an inter-frequency handover. When the mobile is asked to execute inter-frequency measurements on a specific set of cells, it reports the CPICH Ec/No, RSCP and path loss in addition to synchronisation information and observed time differences between radio frames of the different cells. The following events are considered for event-triggered inter-frequency measurement reporting [34]:

- Reporting event 2A: change of the best frequency.
- Reporting event 2B: the estimated quality of the currently used frequency is below a certain threshold and the estimated quality of a non-used frequency is above a certain threshold.
- Reporting event 2C: the estimated quality of a non-used frequency is above a certain threshold.
- Reporting event 2D: the estimated quality of the currently used frequency is below a certain threshold.
- Reporting event 2E: the estimated quality of a non-used frequency is below a certain threshold.
- Reporting event 2F: the estimated quality of the currently used frequency is above a certain threshold.

As in the intra-frequency measurements, hysteresis can be considered for the different thresholds, together with the time to trigger mechanism.

The inter-frequency handover is executed by means of the RRC Hard Handover procedure, removing all the radio links in the current active set and establishing a new active set with the cells of the new frequency.

5.3.3.4 Handover from UTRAN to GSM

In the case of dual UTRAN/GSM terminals, it is possible to execute an inter-RAT handover between UTRAN and GSM. This can be necessary due to the lack of UMTS coverage or to specific operator policies regarding how different services are mapped in the available radio access technologies.

For dual UTRAN/GSM terminals, the RNC configures not only the soft handover events explained in Section 5.3.3.2, but also the conditions necessary to start measurements in GSM according to the thresholds for the events 2D and 2F explained in Section 5.3.3.3. These events specify the minimum levels for the current UTRAN network to initiate or stop GSM measurements by means of the threshold sRATsearch (see page 243). The GSM neighbouring cells to be measured are also indicated by the network.

Depending on how the threshold and parameters are defined, it is very likely that GSM is re-selected, which may turn into a high number of inter-RAT handovers. When the operator prefers to keep the mobile camped on the UTRAN cell by adjusting radio parameters, there are several options:

- Define a low value for the sRATsearch threshold to avoid mobiles measuring GSM. However, if the value is set too low, it might be possible that the mobile loses UTRAN coverage and no GSM cell is measured, thus losing service. Consequently, the setting of this parameter is a trade-off between the number of handovers to GSM and the probability that the service is lost for the mobile.
- Introduce a certain offset in the GSM cells measurements in order to delay the decision of interRAT handover.

In any case, the definition of the GSM cells to be measured is critical in order to have an adequate inter-RAT handover that avoids as much as possible the loss of service situation.

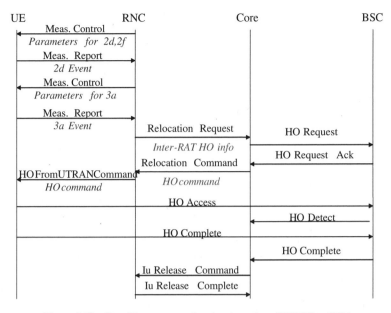

Figure 5.69   Signalling messages in a handover from UTRAN to GSM

The monitoring of GSM cells is controlled by the following events, which are defined for inter-RAT measurement event-triggered reporting [34]:

- Reporting event 3A: the estimated quality of the currently used UTRAN frequency is below a certain threshold, and the estimated quality of the other system is above a certain threshold.
- Reporting event 3B: the estimated quality of the other system is below a certain threshold.
- Reporting event 3C: the estimated quality of the other system is above a certain threshold.
- Reporting event 3D: change of best cell in the other system.

When a handover to GSM is decided, based on the received measurements, the signalling procedure to be executed is as shown in Figure 5.69. The initial messages reflect the configuration that is done by the RNC to configure events 2D and 2F by means of the Measurement Control message. When the quality of UTRAN is below the specified threshold, according to event 2D, the terminal sends the corresponding *Measurement Report* with event 2D and the RNC configures event 3A for inter-RAT measurements. The handover starts when event 3A is reported by the mobile. Then, the RNC sends a *Relocation Request* to the Core Network (i.e. to the MSC), which communicates with the BSC of the GSM access network to set-up the corresponding GSM resources in the requested cell. When the new resources are ready to use, a *Handover from UTRAN* command is sent to the mobile, which initiates the transmission in the corresponding GSM frequency and time slot by means of the *HO Access* message. The procedure is completed by releasing the corresponding resources in the Iu interface that connected the RNC with the MSC.

A relationship should exist between the thresholds to trigger events 2D, 2F and 3A. In particular, event 2D determines the instant to start GSM measurements, which usually involves the use of compressed mode. Therefore, low values of the threshold associated with this event are suitable in order to not trigger the compressed mode unnecessarily, which could cause a certain performance degradation. In any case, this threshold must be above the threshold of event 3A, which specifies the instant when the handover to

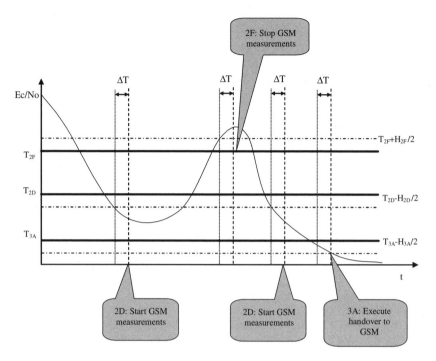

Figure 5.70 Handover from UTRAN to GSM

GSM must start, in order to have GSM measurements available when this occurs. Furthermore, a certain separation between both 2D and 3A associated thresholds is convenient, because a certain delay may exist from the instant when GSM measurements are triggered to the reception of the first measurement, due to the compress mode activation.

Event 2F specifies the instant when GSM measurements are stopped, consequently, a certain separation is necessary between the 2D and 2F associated thresholds to avoid signal variations leading to continuous activations and deactivations of the GSM measurements.

The threshold for event 3A should be set high enough to avoid the mobile losing UTRAN coverage during the handover procedure due to sudden signal degradations. Furthermore, it should be set in accordance with the threshold that triggers the handover from GSM to UTRAN. If both thresholds are very similar, undesirable ping-pong effects between the systems is likely to occur.

Figure 5.70 illustrates an example of the triggering of the different events 2D, 2F and 3A. The considered thresholds are T_{2D}, T_{2F}, T_{3A}, respectively, and the corresponding hysteresis margins are H_{2D}, H_{2F}, H_{3A}.

Directed Retry is a special procedure related to the inter-RAT handover from UTRAN to GSM [36]. It occurs when, during a connection establishment carried out through UTRAN, the RNC decides that the service should be allocated to the GSM network (e.g. because the existing UTRAN load is too high). In this case, the signalling procedure is depicted in Figure 5.71. Note that the RNC answers the RAB Assignment Request message of the RANAP protocol, indicating that directed retry should be executed, and starts the relocation procedure. Then, in the CN, the 3G MSC communicates with the corresponding 2G MSC in order to prepare the inter-RAT handover. After the preparation of the handover in the CN, an RRC *Handover from UTRAN* command is issued to the terminal, thus triggering the procedures to release the UTRAN resources (e.g. signalling radio bearer, radio links, etc.) and to establish the new radio link with the BSC (Base Station Controller) in GSM. The procedure completes with the release of the resources in the Iu interface.

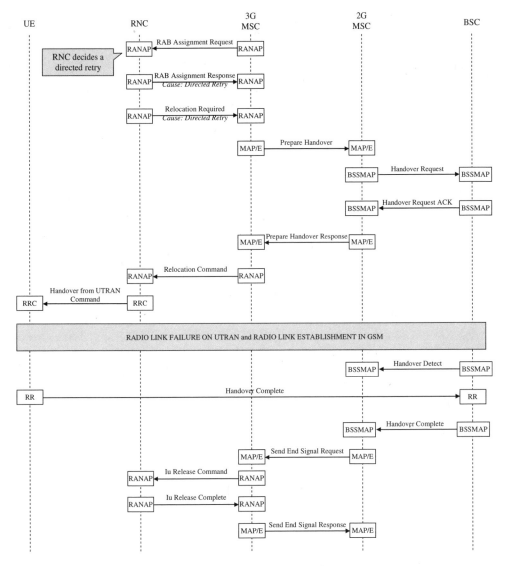

Figure 5.71   Directed retry procedure

5.3.4 NEIGHBOUR CELL LIST DEFINITION

A neighbour cell list is defined for each UTRAN cell. It is referred to as Neighbour Set (NS), and is composed of up to 32 intra-frequency cells, 32 inter-frequency cells and 32 cells from other RATs (e.g. GSM).

As explained in Section 5.3.3, in order to take handover decisions, each terminal must measure and report the cells included in the active set and the monitored set. This list of cells to measure is built from the NS of the cells that are in the active set, up to a maximum of 32, ruling out the repeated ones. The list is configured by the network and sent to mobiles in connected mode in the Cell_Info_list of the RRC measurement control message. Note that when the union of the different neighbouring sets contains more than 32 cells, some of the neighbours will not be measured.

However, the terminal may detect other intra-frequency cells that are not included in either the active or the monitored sets. They constitute the detected set, and it may happen that one of these cells belongs to the NS but has not been included in the monitored set due to the lack of space. In such a case, it is possible to configure the mobile reports so that they alter the cells of the detected set by means of a proper configuration of the triggering condition of events 1A and 1C, allowing the possibility of triggering by cells of the detected set. Nevertheless, under no circumstance it is possible for a cell of the detected set to be added to the active set if it is not included in the NS of the current cells in the active set.

Note that this has an important implication on the interference level. This is because if one mobile detects one cell and it cannot be added to the active set, the mobile will transmit a high power level as long as it enters into this cell's coverage area. This generates a high interference level to other users, which can degrade the quality and the network capacity, as was illustrated in Section 5.2 in the context of admission control. In order to avoid this situation, a proper configuration of the neighbouring cell list is required.

It is advisable not to define a large number of neighbour cells, in order to avoid some cells being included in the monitored set. In that sense, between 15 and 20 is usually acceptable. This requires an appropriate network optimisation of the cell coverage.

It is important to configure events 1A and 1C so that the detected cells are reported, to determine if the received signal level from a cell not included in the NS has exceeded a certain threshold. In such a case, and in order to avoid the excessive interference due to the impossibility of executing handover, it is possible to force the dropping of a call.

With respect to the GSM neighbour cells definition, it is advisable to define other GSM cells apart from those that are co-sited with the corresponding UTRAN cell. The reason is that, in some cases with low load, a mobile terminal could measure a high UTRAN Ec/No even though it is located far from the cell site, thus having a low received signal level from the co-site GSM cells. In such a case, if there is a degradation in the Ec/No (e.g. because of a load increase), the mobile can not switch to the GSM cells because of their low signal strength and the service would be lost.

It is also recommended to define a reduced number of GSM neighbouring cells, because the higher the number of cells to measure, the lower the measurement resolution.

With respect to terminals that are not in connected mode, thus executing cell reselection procedures, the list of cells to measure is indicated in the Cell_Info_List sent through the broadcast channel. This list is not necessarily the same one that is configured for the terminals in connected mode. In any case, the cells included for measurements in connected mode should also be included in the list used for cell reselection.

5.4 CONGESTION CONTROL ALGORITHMS

The traffic generation processes associated with different mobile users are random and statistically independent among them. In cellular radio, mobility and propagation conditions add even more randomness. Due to these uncertainties, it is possible that during some periods overload situations exist in the uplink and/or in the downlink direction, which could prevent some users from achieving their expected requirements due, for example, to an increase in the uplink load factor or in the downlink transmitted power. Congestion or load control mechanisms should be devised to tackle these situations in which the system has reached an overload status and, consequently, the QoS guarantees of the admitted users are at risk due to the evolution of system dynamics.

In contrast to the congestion problem in fixed networks, where a large number of studies have been published [37][38], and despite radio network congestion being a widely recognised and identified problem, not many specific solutions and algorithms for such a wireless environment are available in the open literature [39][40], and few of them are well aligned with 3GPP specifications [41–44]. Also, a general framework for executing load control combined with the admission control phase has been in Reference 45.

When considering different services, the more stringent the required QoS the tighter the Radio Resource Management for that service. In the case of conversational services, since they require a channel of constant quality and very short delay constraints, overload situations may cause a direct impact on performance in terms of, for example, call dropping. However, interactive services require neither channels of constant quality nor tight delay constraints, so that overload situations may have a limited impact and, consequently, the need for congestion control mechanisms may be different depending on whether or not conversational services are involved.

Furthermore, in the case of interactive services, the packet-like traffic generation adds more randomness to the instantaneous load conditions and, therefore, WCDMA scenarios with only interactive traffic will exhibit a natural tendency to auto-recover from overload situations, while at the same time the degradation during overloads may be quite acceptable due to the nature of the considered service, which allows executing retransmissions. To illustrate this effect, Figure 5.72 presents the uplink cell load factor

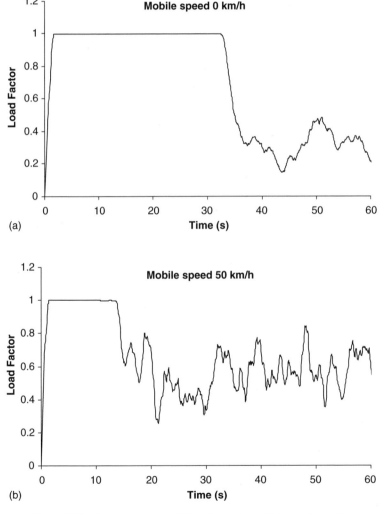

Figure 5.72   Auto-recovery capability of scenarios with interactive traffic

in a situation where a heavy load level of interactive traffic is considered with seven omnidirectional cells of radius 500 m. The www browsing model presented in Appendix A5.3.3 is considered with an average time between pages of 30 s, an average interarrival packet time of 0.125 s and an average of 25 packets per page. The radio bearer allows a maximum instantaneous bit rate of 64 kb/s, and corresponds to the one presented in Section 3.6.2. The maximum UE power is 21 dBm. Figure 5.72(a) shows the case with mobile speed 0 km/h and Figure 5.72(b) corresponds to the 50 km/h case. No congestion control mechanisms are applied, and the average offered load is the same throughout the simulation. It can be observed that at the beginning of the simulation, a high number of users start their sessions, so that the load level increases. This means that the BLER will be higher than the target value and that more packets than normal will need to be retransmitted, with corresponding delay degradation. However, users close to the node B will be able to successfully transmit their packets and eventually reach inactivity periods. Therefore, if some sources turn off, even users far from the node B will be able to successfully transmit their packets. Additionally, users will move around the network in the 50 km/h case and, consequently, in general all users will experience some periods where packets can be correctly received. As a result, after about 10 s the cell load eventually decreases or, in other words, the randomness associated with the scenario has recovered by itself this overload situation. It can also be seen that, even for stationary users (0 km/h), the cell load decreases after about 30 s when the users closer to the node B have successfully carried out their transmissions. Of course, if the load was even higher, this auto-recovery would be more difficult and would last longer, so that taking it to the extreme, a congestion control mechanism would be in any case necessary.

Despite the self-capabilities to bear overload situations when only interactive users are considered, the meaningfulness of the congestion control itself should be considered for such a scenario. That is, if no congestion control actions are taken, the overload will cause packet retransmissions and a delay degradation will follow. This can be particularly problematic for certain connections where the number of retransmissions is high. In this case, congestion control may provide benefits from a connection-by-connection point of view. To illustrate this effect, a static user has been placed far from node B in the previous simulations and the user's statistics in terms of average delay are reported in Table 5.12. Note that, when no congestion control is applied, the average packet delay experienced by the static user is much worse than the average cell performance. However, if a congestion control mechanism like the one that will be presented later in this section is applied, the resulting performance is fairer with this user and the delay is closer to the cell average value. Note, in any case, that the application of congestion control has an impact on the overall cell performance in the sense that the admission probability is decreased and the overall delay is increased because of the actions taken to decrease the uplink cell load factor. These degradations will be dependent on the specific congestion control algorithm and their parameters, as will be explained within this section.

The previous results reveal that in scenarios with only interactive traffic, and due to the robustness of this traffic to handle overload situations, the interest of a congestion control algorithm will depend on the improvement that can be achieved from an individual connection point of view with respect to the degradation that is introduced to the overall performance. However, in scenarios where there is a mix of conversational and interactive users, the situation changes since in this case conversational users are very sensitive to overload situations. To illustrate this effect, Table 5.13 presents the performance obtained in a simulation with both types of traffic. Note that when no congestion control is applied, the BLER of

Table 5.12 Performance figures that devise the need for congestion control mechanisms with only interactive traffic

	Average delay of the static user	Overall average delay	Prob $(\eta > 0.75)$	Admission probability
Without congestion control	1.07 s	0.13 s	0.22	100.0%
With congestion control	0.41 s	0.26 s	0.09	75.6%

Table 5.13 Performance figures that devise the need for congestion control mechanisms with interactive and conversational traffic

	Admission conversational	Admission interactive	BLER conversational	Delay interactive
Without congestion control	100%	100%	3.69%	0.15 s
With congestion control	100%	47%	1.18%	0.65 s

conversational users experiences a significant increase over the target value of 1%, which may be unacceptable for the service quality. In this context, the application of a congestion control strategy may reduce the BLER at the expense of increasing the average delay and decreasing the admission probability of interactive traffic.

In any case, note that the application of a congestion control algorithm always involves introducing some degradations to certain users in order to improve the performance of others. Consequently, the specific algorithm setting must be related to the operator policies with respect to the importance they give to some services or user types compared to others. Taking this into account, this section will devise the principles for congestion control operation and will provide the impact of the different parameters over the performance of the different traffic types.

5.4.1 GENERAL STEPS OF A CONGESTION CONTROL ALGORITHM

The congestion control algorithm must operate to ensure that overload situations can be recovered. This involves the execution of the following steps [46]: detection, resolution and recovery.

5.4.1.1 Congestion Detection

This step intends to determine if the system has reached a congestion situation in order to trigger the correspondent actions in the congestion resolution phase. In the uplink direction, the usual parameter that is monitored to detect congestion is the uplink load factor while in the downlink it is the transmitted power level. Therefore, a criterion to decide that a given node B has entered congestion would be that the uplink load factor or the downlink transmitted power exceeds a given threshold during a certain period of time, for example, if $\eta_{UL} \geq \eta_{CD}$ in uplink (or $P_T \geq P_{CD}$ in downlink), during a certain percentage p of the frames within a period ΔT_{CD}.

Note that the network is planned to operate below a certain maximum load factor $\eta_{\max}$ so the congestion detection threshold η_{CD} should be set in accordance with this maximum planned value. Similarly, the period to decide congestion, ΔT_{CD}, plays a key role, since when it is set too low, the congestion resolution may be triggered too often, even under sporadic load factor fluctuations that could be easily solved without the need of specific actions, thus degrading the performance of certain users unnecessarily. However, high values of the detection period ΔT_{CD} will tend to delay the congestion resolution, which may degrade the performance of those users more sensitive to overload situations.

5.4.1.2 Congestion Resolution

After detecting congestion in the previous step, the congestion resolution algorithm executes a set of actions to lead the system out of the congestion status and avoid system instability. Three general steps are in turn identified in this procedure.

(a) Prioritisation. The actions taken by a congestion resolution algorithm necessarily involve introducing some degradations in the performance observed by certain connections to achieve an overall system improvement or avoid degradations being experienced by certain types of services or users. As a result, the congestion resolution algorithm has to establish some priorities between the different connections depending on the suitability of taking actions over each of them. This process will lead to an ordered list where at the top will be the connections with less priority, which will be the first ones whose performance will be degraded by the congestion control actions.

Several criteria can be considered to build this prioritisation list. They can take into account operator policies (e.g. consumer segment will normally have lower priority than business segment), quality requirements (e.g. interactive users will normally have lower priority than conversational users since their QoS is less stringent) and other technical criteria (e.g. in the case of the downlink, congestion could be caused by certain users that, exceptionally, consume a high amount of power and therefore it becomes beneficial to take the first resolution actions over these users).

(b) Load reduction. This process takes the appropriate actions to reduce either the uplink load factor in the cell or the downlink transmitted power. The following types of actions are considered:

- Blocking new connections requesting admission. The blocking can be executed for all service classes or only in a selective way, for example, by letting some high priority users or traffic classes enter the system even during congestion.
- Limiting the transmission capabilities of certain connections according to the prioritisation list that has been built in the previous step. This is done by limiting the TFCS, which specifies the maximum bit rate that can be used by a certain connection, thus reducing the maximum level of interference generated by this connection. Note that this solution is valid for both the uplink and downlink direction with users operating in DCH channels. The RRC protocol reconfiguration procedures [34], including Transport Channel reconfiguration or TFC reconfiguration, can be used for this purpose (see Section 3.5.3.2). Similarly, transport channel type switching procedures, which involve moving certain users operating in DCH channels to RACH/FACH channels, can also be used. In some cases, it may even be necessary to drop certain connections, for example, if in the downlink one user is exceptionally consuming an excessive amount of power.
- Forcing a handover to certain mobiles. In this case, an inter-frequency handover would usually be the preferred solution, since an intra-frequency handover would still keep the interference generated by the mobile. By means of this strategy, load balancing between different carrier frequencies is possible.

(c) Load check. The load reduction actions will be executed progressively until the load is below a certain threshold (e.g. $\eta_{UL} \leq \eta_{CR}$ in the uplink direction or $P_T \leq P_{CR}$ in the downlink direction). Therefore, it can be decided that the congestion situation has been overcome if, during a certain percentage p of frames within an observation period ΔT_{CR}, the measured load is below the specific congestion resolution threshold η_{CR} or P_{CR}.

5.4.1.3 Congestion Recovery

Once the congestion resolution phase has decided that the congestion situation has been overcome, a congestion recovery algorithm is needed to restore to the different mobiles the transmission capabilities they had before the congestion was triggered (unless, of course, the resolution action has been to drop some calls, which is an irrecoverable decision). It is worth mentioning that such an algorithm is crucial because, depending on how the recovery is carried out, the system could fall again in congestion. Again, RRC protocol reconfiguration procedures will be used to increase the TFCS or to execute transport channel type switching.

This procedure will continue until all the users that have been affected by the resolution mechanisms have restored their transmission capabilities.

Figure 5.73 illustrates the overall steps of the general congestion control procedure. In the following sections, the impact on the system performance of different parameters and algorithms in this procedure will be analysed.

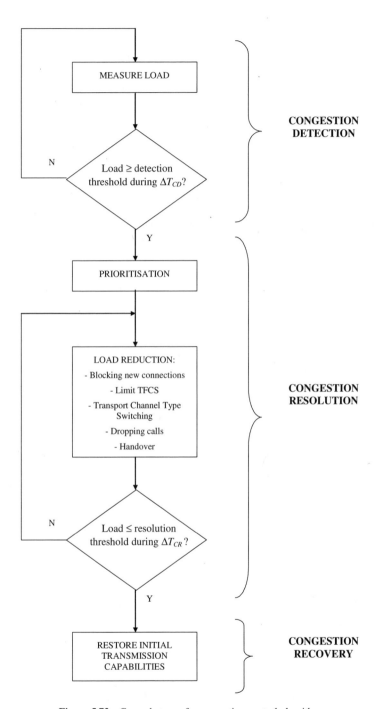

Figure 5.73   General steps of a congestion control algorithm

Table 5.14 Reference performance measurements without congestion control

	BLER conversational (%)	BLER interactive (%)	Delay interactive (s)	Average η	Prob. $(\eta > 0.75)$
Case A	2.40	5.67	0.14	0.78	0.58
Case B	2.84	7.26	0.14	0.82	0.66
Case C	3.69	9.74	0.15	0.86	0.76

5.4.2 CONGESTION RESOLUTION STRATEGIES

This section presents some relevant aspects that affect the performance obtained with different load reduction methods used in the congestion resolution phase. To this end, some simulations are presented for the uplink case assuming a mix of interactive and conversational users in a scenario with seven omnidirectional cells and a radius of 500 m. Conversational users transmit at a constant bit rate of 64 kb/s and interactive users have variable rate transmission with possible bit rates of 0 kb/s, 16 kb/s, 32 kb/s, 48 kb/s and 64 kb/s, according to the radio bearer presented in Section 3.6.2. The system is planned to operate with a load factor of 0.75. The following load situations are considered, corresponding to three different case studies:

- Case Study A: low conversational load, medium interactive load.
- Case Study B: medium conversational load, medium interactive load.
- Case Study C: low conversational load, high interactive load.

As a reference, Table 5.14 presents the obtained performance in the network when no congestion control is considered. In order to decouple the relationship between admission and congestion control, no admission is considered in these results. Note also that in all the cases, the average load factor is higher than the planned value of 0.75 and consequently the BLER of the conversational and interactive users exceeds the target value of 1%. However, this is critical only for conversational users since interactive users have a margin for retransmissions, so they simply experience a certain increase in delay.

When congestion control policies are adopted, the expected effects are a BLER reduction for conversational users, an average packet delay increase for interactive users (because congestion control reduces their bit rate) and, where requests are blocked during congestion periods, a reduction of the admission probability. Taking this into account, two possibilities of load reduction are now explored: executing a selective blocking during congestion and using different TFCS reduction algorithms.

5.4.2.1 Blocking New Calls

Blocking the entrance of new users in the cell is one of the possibilities usually considered to reduce the load after triggering the congestion resolution algorithm. This blocking can affect all the incoming requests or only selected services. Table 5.15 compares the performance obtained with the following two policies [23]:

(a) Block conversational. All the new incoming requests, either conversational or interactive, are blocked during congestion periods.
(b) No block conversational. Only interactive requests are blocked during congestion periods, but conversational users may enter the system.

In all cases, it has been assumed that $\Delta T_{CD} = 0.1$ s, $\Delta T_{CR} = 1$ s, $\eta_{CD} = 0.9$, $\eta_{CR} = 0.75$ and that there is a percentage of time $p = 90\%$ to trigger the different events. Similarly, the high reduction and slow recovery algorithms that will be explained in Sections 5.4.2.2 and 5.4.3, respectively, are considered.

Table 5.15 Impact of blocking conversational calls

	Admission conversational (%)	Admission interactive (%)	BLER conversational (%)	BLER interactive (%)	Delay interactive (s)
Case A					
No block Conv.	100	58	1.18	1.62	0.68
Block Conv.	80	58	1.17	1.67	0.66
Case B					
No block Conv.	100	48	1.20	1.67	1.11
Block Conv.	74	51	1.25	1.72	1.07
Case C					
No block Conv.	100	50	1.24	1.71	0.93
Block Conv.	76	52	1.20	1.70	0.86

Note in Table 5.15 that, compared to Table 5.14, a significant reduction in the BLER is achieved in all cases due to the congestion control algorithm. The improvement is achieved at the expense of a lower admission probability and a higher delay for interactive traffic. In any case, it can be observed that a significant gain is obtained in terms of conversational users' admission probability if conversational users are not blocked during congestion periods. In turn, neither the admission probability for interactive users nor the BLER presents significant variations due to the acceptance of new conversational users during congestion periods. Only the delay of interactive users appears to experience a certain increase due to a slightly higher duration of the resolution periods when conversational users are not blocked. Therefore, as a general comment, it seems to be advisable not to block conversational users during congestion periods in order to preserve this type of traffic.

5.4.2.2 TFCS Limitation Algorithms

The influence of the specific TFCS limitation algorithm during the load reduction process is explored through the two following possibilities:

(a) High reduction algorithm. This algorithm is illustrated in Figure 5.74. In particular, the algorithm acts over the different interactive users according to the order established during the prioritisation phase. For each user i, the maximum transport format $TF_{max,i}$ is reduced by one, thus decreasing the user's bit rate. Afterwards, the algorithm estimates the load factor that would result from the reduction and, if it is still higher than the desired threshold η_{CR}, it continues with the reduction of the same user, until reaching the limit where the maximum transport format is 0 and, thus, the transmission is inhibited for this user. If the estimated load factor is still higher than the threshold, the algorithm will reduce the TFCS of the next user. Note that the corresponding transport channel reconfiguration message with the new TFCS is only sent once the reduction process has ended for each user. The effect of this algorithm will be to execute high TFCS reductions to a reduced number of users, in most cases inhibiting their transmission.

(b) Low reduction algorithm. This algorithm is illustrated in Figure 5.75. In this case, it executes for each user in the prioritised list a reduction of one in the maximum transport format and then it moves to the next user in the table in case that the estimated load factor is still higher than the desired threshold η_{CR}. As a result, this algorithm will lead to a higher number of users affected by the TFCS reductions than the previous algorithm, but the bit rate reduction will be smaller for each user. Similarly, note that a transport channel reconfiguration message will be required each time that the maximum transport format is reduced by one.

Note that both algorithms act progressively, which means that in a given execution a set of users are reduced until the estimated load factor is below the threshold. Therefore, when the real load factor after the reduction is still below the threshold, new reductions will be executed.

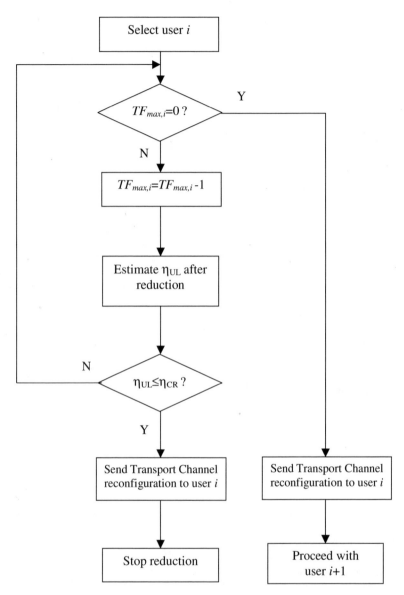

Figure 5.74   High reduction algorithm

Table 5.16 shows the performance results obtained with both algorithms for the three considered case studies. The thresholds $\eta_{CD} = 0.75$, $\eta_{CR} = 0.6$, together with $\Delta T_{CD} = 0.1$ s, $\Delta T_{CR} = 1$ s, and a percentage of time $p = 90\%$ to trigger the different events, have been considered. The recovery algorithm is the slow recovery that will be presented in Section 5.4.3, and conversational calls are not blocked during congestion periods, thus having an admission probability of 100% for conversational users (not shown in the table). It can be observed that the differences between both algorithms are small although the high reduction algorithm seems to reduce the load factor to slightly smaller average values. It should

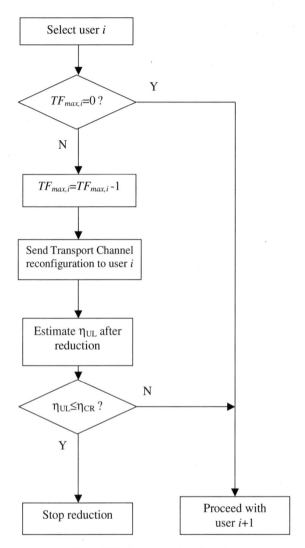

Figure 5.75 Low reduction algorithm

Table 5.16 Impact of the TFCS reduction algorithm

	Admission interactive (%)	BLER conversational (%)	BLER interactive (s)	Delay interactive (s)	Average η	Prob. $(\eta > 0.75)$
Case A						
High	44	1.08	1.29	1.34	0.48	0.10
Low	43	1.11	1.29	1.38	0.50	0.11
Case B						
High	29	1.16	1.31	2.71	0.57	0.16
Low	29	1.12	1.29	2.57	0.59	0.18
Case C						
High	39	1.08	1.30	1.57	0.46	0.10
Low	37	1.08	1.30	1.55	0.51	0.13

be taken into account that both strategies are devised to reduce the same amount of load, which explains why not many differences are perceived from an overall point of view. In any case, it is worth mentioning that the high reduction algorithm requires a somewhat lower signalling due to a smaller number of transport channel reconfiguration messages.

5.4.3 CONGESTION RECOVERY ALGORITHMS

Once it has been decided that the congestion has ended, the congestion recovery algorithm is responsible for restoring the initial transmission capabilities of the mobiles whose bit rate has been reduced in the resolution phase. The algorithm will build a prioritised list of the users that have to be restored. Possible prioritisation criteria would be, for example, the QoS requirements of each connection or the amount of time that a connection has been inhibited. We now explore two different possibilities for increasing the bit rates of the connections in the list:

(a) Slow recovery. This algorithm operates on a user-by-user approach. That is, initially a specific user is again allowed to transmit at the maximum rate. Once this user has emptied the transmission buffer that contained all the packets that had not been allowed to transmit during the congestion, another user recovers the maximum rate and so on, until finishing the list of users. Note that this algorithm can be regarded as a 'time scheduling' algorithm, since the buffers of the users will be emptied one-by-one. In this way, the recovery process is carried out progressively, avoiding sudden interference increases.

(b) Fast recovery. This algorithm increases the transmission bit rate of the users in the list on a frame-by-frame basis. This means that in each frame a new user is allowed to recover its initial transmission capabilities. In this case, the recovery is done in a faster way than with the previous algorithm, but at the risk of having interference increases that could easily return the system to congestion.

Once all the users have been restored, new connections will also be accepted in the system.

Table 5.17 presents the performance results obtained with the two recovery algorithms in the three considered case studies of the previous sections. The congestion detection and resolution thresholds are set to $\eta_{CD} = 0.75$, $\eta_{CR} = 0.6$ together with $\Delta T_{CD} = 0.1$ s, $\Delta T_{CR} = 1$ s and a percentage of time $p = 90\%$ to trigger the different events. Conversational users are not blocked during congestion periods and the low reduction algorithm is considered for congestion resolution. It can be clearly observed that the slow recovery algorithm retains lower values of the average load factor, which becomes a lower BLER for both conversational and interactive services. Also, the average delay of interactive services is smaller with the slow recovery algorithm. The reason is that with the fast recovery algorithm, because of the increase in interference generated when restoring the initial maximum bit rates, it is very likely that the system will fall again into congestion. However, with the slow recovery algorithm, the increase in the

Table 5.17 Impact of the congestion recovery algorithm

	Admission interactive (%)	BLER conversational (%)	BLER interactive (s)	Delay interactive (s)	Average η	Prob. ($\eta > 0.75$)
Case A						
Fast	57	1.11	1.34	1.46	0.53	0.15
Slow	43	1.11	1.29	1.38	0.50	0.11
Case B						
Fast	41	1.16	1.40	3.10	0.65	0.25
Slow	29	1.12	1.29	2.57	0.59	0.18
Case C						
Fast	49	1.16	1.42	2.20	0.55	0.18
Slow	37	1.08	1.30	1.55	0.51	0.13

interference is done in a more controlled way, at the expense of having higher recovery periods that turn into a reduced admission probability for interactive users.

5.4.4 SETTING OF CONGESTION CONTROL PARAMETERS

This section tries to identify the role played by some of the parameters in the congestion control algorithm. To this end, both the observation periods and the congestion thresholds are studied. It should be mentioned that, on the one hand, the setting of these parameters is in some cases dependent on the operator policies with respect to the priorities that are desired for different service classes. On the other hand, the large degree of coupling that exists among the different parameters and algorithms makes a thorough analysis that takes into account all the possible situations difficult. Consequently, the purpose of the following sections is just to identify the main trends that can be derived from the behaviour of each parameter, without trying to provide an optimum setting for all of them.

5.4.4.1 Observation Periods

The observation periods ΔT_{CD} and ΔT_{CR} determine the reaction capability of the algorithm to trigger the execution of the congestion resolution and congestion recovery steps. The congestion detection period ΔT_{CD}, which indicates the time that the load factor should be above the congestion threshold η_{CD}, should be set to relatively short values, to ensure that the algorithm has a fast reaction to the congestion situation. Values below 1 s appear to be adequate for this parameter. In any case, and taking into account that in some cases measurements may exhibit certain fluctuations, it is beneficial to introduce a tolerance margin by specifying that a minimum of $p\%$ of the samples in period ΔT_{CD} must fulfil the detection condition. This tolerance avoids the situation where a low number of samples with $\eta < \eta_{CD}$ within a measurement period indicate that the cell is not in congestion even though $\eta > \eta_{CD}$ during most of the period.

With respect to the congestion resolution observation period ΔT_{CR}, Table 5.18 shows the performance results obtained for three different values of the parameter, namely two low values such as $\Delta T_{CR} = 0.1$ s and $\Delta T_{CR} = 1$ s and a high value such as $\Delta T_{CR} = 10$ s. Case study B (see Section 5.4.2) has been considered. The congestion detection and resolution thresholds are set to $\eta_{CD} = 0.75$, $\eta_{CR} = 0.6$ together with $\Delta T_{CD} = 0.1$ s and a fraction $p = 90\%$ to trigger the different events. High reduction and slow recovery algorithms are considered, without blocking conversational users during congestion.

It can be said that a safe congestion resolution observation period of $\Delta T_{CR} = 10$ s severely penalises the admission rate of interactive users. Note that a high value of ΔT_{CR} makes system-declared congestion situations last longer. If conversational users were also blocked during the congestion period, then a dramatic reduction of the conversational admission probability would also follow. In addition, a safe congestion resolution period of $\Delta T_{CR} = 10$ s severely penalises the average packet delay of interactive traffic, because interactive users are restricted in the transmission rate capabilities during longer periods. At the same time, this period is able to keep the conversational BLER closer to its target value. However, a short congestion resolution period of $\Delta T_{CR} = 0.1$ s provides higher admission rates and a lower average interactive packet delay at the expense of a higher conversational BLER. This BLER increase arises because the congestion situations are not as efficiently controlled. For instance, it might be decided after $\Delta T_{CR} = 0.1$ s that the congestion has been overcome but, after a short period, the algorithm might

Table 5.18 Impact of congestion resolution observation period

	Admission Interactive (%)	BLER Conversational (%)	BLER Interactive (s)	Delay Interactive (s)
$\Delta T_{CR} = 0.1$ s	45	1.22	1.52	0.82
$\Delta T_{CR} = 1$ s	29	1.16	1.31	2.71
$\Delta T_{CR} = 10$ s	14	1.06	1.18	12.05

well trigger congestion once again. Therefore, the final setting of the parameter should be related to operator-related policies, depending on the maximum BLER and delay degradations that can be accepted for conversational and interactive users.

5.4.4.2 Detection and Resolution Thresholds

The congestion detection threshold η_{CD} determines the maximum allowed load level before triggering the congestion resolution procedure. Therefore, it should be set in accordance with the maximum planned load for the corresponding scenario to avoid, as far as possible, the instantaneous load increasing over this planned value. Similarly, the congestion resolution threshold η_{CR} determines the minimum load level for deciding that a congestion situation has been overcome, so it must also be set bearing in mind the maximum planned load.

The difference between the two thresholds corresponds to a hysteresis margin to ensure that the congestion resolution algorithm has brought the load to a low enough level to avoid, in a short period of time, congestion again being triggered. The larger the difference between both thresholds, the lower the probability that the system again enters into congestion after a resolution procedure. However, if the difference is too large, it will be more difficult for the load reduction algorithm to achieve the desired load level and the process will last longer.

Table 5.19 shows the performance measurements that are obtained for two different sets of thresholds, assuming a planned value for the uplink load factor of 0.75, depending on whether the maximum planned value of 0.75 equals the detection or the resolution threshold. In the first set, with $\eta_{CD} = 0.75$ and $\eta_{CR} = 0.6$, the load factor is not allowed to increase more than 0.75 and a more stringent requirement of 0.6 is set for the resolution phase. In the second set, with $\eta_{CD} = 0.9$ and $\eta_{CR} = 0.75$, the detection is more permissive, allowing an increase up to 0.9 before taking actions, when the resolution must ensure that the load factor goes below the planned value of 0.75. Case study B has been considered with the high reduction and slow recovery algorithms. $\Delta T_{CD} = 0.1$ s and $p = 90\%$ have been assumed with different observation periods ΔT_{CR} for the congestion resolution process.

Table 5.19 reveals that, since $\eta_{CD} = 0.9$ and $\eta_{CR} = 0.75$ constitute a late congestion trigger compared to $\eta_{CD} = 0.75$ and $\eta_{CR} = 0.6$, it is found that the conversational BLER degrades more in the former case, with BLER = 1.14%, than in the later, with BLER = 1.06%. This is because when the system triggers congestion, the cell load has already remained at high values for a certain period of time and this has caused some erroneous transmissions. On the other hand, this late detection avoids some interactive users being blocked and, consequently, a higher interactive admission probability is found. In addition, for the late detection case of $\eta_{CD} = 0.9$ and $\eta_{CR} = 0.75$, a lower interactive average packet delay is obtained. This is because the interactive delay is more degraded due to the congestion control actions, which restrict the maximum bit rate, rather than because of packet retransmissions in overload conditions. So, delaying congestion actions is beneficial for interactive traffic. Consequently, and as occurs with the observation periods, the setting of the congestion thresholds will also depend on the maximum degradation that can be tolerated for the considered services in each scenario.

Table 5.19 Impact of congestion detection and recovery thresholds

		Admission interactive (%)	BLER conversational (%)	BLER interactive (s)	Delay interactive (s)
$\eta_{CD} = 0.75$, $\eta_{CR} = 0.6$	$\Delta T_{CR} = 0.1$s	45	1.22	1.52	0.82
	$\Delta T_{CR} = 10$s	14	1.06	1.18	12.05
$\eta_{CD} = 0.9$, $\eta_{CR} = 0.75$	$\Delta T_{CR} = 0.1$s	61	1.34	2.13	0.39
	$\Delta T_{CR} = 10$s	30	1.14	1.41	4.49

5.4.5 MULTI-CELL CONGESTION CONTROL ALGORITHM

Congestion control algorithms can also be devised from a multi-cell perspective in order to capture the coupling that exists between the different cells operating at the same frequency. This allows the reduction of the load in a given reference cell by executing congestion actions over those cells having a major impact on the considered cell. In particular, this can be beneficial when service distribution is not homogeneous in the scenario and, for instance, cells exist that have a higher number of less priority users that allow a better handling of the resolution procedure.

In multi-cell scenarios, the derivatives framework presented in Section 4.6.2 can be useful for devising a multi-cell congestion control algorithm since the load factor and power derivatives provide a clear indication of the cell having the highest impact over the reference cell [47]. A possible multi-cell congestion resolution approach based on this framework is now presented for both the uplink and downlink directions, assuming, without loss of generality, a scenario with conversational and interactive services similar to the previous sections.

5.4.5.1 Uplink Direction

Let us assume a scenario with K cells and let cell 0 be the cell that has triggered congestion resolution. In practice, the cells $k = 1, \ldots, K - 1$ would correspond to the list of neighbouring cells broadcast by cell 0 and operating at the same UTRAN carrier. The derivative-based congestion resolution algorithm operates in the following steps to restore the load factor in cell 0, η_0, below the desired threshold value η_{CR}:

1. Select the cell to execute the load reduction. Assuming that the interactive traffic has less priority than the conversational users, the cell selection must take into account the amount of load reduction that can be achieved in the reference cell 0 by limiting the transmission capabilities of interactive users in the different cells. In particular, if cell k has an amount of load factor devoted to interactive traffic $\eta_{I,k}$, the maximum load reduction achieved in cell 0, if all the interactive users in cell k were inhibited, can be estimated as:

$$\Delta\eta_{\max,k} = \eta_{I,k} \cdot \frac{\partial\eta_0}{\partial\eta_k} \qquad (5.74)$$

Therefore, the selected cell would be the one having the maximum value of $\Delta\eta_{\max,k}$.

2. Compute the required load reduction to be carried out in the selected cell k to achieve the desired load factor η_{CR} at cell 0. In particular, for a $\Delta\eta_0 = \eta_0 - \eta_{CR}$ reduction at cell 0, the reduction in cell k can be estimated as:

$$\Delta\eta_k = \frac{\Delta\eta_0}{\left(\dfrac{\partial\eta_0}{\partial\eta_k}\right)} \qquad (5.75)$$

3. Prioritisation of the interactive users in the selected cell k. In this case, apart from the criteria related to the QoS requirements of each user, a suitable criterion would be to order the users taking into account the influence that the user has over the load in cell 0, which can be measured using the factor $I_{i_k,0}^{UL}$, extracted from Equation (4.37):

$$I_{i_k,0}^{UL} = \frac{L_{i_k,k}}{L_{i_k,0}} \frac{1}{\dfrac{W}{\left(\dfrac{E_b}{N_0}\right)_{i_k} R_{b,i_k}} + 1} \qquad (5.76)$$

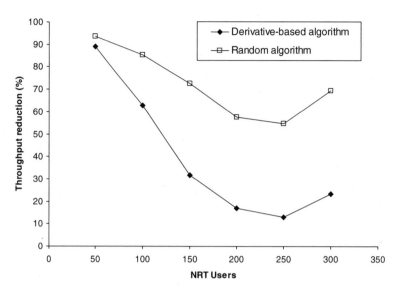

Figure 5.76   Uplink throughput reduction in the neighbouring cells as a function of the number of interactive NRT users

where i_k is the ith user of the k cell, with its corresponding bit rate R_{b,i_k} and requirement $(E_b/N_0)_{i_k}$. $L_{i_k,k}$ is the path loss of the user with respect to its serving cell k and $L_{i_k,0}$ is the path loss with respect to the reference cell. Note that the higher the factor $I_{i_k,0}^{UL}$, the higher the influence of the user over the reference cell 0, and therefore it is reasonable to start the reduction process with the users having higher values of this factor.

4. Reduce the maximum bit rate of the users in cell k until reaching the desired reduction $\Delta\eta_k$ or until having inhibited all the interactive users.

5. Measure load factor in cell 0, η_0, and if it is still higher than η_{CR} return to step 1.

Figure 5.76 shows some performance measurements of the multi-cell algorithm based on the derivative framework in a scenario where the objective is to have the load factor of the reference cell below 0.8. 23 cells have been considered in the scenario. Interactive users have a maximum bit rate of 64 kb/s. A non-homogeneous scenario has been considered, in which two neighbouring cells have twice the interactive traffic compared to the rest of the cells. For comparison purposes, another multi-cell congestion resolution algorithm is considered, in which the cell to be reduced is selected randomly. Figure 5.76 presents the total interactive throughput reduction required in the load reduction process with the two algorithms as a function of the number of interactive users. Note that the algorithm based on derivatives is more efficient and ends the load reduction process with the lowest throughput reduction. It is worth mentioning that the efficiency of the derivative-based algorithm increases in scenarios with non-homogeneous spatial traffic distribution, thanks to the ability of the gradient to identify the cells having the highest influence over the reference cell.

5.4.5.2 Downlink Direction

A similar multi-cell load reduction algorithm can be devised for the downlink direction. In this case, it is assumed that the objective of the algorithm is to limit the power fraction with respect to the maximum available power (i.e. P_{T0}/P_{Tmax0}) of the reference cell to a given bound ϕ_{CR}. Therefore, the steps of the algorithm would be as follows:

1. Select the cell to execute the load reduction. As in the uplink direction, the amount of interactive load in each cell should be considered together with the correspondent derivatives. In particular, let $P_{I,k}$ be the total power devoted to interactive users in cell k. Then, the maximum power reduction that can be achieved in cell 0 when all the interactive users in cell k are inhibited is given by:

$$\Delta P_{max,k} = P_{I,k} \cdot \frac{\partial P_{T0}}{\partial P_{Tk}} \tag{5.77}$$

Consequently, the cell with the highest $\Delta P_{max,k}$ will be selected.

2. Compute the required power reduction in the selected cell k, which will be given by:

$$\Delta P_{Tk} = \frac{\Delta P_{T0}}{\left(\dfrac{\partial P_{T0}}{\partial P_{Tk}}\right)} \tag{5.78}$$

where ΔP_{T0} is the desired reduction in cell 0, given by:

$$\Delta P_{T0} = P_{T0} - \phi_{CR} \cdot P_{T\,max\,0} \tag{5.79}$$

3. Order the interactive users in cell k in decreasing order of the transmitted power.
4. Reduce the bit rate of the users until reaching the desired reduction of ΔP_{Tk} or until having inhibited the transmissions of all the users.
5. Measure $\phi_0 = P_{T0}/P_{T\,max\,0}$ and if it is still higher than ϕ_{CR} return to step 1.

Figure 5.77 presents some performance results of the multi-cell downlink algorithm based on derivatives under the same simulation conditions in the previous section, except that downlink users transmit at 384 kb/s. Again, it can be observed that the multi-cell congestion resolution algorithm based on derivatives executes the load reduction process with the lowest interactive throughput reduction, although in this particular example the differences compared to the uplink case are smaller.

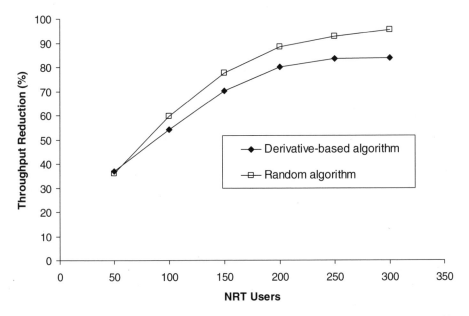

Figure 5.77 Downlink throughput reduction in the neighbouring cells as a function of the number of interactive NRT users

5.5 SHORT TERM RRM ALGORITHMS

Dynamic bandwidth allocation schemes play a relevant role in wireless systems where transmission requirements in terms of bit rate may vary over time depending on the traffic generation patterns for each service. In this sense, variable bit rate services introduce new challenges into the radio resource management problem since the allocated bit rate should match, as far as possible, the required bit rate at each time instant, particularly in the case of stringent delay constraints. Such schemes should provide an efficient use of the radio resources, taking advantage of the inherent traffic multiplexing; that is, a user should ideally get the required service as long as there are idle resources in the network. Furthermore, they should try to be fair, in the sense of guaranteeing to each user the agreed service rate, decoupled from the rest of the users. These requirements need schemes that provide a fast management of the radio resources, at the radio frame time scale, keeping in mind a near real time behaviour. Therefore, in this book, they are referred to as short term RRM algorithms, and include the MAC algorithms and the packet scheduling function for transmission in shared channels.

There is extensive literature concerning dynamic scheduling algorithms in WCDMA systems. Naghshineh and Acampora introduced resource sharing schemes for QoS guarantees into different service classes in microcellular networks [48]. Das *et al.* developed a general framework for QoS provisioning by combining call admission control, channel reservation, bandwidth reservation and bandwidth compaction [49]. However, most of the proposed uplink algorithms are centralised, like the one presented in Reference 50. Another example is the so-called WISPER protocol, which schedules the transmissions according to their BER requirements [51]. Similarly, some scheduling algorithms have been proposed that maximise the total uplink throughput by selecting the most appropriate rates for each transmission [52][53]. These types of centralised approaches allow a higher control on the transmission bit rate, leading to behaviour closer to the optimum, at the expense of requiring a higher amount of signalling to indicate the allocated bit rates. Nevertheless, current WCDMA systems like UTRAN operate in a decentralised way, since the instantaneous bit rate (i.e. the transport format combination) is selected at the MAC layer of the UE, according to the maximum bit rate that is signalled by the network, by limiting the set of allowed transport format combinations (i.e. the TFCS). This mechanism avoids the need to signal continuously the allocated bit rate since modifications of TFCS are expected to occur only sporadically (e.g. during congestion periods). In this respect, few studies aligned to the 3GPP specifications are available in the open literature that feature TFC selection algorithms at the UE-MAC layer [46][54].

With respect to the downlink direction, scheduling algorithms can be implemented more easily than in the uplink due to the inherent centralised nature of the downlink. Therefore, the scheduler may have control over all the resources available at a given instant, and both the instantaneous rate and the power can be allocated to each transmission [55]. Compared with TDMA networks, where the system can only serve a packet at a time [56], in a WCDMA network multiple packets can be transmitted by the base station simultaneously, even with different spreading factors. This raises the question of which is the best way to perform scheduling in a wireless network with multiple servers [57]. In this case, the scheduling algorithm may operate with a predominant TDMA component (i.e. allocating few simultaneous users with high bit rates) or with a predominant CDMA component (i.e. allocating a high number of users with low bit rates).

In the rest of this section, some proposed decentralised uplink algorithms and downlink scheduling algorithms in the framework of the current systems are described so as to reveal possible real solutions to the short term RRM in UTRAN FDD.

5.5.1 UPLINK UE-MAC ALGORITHMS

UE-MAC algorithms in the uplink direction aim at selecting the instantaneous Transport Format (TF) or Transport Format Combination (TFC) for a transport channel or combination of transport channels, respectively, in each TTI. Equivalently, this corresponds to selecting the instantaneous transmission bit rate autonomously at the MAC layer of each UE. The network limits the maximum bit rate that can be selected by indicating the allowed transport formats (i.e. the TFS or TFCS) during the Radio Bearer set-up procedure or as a result of, for example, a reconfiguration action taken by congestion control.

For CBR services, radio access bearers typically contain two transport formats: TF1, corresponding to transmission at a given bit rate; and TF0, corresponding to no transmission. In this case, the MAC selection is trivial and simply depends on whether the source is active or not (i.e. if the source is active and there are transport blocks in the buffer, TF1 is selected, otherwise TF0 is used). Consequently, UE-MAC algorithms are mainly targeted to variable bit rate services (e.g. interactive and background) usually with certain delay tolerance. In this case, the TF selection can be carried out according to different criteria. Some examples will now be presented.

It is assumed that the different transport formats of the TFS are ordered from TF1 (lowest bit rate) to TFmax (highest bit rate), and TF0 corresponds to no transmission. In particular, three different algorithms are described [46]:

5.5.1.1 Maximum Rate Strategy (MR Algorithm)

This is the simplest approach for UE-MAC operation, and consists of selecting the TF that allows the highest transmission bit rate according to the amount of bits to be transmitted. In particular, the number of transport blocks to be transmitted in the next TTI will be given by:

$$numTB = \min\left(TBmax, \left\lceil \frac{L_b}{TBsize} \right\rceil\right) \tag{5.80}$$

where $\lceil x \rceil$ denotes the lowest integer value higher than or equal to x and $TBmax$ corresponds to the number of transport blocks that can be transmitted with the highest transport format TFmax. $TBsize$ is the number of payload bits (i.e. without including MAC/RLC headers) per transport block and L_b is the total amount of bits waiting for transmission in the buffer. Therefore, the selected TF would be the one allowing the transmission of $numTB$ transport blocks.

Note that although this approach will tend to reduce the delay, it will also tend to increase the generated interference because the higher bit rates require the higher transmission powers.

5.5.1.2 Time-oriented Strategy (TO Algorithm)

Taking into account that certain delay bounds should be guaranteed, the possibility remains of selecting the TF that allows the transmission of the transport blocks in the buffer within a specified delay bound.

Let us assume that the maximum delay bound is TO ms, and that a total of L_b bits are to be transmitted within this delay bound. Furthermore, let's assume that $TBmax$ is the maximum number of transport blocks transmitted in a TTI, corresponding to TFmax. In order to transmit these bits in a maximum of TO ms, the number of bits to be transmitted per TTI would be:

$$L = \frac{L_b}{TO} \cdot TTI \tag{5.81}$$

So, the number of transport blocks to be transmitted in the next TTI will be:

$$numTB = \min\left(TB \max, \left\lceil \frac{L}{TBsize} \right\rceil\right) \tag{5.82}$$

$TBsize$ being the number of payload bits in a transport block for the considered RAB. The selected TF would then be the one allowing the transmission of $numTB$ transport blocks.

5.5.1.3 Rate-oriented Strategy (SCr Algorithm)

When a certain average bit rate needs to be guaranteed, a new possibility arises that makes use of the 'service credit' (SCr) concept [58] as an adaptation of token bucket algorithms used for flow control in fixed networks. The SCr of a connection accounts for the difference between the obtained bit rate and the bit rate expected by this connection. Essentially, if $SCr < 0$, this means that the connection has obtained a higher bit rate than expected, while $SCr > 0$ means that the connection has obtained a lower bit rate than expected.

In each TTI, the SCr of a connection will be updated as follows:

$$SCr(n) = SCr(n-1) + (GBR/TBsize) - TrTB(n-1) \qquad (5.83)$$

where $SCr(n)$ accounts for the service credit in the nth TTI, GBR corresponds to the number of bits per TTI that would be transmitted at the Guaranteed Bit Rate or target bit rate, $TBsize$ is the number of payload bits (i.e. without including MAC/RLC headers) per transport block for the considered RAB and $TrTB(n-1)$ accounts for the number of successfully transmitted transport blocks in the $(n-1)$th TTI.

At the beginning of the connection, the algorithm starts with $SCr(0) = 0$.

The ratio $GBR/TBsize$ reflects the mean number of transport blocks that should be transmitted per TTI according to the target bit rate. As a result, $SCr(n)$ is a measure of the number of transport blocks that have to be transmitted in the current TTI to maintain the guaranteed bit rate.

Assuming that in the buffer there are L_b bits and that $TBmax$ is the maximum number of Transport Blocks that can be transmitted per TTI, corresponding to TFmax, the number of transmitted transport blocks in the nth TTI is given by:

$$numTB = \begin{cases} \min\left(\left\lceil \dfrac{L_b}{TBsize} \right\rceil, SCr(n), TBmax \right) & \text{if } SCr(n) > 0 \\ 0 & \text{if } SCr(n) \leq 0 \end{cases} \qquad (5.84)$$

5.5.1.4 Comparison of UE-MAC Strategies

This section shows the comparison between the three UE-MAC strategies presented in the previous sections. An interactive service with a maximum uplink bit rate of 64 kb/s is considered. The TTI is 20 ms, the transport block size is 336 bits, including a MAC/RLC header with 16 bits, thus the payload is $TBsize = 320$ bits. The TFS is shown in Table 5.20. Some simulations have been executed considering a www traffic model like the one presented in Appendix A5.3.3 with a reading time between www pages of 30 s, an average interarrival time between packets in a page of 125 ms and an average of 5 pages per session. The average packet length is 366 bytes.

One important measurement to understand the behaviour of the different UE-MAC strategies is the transport format distribution being used. Furthermore, it allows the characterisation of the activity of

Table 5.20 Transport format set for the considered RAB

Transport format	Number of transport blocks	Bit rate (kb/s)	Spreading factor
TF0	0	0	N/A
TF1	1	16	64
TF2	2	32	32
TF3	3	48	16
TF4	4	64	16

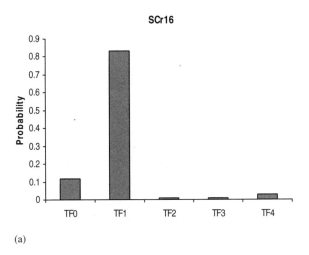

(a)

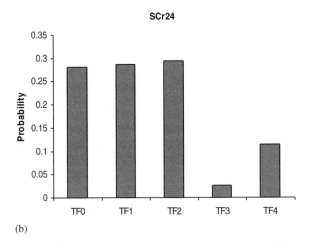

(b)

Figure 5.78   TF distribution for SCr strategy with a target bit rate of (a) 16 kb/s and (b) 24 kb/s

each traffic flow at the radio interface, which is required to do a proper admission control that takes into account UE-MAC behaviour.

Some examples of how this selection is done depending on the specific UE-MAC algorithm are given in Figure 5.78(a) and (b), for the cases SCr16, (SCrX standing for a service credit strategy with a target bit rate of X kb/s) and SCr24, respectively, Figure 5.79 for the MR strategy and Figure 5.80 for the TO12 case, where TOX stands for the delay oriented strategy with a target delay for each packet of X radio frames $(10 \cdot X \text{ ms})$. In all these cases, the distribution is only measured during activity periods of the traffic source (i.e. during www pages or packet calls), so the reading time that would lead to the selection of TF0 is not accounted for in the distributions.

It is observed that, for the SCr24 case (see Figure 5.78(b)) when transmitting, TF1 and TF2 are used most of the time because the UE buffer queues several packets and so tends to transmit the information at 24 kb/s, which falls between TF1 = 16 kb/s and TF2 = 32 kb/s as shown in Table 5.20. In turn, in the

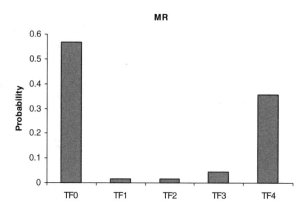

Figure 5.79   TF distribution for MR strategy.

periods when the UE buffer is empty and TF0 is selected, the UE is gaining service credits and, consequently, when a new packet arrives the instantaneous transmission rate is increased by selecting TF3 and TF4 to keep the average bit rate around the target value. For the SCr16 case (see Figure 5.78(a)), it is observed that TF1 (i.e. 16 kb/s) is most often selected, and there are very few inactivity periods at the radio interface (i.e. TF0 is seldom used). The reason is that, due to the lower target bit rate, most of the time there are packets waiting for transmission in the buffer. As a result, in very few occasions are service credits gained and TF4 is rarely used. Note that in this case the interference generated by the user will be approximately constant during the whole activity period since most of the time the same bit rate is used.

When MR strategy is applied (see Figure 5.79), UE-MAC chooses the TF according to the buffer occupancy and tries to transmit the information as fast as possible. Consequently, TF4 is used for most of the transmitting time but there are also many time periods where the radio interface is unused (i.e. TF0 is selected) because the buffer is empty. In this case, the interference pattern generated to the rest of users will alternate high interference periods associated with the highest bit rate of TF4 and periods where no interference is generated.

Finally, for the time-oriented strategy TO12 (see Figure 5.80), the TF selection is highly dependent on the instant when packets arrive at the buffer in order to achieve a total delay closer to the target value.

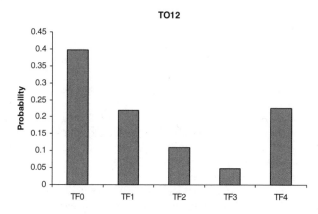

Figure 5.80   TF distribution for TO strategy with a target delay of 12 radio frames.

Table 5.21 Radio interface usage statistics for different UE-MAC algorithms

	Source activity factor	Activity factor at the radio interface	Average spreading factor
MR	0.10	0.04	18
SCr24	0.10	0.06	40
SCr16	0.12	0.12	62
TO12	0.10	0.05	36
TO18	0.10	0.06	41

When there are few packets in the buffer, the selected TF is lower (i.e. TF1 or TF0) in order to accommodate the delay to the desired value, while when the buffer contains a high number of packets, TF4 is selected to avoid long queuing delays.

By making use of the previous results and the traffic model parameters, some radio interface usage statistics can be derived that are of interest, for example, for an ulterior setting of the admission control in terms of estimating the activity factor and the average bit rate of each connection (see Section 5.2.1.3). They are presented in Table 5.21. Specifically, the provided statistics are the activity factor at a source level (i.e. the ratio between the amount of time that the source is in an activity period with respect to the total session time), the activity factor at the radio interface (i.e. the ratio between the amount of time that a TF different from TF0 is selected with respect to the total session time) and, finally, the average spreading factor that is used when transmitting. In terms of generated interference, it should be taken into account that the average transmitted power is approximately proportional to the ratio between the activity factor at the radio interface and the spreading factor, which is very similar for all the strategies, and therefore the average interference will also be very similar in all the cases. However, the interference distribution can be very different because the sources with the shortest activity periods will transmit higher instantaneous power levels.

With respect to the performance observed by different users from an overall point of view, the packet delay cumulative distribution functions for the considered strategies are presented in Figure 5.81. These

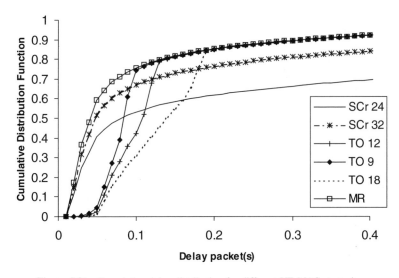

Figure 5.81 Cumulative delay distribution for different UE-MAC strategies.

Table 5.22 Delay and rate for different UE-MAC strategies.

	Average packet delay (s)	Packet delay jitter (s)	Rate per page (kb/s)	Rate per page deviation (kb/s)
SCr16	1.80	2.28	14.2	2.1
TO18	0.18	0.16	21.0	12.1
SCr24	0.54	0.95	19.0	5.0
TO12	0.16	0.16	22.1	11.0
MR	0.12	0.18	23.6	11.3

reveal that the delay distribution is quite different for TO and SCr strategies. Specifically, few packets experience low delay for TO because the strategy tends to transmit the packet information in the specified delay bound, which is reflected in the CDF by a sudden increase around the timeout value. In turn, with SCr some packets can be transmitted with a very low delay, for example, when the traffic source has been off for some time and the terminal is gaining SCr until the arrival of a new packet.

Table 5.22 presents some overall performance measurements in terms of average packet delay, jitter of the packet delay, the average bit rate along a page (i.e. the total number of bits transmitted per page with respect to total page duration) and bit rate deviation around the average value when several users are considered. It can be observed that the MR strategy provides the highest bit rate per page. In turn, TO is revealed to be quite insensitive to the specific target delay value in terms of bit rate. This is because this strategy takes into account the buffer occupancy to keep the total packet delay including buffering and transmission time around TO, which results in a lower delay jitter when compared to the SCr strategy. However, since SCr strategy does not take into account the buffer occupancy, it provides a better control of the transmission rate reflected in a low rate per page deviation.

The above results reveal that, from the user's perspective, the performance achieved with the different UE-MAC algorithms is quite different. Consequently, depending on the targets to be provided on a connection basis, one algorithm could be more suitable than another. The next point of interest is to devise whether, from the system perspective, the different possible algorithms applied at UE-MAC level lead to different interference situations. Table 5.23 compares several parameters at system level, in terms of the uplink load factor and BLER for two different load levels. Quite similar values are obtained in all cases. Thus, despite the fact that every user may apply very different patterns to the spreading factor usage and therefore to the generated interference, the fact that the system performance is the result of the average behaviour of many individual sources and of the time-varying and user-independent propagation conditions leads to the conclusion that the specific algorithm applied at UE-MAC level does not provide very different overall interference patterns in the network.

Table 5.23 System level performance for different UE-MAC strategies

	Average η (200 users)	Average η (500 users)	BLER (%) (200 users)	BLER (%) (500 users)	$P(\eta > 0.75)$ (200 users)	$P(\eta > 0.65)$ (500 users)
MR	0.21	0.53	1.18	2.40	0.03	0.22
SCr16	0.26	0.58	1.12	2.43	0.03	0.24
SCr24	0.23	0.54	1.18	2.57	0.03	0.21
TO12	0.24	0.54	1.11	2.09	0.03	0.21

5.5.2 PACKET SCHEDULING ALGORITHMS IN THE DOWNLINK

The task of the downlink scheduling is to decide the instantaneous transmissions that are carried out in the downlink direction of each node B. They are executed at the MAC layer on the network side depending on the specific transport channels that are being used.

(a) DCH. For dedicated channels, a MAC-d entity exists for each UE having a DCH allocated in a cell [59]. In this case, the scheduling function is executed among the different transport channels that are multiplexed over the same coded composite transport channel according to the specific TFCS assigned by the RRC. In principle, this function is executed independently for each dedicated channel, so similar strategies such as the UE-MAC algorithms applied in the uplink direction could be considered (see Section 5.5.1) together with priority criteria between the different transport channels.

(b) DSCH and FACH. For DSCH and FACH channels, the scheduling algorithm is executed at the MAC-c/sh entity, located at the controlling RNC for a set of nodes B. In this case, a single entity exists that receives the non real time packet flows from different users so the objective of the packet scheduling is to time and code multiplex the different flows over the available resources at the DSCH and the FACH, thus deciding for each user the transport format and the allocated code sequence in the OVSF tree depending on the spreading factor. Allocations may be changed on a radio frame-by-frame basis. Note that, since packet scheduling is executed at the RNC, decisions may be taken considering information from the users of different cells.

Some studies have been presented [60] comparing the allocation of different data services in DSCH channels or in DCH, including switching mechanisms between both transport channels. Also, a comparison between DCH and DSCH transmission, taking into account the possibility of having multiple scrambling codes, has been presented [61]. Some performance evaluation results of www browsing services over DSCH making use of a round-robin scheduling algorithm have been provided [62]. A scheduling algorithm for throughput maximisation has been proposed [63] as well as a joint power and rate adaptation algorithm [64]. Finally, performance evaluations of DSCH have been presented including signalling aspects and channelisation code tree usage [65].

(c) HS-DSCH. For HS-DSCH, the packet scheduling is executed at the MAC-hs entity, which is located at the node B. Consequently, it operates having only information about the users in one cell. In this case, the spreading factor is fixed and equal to 16, and several codes up to 15 can be allocated simultaneously. Then, the scheduling algorithm should decide which users are allocated in the available codes in each sub-frame of 2 ms.

Another differential aspect with respect to the packet scheduling in DSCH is that in HS-DSCH the modulation may be changed for each transmission between QPSK or 16-QAM depending on the channel quality, periodically reported by each mobile in the CQI (Channel Quality Information), so some link adaptation mechanisms need to be included to decide the appropriate modulation in each case. For example, when the CQI indicates that quality is bad, the preferred modulation scheme will be QPSK in order to have less signal-to-interference ratio requirements, while if the channel quality is good, the preferred modulation scheme will be 16-QAM in order to benefit from a higher transmission bit rate. The CQI can also be used to prioritise some users over others, for example, by allocating resources mainly to those users having a good channel quality.

HS-DSCH also allows the use of hybrid ARQ, which involves packet combining or incremental redundancy of successive retransmissions of a given packet. This process should be taken into account by the scheduling algorithm to decide whether a retransmission or an original packet is required by a certain user.

Some studies regarding packet scheduling algorithms for HS-DSCH have been presented [66]. Also, analysis of hybrid ARQ for HS-DSCH has been done [67–69].

5.5.2.1 Example of Packet Scheduling Algorithm for DSCH

This section describes a packet scheduling algorithm that carries out the time and code multiplexing of transmissions over the DSCH. It is clear that many possible packet scheduling algorithms could be proposed and many service mix scenarios considered. Furthermore, the scheduling algorithm may operate according to different policies, for example, maximizing the total throughput from an overall point of view [52] or trying to retain the specific requirements of the different users [70]. Nevertheless, and to gain insight into the DSCH management, a specific scheduling algorithm suitable for layered streaming video and the interactive traffic is now considered to illustrate, with a particular example, the main aspects that DSCH scheduling should take into account.

In general, a packet scheduling algorithm for transmissions in the DSCH channel operates on a frame by frame basis although each granted transmission is allowed to transmit during the whole TTI duration. The input of the algorithm will be the set of flows of users transmitting at the DSCH channel and the general procedure will be as follows.

1. Prioritisation. This step consists of ordering the different users' requests for DSCH transmission (i.e. the different flows of transport blocks to be transmitted for each user) according to specific criteria that should take into account service differentiation (i.e. priorities between services) and QoS aspects (e.g. timeouts, amount of service that each flow has received, etc.). In particular, in the presented example, the following criteria are considered.

   - Service Class. It is worth noting that this decision is more related to commercial operator policies rather than to technical aspects, depending on how important the different service classes are for the operators.
   - Users belonging to the same service class. In this case the priority is established according to the service credit concept, in the same sense as explained in Section 5.5.1.3. In particular, the higher the service credit the higher the priority will be since this means that the flow has received less bit rate. Note that other criteria could be used such as the time that one packet has remained in the buffer or the amount of information waiting for transmission for a given user [70].

2. Resource allocation. Once requests are ordered, the next step consists of deciding whether or not they are accepted for transmission in the DSCH channel and which is the accepted transport format. For each request the algorithm executes the following steps:

   2.1. Initial TF selection. The algorithm selects a preferred value for the TF. As in the UE-MAC case, different policies could be explored, like selection according to the SCr algorithm, the TO or the MR algorithm (see Section 5.5.1.3). The selection made here is not definitive and can be modified in successive steps in the scheduling algorithm. In any case, it has influence over the number of users that will finally be accepted. For instance, when the selection is to transmit with the highest transport format, the algorithm will tend to allocate few users with high bit rates. However, if the initially selected transport format is lower, the algorithm will tend to allocate more users with more reduced bit rates.

   2.2. Availability check. After the initial TF selection, the algorithm checks whether or not this selection is feasible depending on the available resources at the DSCH and modifies the transport format accordingly. Essentially, there are two conditions that should be fulfilled to grant the transmission of a given request. They deal with the availability of OVSF codes and transmitted power.

With respect to OVSF code availability, Kraft's inequality is the condition to be checked but taking into account only the part of the code tree that is occupied by the DSCH. This part of the code tree starts at a given root code whose spreading factor is denoted as SF_{root} (see Figure 5.82). Then, assuming that a total of n_s users have already been allocated in the DSCH channel with their corresponding spreading factors SF_i, the algorithm should check the acceptance of the $(n+1)$th request according to the following inequality:

$$\sum_{i=1}^{n_s} \frac{1}{SF_i} + \frac{1}{SF_{n+1}} \leq \frac{1}{SF_{root}} \tag{5.85}$$

where SF_{n+1} is the spreading factor associated with the TF selected by the $(n+1)$th request.

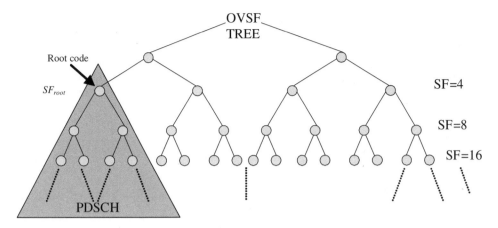

Figure 5.82 OVSF code tree

If the code availability condition holds, the algorithm will check the power availability. Otherwise, the selected transport format will be reduced by one in order to increase the spreading factor and the code availability condition will be checked again. The process is represented graphically in Figure 5.83.

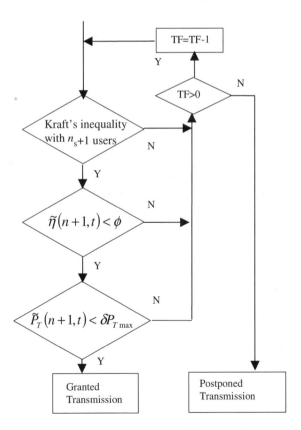

Figure 5.83 Resource allocation process in the packet scheduling algorithm

With respect to power availability, for simplicity the check can be done with two different conditions, although it is possible to join both under a single condition. They respond to the power availability conditions that were explained in the context of downlink admission control. Specifically, the two conditions to check are the estimated downlink load factor and then the required transmission power, as depicted in Figure 5.83. A transmission is granted only if the estimated load factor after its acceptance is below a certain threshold ϕ and the estimated transmitted power level is below a fraction δ of the maximum available power. Otherwise, the transport format is reduced by one, or equivalently, the transmission bit rate is reduced and the conditions are checked again. If this is not possible because TF = 0, the request should wait for the next frame.

The estimated load factor whenever there are n transmissions in the system in frame t (including both DCH and DSCH transmissions) is given by:

$$\tilde{\eta}(n, t) = \sum_{i=1}^{n} \frac{\rho + f_{DL,i}(t-1)}{\left(\dfrac{E_b}{N_0}\right)_i R_{b,i}} + \rho \qquad (5.86)$$

$R_{b,i}$ being the bit rate of request i and $f_{DL,i}(t-1)$ the other-to-own cell interference factor for the user according to estimations or measurements done up to frame $t-1$. Due to the difficulties in obtaining measurements of this factor, it is possible to take an average or typical value assumed equal for all the users or to make an estimation based on the path loss measurements reported by the mobile. It should be pointed out that the differences between the expected load factor and the real value can be due to the inaccuracies in the measurement of the other-to-own-cell interference factor $f_{DL,i}$ and the path loss.

Similarly, the expected power is given by:

$$\tilde{P}_T = \frac{P_p + P_N \displaystyle\sum_{i=1}^{n} \frac{L_{p,i}}{\left(\dfrac{E_b}{N_0}\right)_i R_{b,i}} + \rho}{1 - \tilde{\eta}(n, t)} \qquad (5.87)$$

P_p being the power devoted to common control channels and P_N the background noise.

It should be mentioned that control parameters ϕ and δ (both <1) should be appropriately set in order to leave a certain margin for possible fluctuations between the expected values and the real measurements.

Impact over a Two-layered Video Streaming To illustrate the impact of the setting of different parameters, the previous algorithm is now evaluated in a sample scenario with streaming video users. Streaming video service is one of the expected interests in 3G systems. Quality requirements deal with the achieved bit rate, the percentage of lost packets and the delay jitter, rather than the end-to-end delay. A streaming service allows an initial set-up delay that gives room to some packet transmissions before the video is reproduced. These packets can be stored in the mobile terminal buffer. Therefore, with a proper buffer dimensioning, the user will be unaware of possible packet retransmissions because the stored buffer allows for a continuous packet flow reproduction. Thus, this property gives some more room for scheduling the streaming service as packet retransmissions may play a role [71].

In order to differentiate quality levels, for this service a two layered video application is assumed that is characterised by two different flows: a basic layer, with the minimum requirements for a suitable visualisation, and an enhancement layer, which contains additional information to improve the quality of the received images. Taking into account that the DSCH is always associated with a DCH through which control information and DSCH allocations can be transmitted, it will be assumed that the basic layer of the video service will be transmitted through the DCH channel while the enhancement layer will be

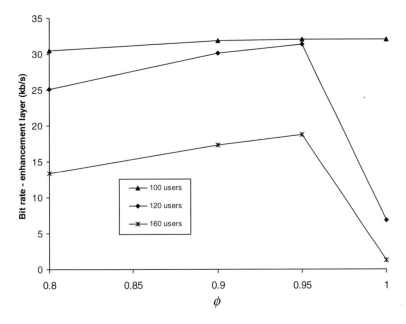

Figure 5.84   Average bit rate of the enhancement layer

transmitted only if there is capacity in the DSCH channels. A basic layer with 32 kb/s and an enhancement layer of up to 64 kb/s will be considered in the following.

One of the most relevant parameters in the design of the packet scheduling algorithm is the threshold ϕ. Figure 5.84 presents the average bit rate obtained during a streaming session in the enhancement layer for different ϕ values and numbers of users in the system. $\delta = 1$ has been assumed. It can be concluded that the selection $\phi = 0.95$ provides the best behaviour, since the enhancement layer gets the highest possible bit rate for the different load levels. In turns, Figure 5.85 plots the delay jitter achieved for different ϕ and for 120 users. Again $\phi = 0.95$ is revealed as a suitable value because it provides the lowest possible delay variation.

DSCH Dimensioning: Root Code Selection   We now present some performance indicators of the previously described packet scheduling algorithm in a scenario with a mix of conversational users that transmit in DCH channels and interactive users that transmit in DSCH, to examine the effects of the selection of the spreading factor of the DSCH root code SF_{root}. This spreading factor determines the part of the OVSF code tree that is used by the DSCH and the part that remains for the rest of channels such as DCH (see Figure 5.82). Consequently, the selection of SF_{root} must be done depending on the specific needs of the provided services.

Specifically, when mixing real time conversational users through DCH and non real time interactive users through DSCH, a trade-off arises in the suitable selection of SF_{root}. This trade-off is illustrated in Figures 5.86, 5.87 and 5.88, which present several performance figures for both types of traffic. In particular, Figure 5.86 plots the average page delay of interactive users for different DSCH allocations given by the value of SF_{root}, while Figures 5.87 and 5.88 show the dropping and admission probabilities, respectively, of conversational users. A constant conversational offered load of 40 Erlangs has been considered in the whole scenario, composed of seven omnidirectional cells, while the interactive offered load has been progressively increased.

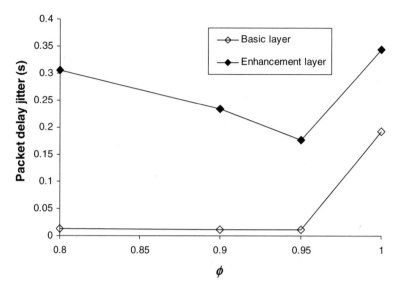

Figure 5.85   Performance in terms of delay jitter for the basic and enhancement layers

Note in Figure 5.86 that the increase in the number of interactive users in the system causes a corresponding page delay increase because the requests of the different users need to queue while waiting for transmission before the packet scheduling algorithm can allocate them. Furthermore, the lower the SF_{root} value, the lower the packet delay for the same interactive load level because of a higher capacity of the DSCH channel leading to a higher number of users that can be simultaneously allocated. However, from the point of view of conversational traffic, the lower the SF_{root}, the lower number of codes available for DCH channels. Consequently, the admission probability is reduced when SF_{root} is decreased, as seen in Figure 5.88. Furthermore, call droppings are caused when a user must handoff a call to a cell where there are no available OVSF codes, as seen in Figure 5.87, which reveals that dropping probability is not negligible for high interactive loads supported by low SF_{root} values.

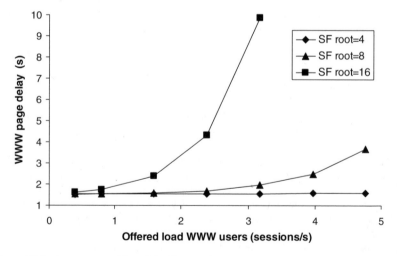

Figure 5.86   Average page delay of WWW browsing users for different loads and SF_{root} values

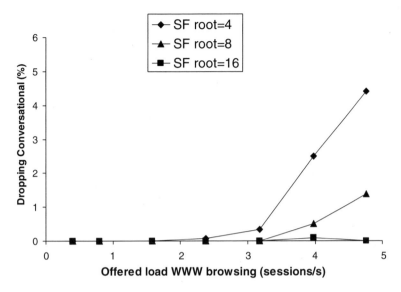

Figure 5.87 Dropping probability of conversational users for different loads and SF_{root} values

Keeping all the above in mind, and defining some performance QoS targets for both conversational and interactive traffic, it is possible to define a feasible region of operation for each DSCH dimensioning depending on the existing mix of conversational and interactive traffic, as shown in Figure 5.89. In particular, the considered QoS figures are a dropping probability lower than 1% for conversational users and an average page delay lower than 4 s for interactive users. It can be observed that, as the conversational traffic increases, and for the $SF_{root} = 16$ case, the interactive traffic is limited to no more than 2.1 sessions/s, otherwise the limited DSCH capacity does not allow assuring the target delay since interactive packets should be queued too long before getting access to the radio channel. In this

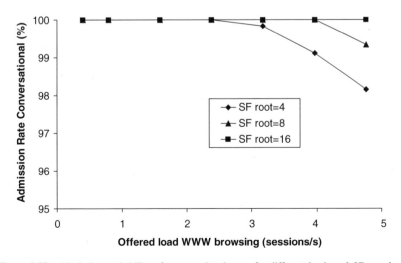

Figure 5.88 Admission probability of conversational users for different loads and SF_{root} values

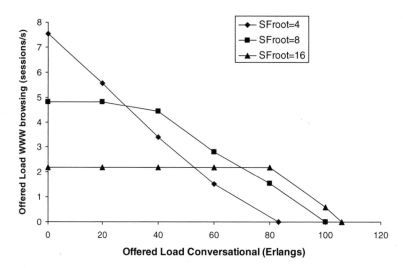

Figure 5.89 DSCH dimensioning regions

case, for conversational loads higher than 80 Erlangs, it would be necessary to reduce the interactive load, otherwise the conversational dropping probability could not be provided because of the OVSF code scarcity. For the $SF_{root} = 8$ case, the supported interactive traffic when no conversational traffic is present rises to about 5 sessions/s because more capacity is devoted to the DSCH channel and facilitates the satisfaction of the interactive packet delay bound. When conversational traffic is present, it is necessary to progressively reduce interactive load, otherwise the dropping criteria cannot be met. Similar conclusions can be drawn for the $SF_{root} = 4$ case.

5.6 POWER CONTROL

Power control is one of the most important procedures in a WCDMA environment such as UTRAN FDD, due to its interference-limited nature. As a result, it is important that each transmission is carried out with the minimum required power to ensure the quality requirements for the considered service. As depicted in Figure 5.90 the power control procedure operates in two steps [72][73], denoted as outer and inner loops, respectively.

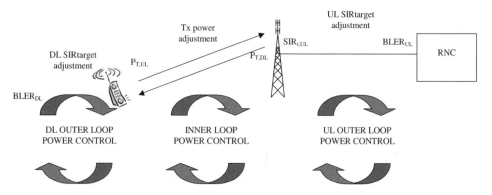

Figure 5.90 Inner and outer loop power control

The function of the outer loop power control is to map the required quality (e.g. in terms of block error rate, BLER) into an appropriate value of the signal-to-noise-and-interference ratio target (SIR). This value depends on the radio channel characteristics and has a long-term variation as the environment changes (e.g. when varying mobile speed or when changing from Line Of Sight, LOS, to Non Line Of Sight, NLOS, situations). In the uplink direction, the outer loop power control algorithm is executed at the RNC by monitoring the transport block error rate for a given radio link. If the measured BLER is higher than the requirement for the radio link, the SIR target is increased and, on the contrary, if the measured BLER is below the requirement, the SIR target should be decreased in order to save energy and to reduce interferences. The time scale for outer loop power control operation is in the order of seconds, so that a significant number of transport blocks have been processed to obtain accurate BLER measurements. In the downlink direction, the procedure is executed internally by the mobile terminal in order to meet the BLER criterion that has been set by UTRAN during the establishment or reconfiguration procedures.

Once the SIR target has been set either in the uplink or in the downlink direction, the inner loop power control is responsible of adjusting the appropriate transmission power level in order to meet this SIR target requirement in each direction. The inner loop power control can operate either in open loop or in closed loop mode. The open loop power control is used only when no feedback channel is available in the opposite direction, as would be the case of RACH and FACH transport channels. In this case, the transmitted power is set depending on power measurements in the opposite link (e.g. the uplink power is set depending on the downlink measurements). Due to the different channel behaviour at the different uplink and downlink frequencies, this mechanism is only able to compensate slow fading variations while the fast fading remains present. Therefore, closed loop power control is the preferred solution whenever possible. In particular, DCH, DSCH and CPCH channels make use of closed loop power control while RACH and FACH make use of open loop power control.

In the uplink closed loop power control, the node B measures the received SIR and if it is below the SIR target requirement a Transmit Power Control (TPC) command with value '1' is sent in the downlink indicating that the UE must increase the transmitted power. However, if the measured SIR is higher than the target requirement a TPC command with value '0' is sent indicating that the UE must decrease the transmitted power. The operation in the downlink is equivalent but in this case the UE executes the SIR measurement and feeds back the TPC commands to the node B. One TPC command is sent in each time slot of the 10 ms radio frame, which leads to a power control rate of 1500 Hz. The transmitted downlink TPC command can have 2, 4 or 8 bits, depending on the time slot format. A repetition code is applied, so that if the TPC command is set to '1' all the bits of the TPC field are set to this value. In turn, the transmitted uplink TPC commands can have either 1 or 2 bits, and again a repetition code is used.

The amount of power increase or decrease when receiving a TPC command is signalled by higher layers by the magnitude TPC step size, which can take values 1 or 2 dB for the uplink. In the downlink, the TPC step size may be take the values 0.5, 1, 1.5 or 2 dB [72][73]. The maximum transmission power for a radio link in both the uplink and downlink directions is signalled by UTRAN. The power increase or decrease applies to both the DPDCH and the DPCCH, whose relative power is set by higher layers according to parameters β_c and β_d as has been described on page 75.

During the establishment of a dedicated physical channel, an initial period named the power control preamble of up to 7 frames can be left in which only the DPCCH channel is transmitted in order to converge the power control to an appropriate initial value [34][72]. After this period, the DPDCH can be transmitted.

A special situation for uplink power control occurs when the mobile is in soft handover, simultaneously connected to more than one cell. In this case, different and contradictory TPC commands can be transmitted by the different cells simultaneously so the mobile terminal must combine them appropriately to take the decision about increasing or decreasing the transmitted power. To this end, two possible algorithms are defined in the specifications [72].

On the other hand, when compressed mode is used in either the uplink or the downlink direction, this has an important implication over power control, since the closed loop power control is interrupted and some TPC commands are not transmitted during transmission gaps. In order to compensate, the power control step can be increased up to 3 dB for some time after the gap, so that a faster power control adjustment is achieved.

REFERENCES

[1]   H. Holma, A. Toskala, *WCDMA for UMTS*, John Wiley & Sons Ltd, 2nd edition, 2000
[2]   J. Laiho, A. Wacker, T. Novosad, *Radio Network Planning and Optimisation for UMTS*, John Wiley & Sons Ltd, 2002
[3]   O. Sallent, S. Ruiz, R. Agustí, F. Adelantado, M.A. Díaz-Guerra, J.L. Miranda, J. Montero, 'Using Data Extracted from a GSM Network for 3G Planning and RRM Evaluation', *IEEE Wireless Communications and Networking Conference (WCNC'03)*, New Orleans (EEUU), March, 2003
[4]   3GPP TS 05.08 v8.20.0 'Technical Specification Group GSM/EDGE Radio Access Network; Radio Subsystem Link Control (Release 1999)'
[5]   ARROWS IST-2000-25133 project http://www.arrows-ist.upc.es/
[6]   J.J. Olmos, S. Ruiz, 'Transport Block Error Rates for UTRA-FDD Downlink with Transmission Diversity and Turbo Coding', *13th IEEE International Symposium on Personal, Indoor and Mobile Radio Communications PIMRC-2002*, **1**, pp. 31–35
[7]   S.A. Kyriazakos, G.T. Karetsos, *Practical Radio Resource Management in Wireless Systems*, Artech House, 2004
[8]   Z. Lui, M. El Zarki, 'SIR Based Call Admission Control for DS-CDMA Cellular Systems', *IEEE Journal on Selected Areas in Communications*, **12**, 1994, pp. 638–644
[9]   L. Badia, M. Zorzi, A. Gazzini, 'On the Impact of User Mobility on Call Admission Control in WCDMA Systems', *56th IEEE VTC Fall Conference*, 2002, Vancouver, pp. 121–126
[10] S. Redana, A. Capone, 'Received Power-Based Call Admission Control Techniques for UMTS Uplink', *56th IEEE VTC Fall Conference*, 2002, Vancouver, pp. 2206–2210
[11] H. Holma, J. Laakso, 'Uplink Admission Control and Soft Capacity with MUD in CDMA', *IEEE Vehicular Technology Conference in Fall 1999*, 1999, Amsterdam, pp. 431–435
[12] F. Gunnarsson, E. Geijer Lundin, G. Bark, N. Wiberg 'Uplink Admission Control in WCDMA Based on Relative Load Estimates,' *IEEE International Conference on Communications*, ICC-2002, pp. 3091–3095
[13] N. Dimitriou, G. Sfikas, R. Tafazolli, 'Call Admission Policies for UMTS', *51st IEEE VTC Spring Conference*, 2000, Tokyo, pp. 1420–1424
[14] A. Capone, S. Redana, 'Call Admission Control Techniques for UMTS', *54th IEEE VTC Fall Conference*, 2001, Atlantic City, pp. 959–929
[15] C.J. Ho, J.A. Copeland, C.T. Lea, G.L. Stuber, 'On Call Admission Control in DS/CDMA Cellular Networks', *IEEE Transactions on Vehicular Technology*, **50**(6), November 2001, pp. 1328–1343
[16] V. Phan-Van, S. Glisic, 'Radio Resource Management in CDMA Cellular Segments of Multimedia Wireless IP Networks', *The 4th International Symposium on Wireless Personal Multimedia Communications (WPMC)*, Aalborg, Denmark, September, 2001
[17] E.G. Lundin, F. Gunnarsson, F. Gustafsson, 'Uplink Load Estimation in WCDMA,' *IEEE Wireless Communications and Networking*, 2003, WCNC, pp.1669–1674
[18] A. Hämäläinen, K. Valkealahti, 'Adaptive Power Increase Estimation in WCDMA', *13th IEEE International Symposium on Personal, Indoor and Mobile Radio Communications PIMRC*, 2002, Lisbon, pp. 1407–1411
[19] N. Dimitriou, G. Sfikas, R. Tafazolli, 'Call Admission Policies for UMTS', *51st IEEE VTC Spring Conference*, 2000, Tokyo, pp. 1420–1424
[20] O. Sallent, J. Pérez-Romero, R. Agusti, 'Optimizing Statistical Uplink Admission Control for W-CDMA', *57th IEEE VTC Fall Conference*, 2003, Orlando
[21] F. Adelantado, O. Sallent, J. Pérez-Romero, R. Agustí, 'Time Correlation of the Intercell to Intracell Interference Ratio in a W-CDMA Network', *IEE Electronics Letters*, December, 2002, **38**(25), pp.1735–1737
[22] W. Ying, Z. Jingmei, W. Weidong, Z. Ping, 'Call Admission Control in Hierarchical Cell Structure', *IEEE 55th VTC Spring Conference*, 2002, Birmingham, pp. 1955–1959
[23] 3GPP 25.922 v6.0.1 'Radio resource management strategies (release 6)'
[24] S. Akhtar, S.A. Malik, D. Zeghlache, 'Prioritized Admission Control for Mixed Services in UMTS WCDMA Networks', *12th IEEE International Symposium on Personal, Indoor and Mobile Radio Communications PIMRC*, 2001, San Diego, pp. 133–137

[25] J. Outes, L. Nielsen, K. Pedersen, P. Mogensen, 'Multi-Cell Admission Control for UMTS', *IEEE 53rd Vehicular Technology Conference Spring*, May 2001, Rhodes, Greece

[26] T. Minn, K.Y. Seu, 'Dynamic Assignment of Orthogonal Variable Spreading Factor Codes in W-CDMA', *IEEE Journal on Selected Areas in Communications*, August, 2000, pp.1429–1440

[27] J. Knutsson, P. Butovitsch, M. Persson, R.D. Yates, 'Downlink Admission Control Strategies for CDMA Systems in a Manhattan Environment', *IEEE Vehicular Technology Conference VTC*, 1998, pp. 1453–1457

[28] M. Kazmi, P. Godlewski, C. Cordier, 'Admission Control Strategy and Scheduling algorithms for Downlink Packet Transmission in WCDMA', *52nd IEEE Vehicular Technology Conference Fall*, 2000, Boston, pp. 674–680

[29] J. Pérez-Romero, O. Sallent, R. Agusti, G. Parés, 'A Downlink Admission Control Algorithm for UTRA-FDD', *4th IEEE Conference on Mobile and Wireless Communications Networks MWCN*, Stockholm, 2002, pp. 18–22

[30] J. Pérez-Romero, O. Sallent, R. Agusti, 'Impact of User Location in W-CDMA Downlink Resource Allocation', *IEEE International Symposium on Spread Spectrum Techniques and Applications (ISSSTA)*, 2002, Prague, **2**, pp. 420–424

[31] J. Pérez-Romero, O. Sallent, D. Ruiz, R. Agustí, 'An Admission Control Algorithm to Manage High Bit Rate Static Users in W-CDMA', *IST Mobile & Wireless Telecommunications Summit*, Lyon, 2004

[32] 3GPP TS 23.122 'NAS Functions Related to Mobile Station (MS) in Idle Mode'

[33] 3GPP TS 25.304 'User Equipment (UE) procedures in Idle Mode and Procedures for Cell Reselection in Connected Mode'

[34] 3GPP TS 25.331 'Radio Resource Control (RRC); Protocol Specification'

[35] 3GPP TS 04.18 v8.24.0 'Technical Specification Group GSM/EDGE Radio Access Network; mobile radio interface layer 3 specification; radio resource control protocol (release 1999)'

[36] 3GPP TS 25.931 v5.1.0 'UTRAN Functions, Examples on Signalling Procedures (release 5)'

[37] W.R. Stevens, *TCP/IP Illustrated*, Vol.1, Addison-Wesley Professional Computing Series, 1994

[38] S.H. Low, F. Paganini, J.C. Doyle, 'Internet congestion control', *IEEE Control Systems Magazine*, **22**(1), February, 2002, pp. 28–43

[39] T.K. Liu, J.A. Silvester, 'Joint Admission/congestion Control for Wireless CDMA Systems Supporting Integrated Services', *IEEE Journal on Selected Areas in Communications*, **16**(6), August, 1998, pp. 845–857

[40] N. Passas, L. Merakos, 'A Graceful Degradation Method for Congestion Control in Wireless Personal Communication Networks', *Proceedings of the IEEE Vehicular Technology Conference VTC*, 1996, pp. 126–130

[41] R. De Bernardi, D. Imbeni, L. Vignali, M. Karlsson, 'Load Control Strategies for Mixed Services in WCDMA', *51st IEEE Vehicular Technology Conference (VTC) Spring*, 2000, Tokyo, pp. 825–829

[42] W. Rave, T. Kohler, J. Voigt, G. Fettweis, 'Evaluation of Load Control Strategies in an UTRA/FDD Network', *IEEE 53rd Vehicular Technology Conference Spring*, May 2001, Rhodes, Greece, pp. 2710–2714

[43] J. Pérez-Romero, O. Sallent, R. Agustí, J. Sánchez, 'Managing Radio Network Congestion in UTRA FDD', *IEE Electronics Letters*, October, 2002, **38**(22), pp. 1384–1386

[44] J. Sachs, T. Balon, M. Meyer, 'Congestion Control in WCDMA with Respect to Different Service Classes', *European Wireless Conference*, Munich, 1999

[45] J. Muckenheim, U. Bernhard, 'A Framework for Load Control in 3G CDMA Networks', *IEEE GLOBECOM*, San Antonio, 2001, pp. 3738–3742

[46] O. Sallent, J. Pérez-Romero, R. Agustí, F. Casadevall, 'Provisioning Multimedia Wireless Networks for Better QoS: RRM Strategies for 3G W-CDMA', *IEEE Communications Magazine*, February, 2003, **41**(2), pp. 100–106.

[47] J. Pérez-Romero, O. Sallent, R. Agustí, 'A Novel Approach for a Multicell Load Control in W-CDMA', *5th International Conference on 3G Mobile Communication Technologies (3G 2004)*, London, UK, October, 2004

[48] M. Naghshineh, A. S. Acampora, 'Design and Control of Microcellular Networks with QoS Provisioning for Data Traffic', *Wireless Networks 3*, 1997, pp. 249–256

[49] S. K. Das *et al.*, 'A Call Admission and Control Scheme for QoS Provisioning in Next Generation Wireless Networks', *Wireless Networks 6*, 2000, pp. 17–30

[50] L. Xu, X. Shen, J. Mark, 'Dynamic Bandwidth Allocation with Fair Scheduling for WCDMA', *IEEE Wireless Communications*, **9**(2), April 2002, pp. 26–32

[51] I. F. Akyldiz, D. A. Levine, I. Joe, 'A Slotted CDMA Protocol with BER Scheduling for Wireless Multimedia Networks', *IEEE/ACM Transactions on Networking*, April 1999, **7**(2), pp. 146–158

[52] S.J. Oh, D. Zhang, K.M. Wasserman, 'Optimal Resource Allocation in Multiservice CDMA Networks', *IEEE Transactions on Wireless Communications*, July 2003, **2**(4), pp. 811–821

[53] S.A. Jafar, A. Goldsmith, 'Adaptive Multirate CDMA for Uplink Throughput maximization', *IEEE Transactions on Wireless Communications*, March 2003, **2**(2), pp. 218–228

[54] K. Dimou, P. Godlewski, 'MAC Scheduling for Uplink Transmission in UMTS W-CDMA', *IEEE 53rd Vehicular Technology Conference Spring*, Rhodes, Greece, May 2001

[55] D.I. Kim, E.Hossain, V.K. Bhargava, 'Downlink Joint Rate and Power Allocation in Cellular Multirate WCDMA Systems', *IEEE Transactions on Wireless Communications*, January, 2003, **2**(1), pp. 69–80

[56] H. Fattah, C. Leung, 'An Overview of Scheduling Algorithms in Wireless Multimedia Networks', *IEEE Wireless Communications*, October 2002, **9**(5), pp. 76–83.

[57] Y. Cao, V.K. Li, 'Scheduling Algorithms in Broadband Wireless Networks', *Proceedings of the IEEE*, January, 2001, **89**(1), pp. 76–87

[58] L. Alonso, 'Técnicas de Acceso y Gestión de Recursos para Garantizar Calidad de Servicio en Sistemas de Comunicaciones Móviles basados en CDMA', PhD Thesis at the Universitat Politecnica de Catalunya (UPC), May, 2001

[59] 3GPP TS 25.321 'Medium Access Control (MAC), protocol specification'

[60] K.W. Helmersson, G. Bark, 'Performance of Downlink Shared Channels in WCDMA Radio Networks', *IEEE 53rd Vehicular Technology Conference (VTC) Spring*, May 2001, pp. 2690–2694

[61] L. Oliveira, A. Rodrigues, 'Performance of Downlink Packet Data Transmission in a UMTS Network', *IEEE 54th Vehicular Technology Conference (VTC) Fall*, October, 2001, Atlantic City, pp. 1857–1860

[62] P. Giacomazzi, L. Musumeci, G. Verticale, 'Performance of Web-Browsing Services over the WCDMA-FDD Downlink Shared Channel', IEEE GLOBECOM, 2001, pp.3514–3518

[63] B. Maskarevitch, 'Learning Rate Control for Downlink Shared Channel in WCDMA', *14th IEEE International Symposium on Personal, Indoor and Mobile Radio Communications PIMRC*, 2003, Beijing, pp. 2919–2922

[64] X. Qiu, L. Chang, Z. Kostic, T. M. Willis III, N. Mehta, L. J. Greenstein, K. Chawla, J. F. Whitehead, J. Chuang, 'Some Performance Results for the Downlink Shared Channel in WCDMA' *IEEE International Conference on Communications* (ICC 2002), April–May, 2002, pp. 376–380

[65] S. Kourtis, R. Tafazolli, 'Downlink Shared Channel: An Effective Way for Delivering Internet Services in UMTS', *IEE 3G Mobile Communication Technologies Conference*, 2002, London, pp. 479–483

[66] Y. Ofuji, S. Abeta, and M. Sawahashi, 'Comparison of Packet Scheduling Algorithms Focusing on User Throughput in High Speed Downlink Packet Access,' *IEICE Trans. Commun.*, 2003, **E86-B**(1), pp. 132–141

[67] P. Frenger, S. Parkvall, and E. Dahlman, 'Performance Comparison of HARQ with Chase Combining and Incremental Redundancy for HSDPA,' *54th IEEE Vehicular Technology Conference in Fall*, October, 2001, Atlantic City, pp. 1829–1833

[68] N. Miki, H. Atarashi, S. Abeta, and M. Sawahashi, 'Comparison of Hybrid ARQ Packet Combining Algorithm in High Speed Downlink Packet Access in a Multipath Fading Channel', *IEICE Trans. Commun.*, July 2002, **E85-A**(7), pp. 1557–1568

[69] F. Frederiksen, T.E. Kolding, 'Performance and Modeling of WCDMA/HSDPA Transmission/H-ARQ Schemes,' *56th IEEE Vehicular Technology Conference in Fall*, Vancouver, Sept. 2002, **1**, pp. 472–476

[70] O. Sallent, J. Pérez-Romero, F. Casadevall, R. Agustí, 'An Emulator Framework for a New Radio Resource Management for QoS Guaranteed Services in W-CDMA Systems', *IEEE Journal on Selected Areas in Communications*, October, 2001, **19**(10), pp. 1893–1904

[71] J. Pérez-Romero, O. Sallent, R. Agusti, 'Downlink Packet Scheduling for a Two-Layered Streaming Video Service in UMTS', *IST Mobile & Wireless Telecommunications Summit*, Thessaloniki, 2002, pp. 212–216

[72] 3GPP TS 25.214 'Physical Layer Procedures (FDD)'

[73] 3GPP TS 25.211 'Physical Channels and Mapping of Transport Channels onto Physical Channels (FDD)'

[74] UMTS 30.03 v3.2.0 TR 101 112 'Selection Procedures for the Choice of Radio Transmission Technologies of the UMTS', ETSI, April, 1998

[75] 3GPP TR 25.942 v5.3.0 'Radio Frequency (RF) System Scenarios'

[76] P. Karlsson (editor) *et al.*, 'Target Scenarios Specification: vision at project stage 1', Deliverable D05 of the EVEREST IST-2002–001858 project, April, 2004. Available at http://www.everest-ist.upc.es/

[77] P. Goria, C. Guerrini, R. Agustí, F. Casadevall, J. Pérez-Romero, O. Sallent, 'System specification. Radio Resource Management Algorithms: Identification and requirements', Deliverable D04 of the ARROWS IST-2000–25133, April, 2001. Available at http://www.arrows-ist.upc.es/

[78] 3G Offered Traffic Characteristics, Final Report, UMTS Forum, November 2003

[79] 3GPP TSG GERAN#1 GP-000042, 'Packet Data Traffic Models for e-Mail (with and without attachments) and Streaming Multimedia Applications'

[80] J. Ho, Y. Zhu, and S. Madhavapeddy, 'Throughput and Buffer Analysis for GSM General Packet Radio Service (GPRS)', *IEEE Wireless Communication and Networking Conference (WCNC)*, **3**, pp 1427–1431, 1999

APPENDIX - SIMULATION MODELS

This appendix presents a survey of simulation models that are used in system-level simulations for RRM evaluation. This includes propagation, mobility and traffic generation models. These topics have been extensively covered in the literature and therefore the objective of this appendix is not to provide a comprehensive description of all the different possibilities but to give some examples that are used by standardisation organisms such as 3GPP. For further details on the presented models the reader is referred to References 74 and 75.

A5.1 PROPAGATION MODELS

Propagation models try to capture the effects arising from the electromagnetic wave propagation in mobile environments. In general, these effects lead to three different models to characterise the average path loss depending on the distance between transmitter and receiver, the slow variation around the average value due to shadowing and scattering, and finally the fast signal variations due to multi-path effects.

Propagation models differ significantly depending on the considered environment. In the following, models for macrocell and microcell environments are presented.

A5.1.1 MACROCELL PROPAGATION

Macrocell environments are characterised by large cells and high transmitted powers in outdoor urban and suburban areas with antennas over the roof top levels. The path loss model used in this case is given by:

$$L_p(dB) = 40 \cdot \left(1 - 4 \cdot 10^{-3} \Delta h_b\right) \log_{10} d - 18 \log_{10} \Delta h_b + 21 \log_{10} f + 80 \tag{5.88}$$

where d is the distance in km between node B and the mobile terminal, f is the carrier frequency in MHz and Δh_b is the node B antenna height, measured in metres from the average roof top level. Typical values for these parameters are $\Delta h_b = 15$ m and $f = 2000$ MHz, leading to:

$$L_p(dB) = 128.1 + 37.6 \log_{10} d \tag{5.89}$$

The path loss L shall in no circumstances be less than the free space loss given by:

$$L_f(dB) = 20 \log_{10} d + 20 \log_{10} f + 32.44 \tag{5.90}$$

Furthermore, in order to avoid very low path loss values, a minimum coupling loss (MCL) of 70 dB is usually assumed [75]. Therefore, the path loss is given by:

$$L(dB) = \max(L_p, L_f, MCL) \tag{5.91}$$

The shadowing effects are modelled by means of a log-normal variation. This corresponds to the addition of a Gaussian random variable S with average 0 and standard deviation σ to the path loss value in dB. A typical standard deviation of $\sigma = 10$ dB is usually considered. The total path loss is then given by:

$$L_{tot}(dB) = L(dB) + S(dB) \tag{5.92}$$

Table 5.24 Example of channel impulse responses for a vehicular environment

Tap	Channel A Delay (ns)	Channel A Power (dB)	Channel B Delay (ns)	Channel B Power (dB)	Doppler Spectrum
1	0	0.0	0	−2.5	CLASSIC
2	310	−1.0	300	0	CLASSIC
3	710	−9.0	8900	−12.8	CLASSIC
4	1090	−10.0	12900	−10.0	CLASSIC
5	1730	−15.0	17100	−25.2	CLASSIC
6	2510	−20.0	20000	−16.0	CLASSIC

It is assumed that the shadowing values are correlated with the distance. In particular, an exponential normalised decorrelation function is considered, for two points separated x m:

$$R(x) = e^{-\frac{x}{d_c}\ln 2} \tag{5.93}$$

where d_c is the decorrelation distance, 20 m being a typical value.

The fast fading is modelled by means of channel impulse responses based on a tapped-delay line model characterised by the number of taps (i.e. the number of channel paths) and, for each tap, the relative delay with respect to the first path, the average power with respect to the strongest path and the Doppler spectrum. For each environment, two different impulse responses are considered, each occurring with a certain probability [74]. As an example, Table 5.24 presents the two channel impulse responses A and B considered for a vehicular macrocell environment.

A5.1.2 MICROCELL PROPAGATION

Typically, the microcell propagation models are used for the path loss computation in urban environments, with small cells, low transmitted powers and antennas below roof top levels. Manhattan environments with different buildings and street crossings are considered. An example of this environment is shown in Figure 5.91

The path loss is given by:

$$L_p(dB) = 20 \log_{10} \left(\frac{4\pi d_n}{\lambda} D\left(\sum_{j=1}^{n} s_{j-1} \right) \right) \tag{5.94}$$

where λ is the wavelength in metres, n is the number of straight street segments through the shortest path between UE and node B (i.e. $n = 3$ in the example shown in Figure 5.91), s_j is the length in metres of the jth segment and d_n is called the 'illusory' distance between the UE and node B. It is computed in metres taking into account the n street segments by means of the following recursive expressions:

$$k_n = k_{n-1} + d_{n-1}a \tag{5.95}$$
$$d_n = k_{n-1}s_{n-1} + d_{n-1} \tag{5.96}$$

In these expressions a is a function of the street crossing angle, being 0.5 for 90 degree crossings. The initial values are set to $k_0 = 1$ and $d_0 = 0$ and the computation stops when the last segment has been added (in the example, it ends with d_3 after the inclusion of segment s_2).

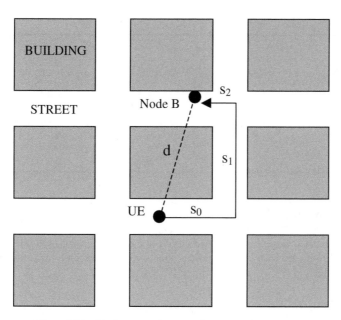

Figure 5.91   Manhattan environment for microcell propagation

The function $D(x)$ incorporates the dual slope behaviour that characterises the propagation in microcell environments. It is defined as:

$$D(x) = \begin{cases} x/x_b & x > x_b \\ 1 & x \leq x_b \end{cases} \qquad (5.97)$$

In this case, the path loss dependence with the distance changes at the so-called break point at distance x_b (typically set to 300 m). Therefore, the slope is 2 before the break point and 4 after it.

On the other hand, the model also takes into account the above roof top propagation by means of the COST Walfish-Ikegami model, dependent of the geographical distance d in metres between node B and the UE (see Figure 5.91). According to this model, the path loss is given by:

$$L_r(dB) = 24 + 45 \log_{10}(d + 20) \qquad (5.98)$$

Finally, the path loss is given by the minimum of the illusory distance and the above roof top propagation:

$$L(dB) = \min(L_p, L_r) \qquad (5.99)$$

Furthermore, as in the macrocell case, a minimum coupling loss MCL of 53 dB is assumed in microcell environments [75].

With respect to the slow fading, a lognormal shadowing with standard deviation 10 dB is usually considered as in the macrocell case.

Table 5.25 presents the channel impulse responses for a microcellular environment.

Table 5.25 Example of channel impulse responses for a microcellular environment

| Tap | Channel A | | Channel B | | Doppler Spectrum |
	Delay (ns)	Power (dB)	Delay (ns)	Power (dB)	
1	0	0.0	0	0.0	CLASSIC
2	110	−9.7	200	−0.9	CLASSIC
3	190	−19.2	800	−4.9	CLASSIC
4	410	−22.8	1200	−8.0	CLASSIC
5			2300	−7.8	CLASSIC
6			3700	−23.9	CLASSIC

A5.2 MOBILITY MODELS

Mobility models are used to determine the trajectories that mobile terminals follow in dynamic system-level simulations. As in the case of propagation models, different mobility models are considered for different environments.

A5.2.1 MOBILITY MODEL FOR MACROCELL ENVIRONMENTS

Macrocell environments corresponding typically to urban and suburban areas use a pseudo-random mobility model with semi-directed trajectories [74]. In this model, the mobiles are characterised by speed and direction, and the position is updated every d_c metres following the corresponding direction, where d_c is the decorrelation length, typically of 20 m. At each position update, the direction is changed with probability 0.2 and in this case the angle for the direction change is selected randomly in the range [−45, 45] degrees with uniform distribution. The process is illustrated in Figure 5.92. At initialisation, the mobiles are scattered in the scenario and the mobile direction is chosen randomly. Mobile speed is kept constant, with typical values of 50 km/h and 120 km/h for urban and suburban areas, respectively.

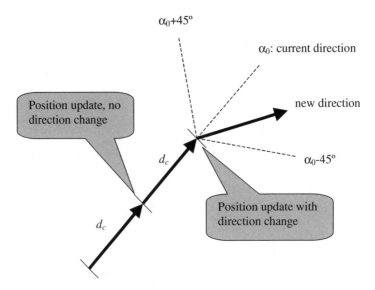

Figure 5.92 Mobility model for macrocell environments

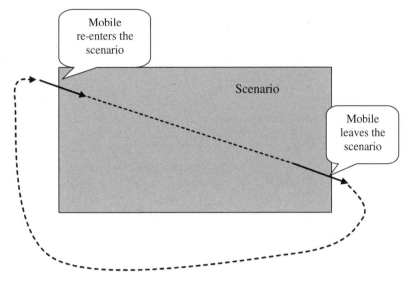

Figure 5.93   Wrap-around technique

When a mobile reaches the edge of the scenario, it is usual to apply the 'wrap-around' technique in which the mobile enters through the opposite side of the edge that it is leaving, as if the scenario wrapped around on itself (see Figure 5.93). By using the wrap-around technique, it is possible to collect statistics from all the cells in the scenario, reducing the simulation time needed to obtain statistically acceptable results. Instead, without wrap-around, due to the border effect it is necessary to collect statistics only from central cells in the scenario, thus increasing the simulation time.

A5.2.2 MOBILITY MODEL FOR MICROCELL ENVIRONMENTS

In the case of microcell environments, the typical structure is the Manhattan grid shown in the example in Figure 5.91. In this case, initially, the mobiles are distributed randomly in the streets and move along them following straight trajectories. When a mobile reaches a cross street, it turns with probability 0.5 and stays in the same street also with probability 0.5. In the case of turning, the new direction is selected with equal probability (see Figure 5.94) [74]. The mobile can only turn at the central point of the cross street. The mobile position is updated every 5 metres and the speed can be changed at each position update with probability 0.2. The average speed is 3 km/h and a normal distribution with standard deviation 0.3 km/h is assumed for speed changes with the constraint that the speed must be higher than 0 km/h.

A5.3 TRAFFIC MODELS

Traffic models are used in system-level simulations in order to emulate the behaviour of a given application. This behaviour is considered at different levels, including the session generation process (i.e. the rate at which sessions are generated and the session duration) and the data generation within a session (i.e. how the different information packets are generated and how the source activity varies). Normally, traffic models are characterised by random generation processes according to specific statistical distributions whose parameters (e.g. first and second order statistics) are adjusted to fit the real generation patterns. Note that during a system-level simulation the traffic models determine the

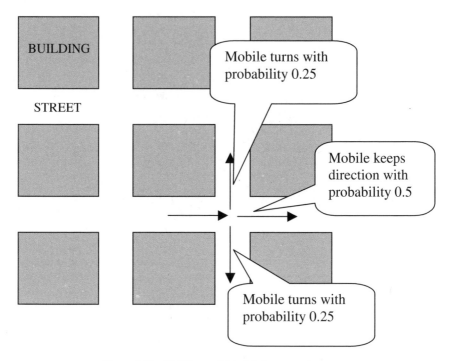

Figure 5.94 Mobility model in Manhattan environments

number of active and simultaneous users at each instant (see Chapter 4, Section 4.4.1). Consequently, it is critical to obtain accurate simulation results in the evaluation of WCDMA systems to have adequate models that fit the real source behaviour as much as possible.

In the following, details are given regarding the traffic models for some sample representative services corresponding to the four different service classes existing in UMTS, namely conversational, streaming, interactive and background. Essentially, the provided models are those that were considered for algorithm evaluation in the framework of the ARROWS and EVEREST IST projects. For details, see References 76 and 77.

A5.3.1 VIDEO-TELEPHONY TRAFFIC MODEL

Video-telephony service belongs to the conversational service class whose required characteristics are strictly given by human perception. Typically, a video-telephony flow consists of a continuous sequence of data blocks that shall be presented to the user in the right sequence at pre-determined instants. Video-telephony implies a full-duplex system, carrying both video and audio. A CBR generation model is usually assumed for traffic generation.

With respect to session generation, it is usual to characterise video-telephony applications in a similar way to voice services. This corresponds to a call generation rate based on a Poisson arrival scheme and exponential call duration.

A5.3.2 VIDEO-STREAMING TRAFFIC MODEL

The video-streaming service corresponds to applications with audio or video one-way real time data flows. Packet video sources may be CBR or VBR sources. For VBR, the model generates video frames with a variable size and forwards them to the network according to a fixed frequency (i.e. 30 frames every second).

An accurate video streaming representation could be very difficult due to the tight dependence of the bit rate with respect to the particular video sequence used. Therefore, the model needs to be able to capture the statistical properties and traffic characteristics of the envisaged video sequences and compression schemes.

The most popular standards defining compression mechanisms today are ISO MPEG and ITU H.26x series, with MPEG-4 and H.263 Version 2 being the move record versions. These standards allow three different kinds of coding schemes for a video frame, in order to improve the coding efficiency: Intra-frame ('I'), Predictive frame ('P') or Bidirectionally-predictive ('B'). An I-frame is coded in isolation from other, using transform/quantisation/entropy techniques. A P-frame is predictively coded using the previously coded frame, so that only the difference between the prediction and the actual frame is coded. A B-frame is predicted bidirectionally, using both its previous frame and the successive one. According to this, an I-frame is used to code frames efficiently corresponding to scene changes, i.e. frames that are different from preceding ones and cannot be easily predicted. Within a scene, frames are similar and may be coded predictively as P or B. Groups of frames between two successive I frames are called a 'group of pictures' (GOP).

A video traffic model for VBR encoder based on MPEG/H.263 has been Referenced 78. The model does not assume a fixed GOP structure, since a generic video content does not necessarily follow a predefined pattern. The model considers two states, corresponding to the I and P frames. The transitions between these two states occur according to certain probabilities, as represented by the Markov chain in Figure 5.95. The specific transition probabilities are set according to the video sequence that is being characterised.

Inside each state, the bit rate during a frame period (i.e. the frame size) is generated using an autoregressive AR(1) process, in order to model the long-term temporal correlation between a frame and the previous one. This is given by:

$$X_n = a + bX_{n-1} + \varepsilon \tag{5.100}$$

where X_n is the size of the nth frame of the generated sequence, a and b are real numbers, and ε is a normal distributed random variable component with a zero mean and a σ^2 variance. The first value of the sequence (X_0) can also be assumed to be a normal random variable. The specific parameters of the AR(1) process have to be estimated from the video sequence that must be characterised.

The above equation makes clear how to generate randomly the next element in the sequence from the previous one, modelling the autocorrelated video traffic. According to the model, the output bit rate within a frame period is constant and changes from frame to frame according to the AR(1) model. The underlying Markov chain is considered in order to capture the abrupt changes in the frame bit rates that occur due to scene changes or visual discontinuities.

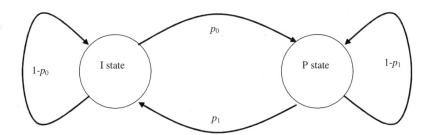

Figure 5.95   Two state model for the video encoder

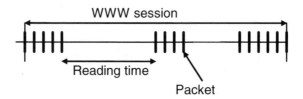

Figure 5.96   WWW browsing session model

Even if the proposed model is quite simple, it is able to generate data with a good autocorrelation function and power spectral density with respect to the sequence that must be characterised.

Besides the model for the bit rate, it is also necessary to define a model for the generation of the video streaming sessions over time. The total duration of a video streaming session can be assumed to be an exponentially distributed variable. Also, the inter-arrival time between two sessions can be assumed to be exponentially distributed

A5.3.3 WWW BROWSING TRAFFIC MODEL

The WWW browsing traffic model applies to interactive services in which the end user requests online data from remote equipment. The traffic model need here is described in Reference 74. This consists of a WWW browsing session, in which the user downloads some web pages, called 'packet calls'. Each of these packet calls is formed by different packets typically representing the HTML objects in a web page. Between two consecutive packet calls, there is a 'reading time', which models the time that the user would spend reading the content of the downloaded page. The overall WWW browsing session model is shown graphically in Figure 5.96. In the case of File Transfer Protocol (FTP), the same model can be applied simply by considering a single packet call in the session.

The model considers the following statistical characterisation:

- Session arrival process, defined as a Poisson arrival process.
- Number of packet calls per session, defined as a geometrically distributed random variable.
- Reading time between packet calls, defined as an exponentially distributed random variable. The reading time starts when the last packet of the packet call is completely received by the user, and ends when the user makes a request for the next packet call.
- Number of packets within a packet call. Different statistical distributions can be used to generate the number of packets. For example, the number of packets can be modelled as a geometrically distributed random variable.
- Inter–arrival time between packets within a packet call, defined as an exponentially distributed random variable.
- Packet size. The traffic model can use the packet size distribution that best suits the traffic case under study. Pareto distribution with cut-off is often used [74] with a typical average value of 480 bytes.

A5.3.4 EMAIL TRAFFIC MODEL

A suitable traffic model for the email service considering the cases with and without attachments has been proposed [79][80]. The model assumes that for most Internet based email services, the incoming messages of a user are stored at a dedicated email server. This email server keeps safe all messages of the user in a mailbox until he/she logs on to the network, initiates the email application and retrieves the messages. In general, when the user runs the email application, the headers of all the available messages are downloaded to the computer from the email server. Then the user will scan through the headers and download the required messages one after another. After downloading each message, the user may read it and compose a reply to the sender. When the user finishes with the current message, he/she will download the next message, and so on.

Consequently, an individual email user generates an ON-OFF traffic pattern in the downlink during an email session. An ON period represents the time interval when a message is being downloaded from the email server to the mobile terminal. An OFF period is the time interval between the completion of a message download and the beginning of the next message download. It represents the message reading time of the user.

The number of email messages to be downloaded, or equivalently the number of ON periods, during a session depends on the number of messages arrived at the email server since the last email session, and the number of messages that the user selects to read during the email session. The arrival of messages to the mailbox can be approximated by a Poisson process with arrival rate λ_e messages per user per hour. Furthermore, let the average time between two email sessions of a user be T_e and let p_e be the probability that a newly arrived message will be read by the user. Then, the random variable m_e representing the number of messages to be read by the user during the email session follows approximately the Poisson distribution as:

$$\Pr(m_e = n) = \frac{(p_e \lambda_e T_e)^n}{n!} \cdot e^{-p_e \lambda_e T_e} \tag{5.101}$$

The email size x_e, based on an empirical analysis of actual email messages, follows a two-stage Weibull distribution given by:

$$F(x_e) = \begin{cases} 1 - e^{-e^{-k_1 x_e^{c_1}}} & \text{if } F(x_e) \leq 0.5 \\ 1 - e^{-e^{-k_2 x_e^{c_2}}} & \text{if } F(x_e) > 0.5 \end{cases} \tag{5.102}$$

where 0.5 represents the probability that the email contains an attachment. c_1 and k_1 model the message length with attachment, where c_1 ranges from 1.2 to 3.2 with mean 2.04, and k_1 from 14 to 21 with mean 17.64. Similarly, c_2 and k_2 model the message length without attachment, where c_2 ranges from 0.31 to 0.6 with mean 0.3, and k_2 ranges from 2.8 to 3.4 with mean 3.61. The length of the ON period is a function of the message size and the instantaneous throughput available to the user. In turn, the OFF period t_e follows the Pareto distribution with the probability distribution function:

$$F(t_e) = 1 - \left(\frac{k_e}{t_e}\right)^{\alpha_e} \tag{5.103}$$

According to the Pareto distribution, the parameter k_e is the minimum duration of the OFF period. The parameter α_e relates to the degree of the heavy-tail behaviour of the OFF period. Typical values are $k_e = 30 \sim 60$ s and $\alpha_e = 0.5 \sim 1.5$.

6

CRRM in Beyond 3G Systems

This chapter covers the radio resource management issues arising in heterogeneous network scenarios, where different RATs coexist and operate in a coordinated way, including reconfigurability capabilities at different levels of the network and terminals and providing broadband radio access for high bit rate services. Such scenarios are often referred to as 'beyond 3G' systems and pave the way for the extension to include new 4G radio access technologies. When jointly considering different RATs, a more efficient use of the available radio resources can be achieved with the introduction of CRRM (Common Radio Resource Management) algorithms that take into consideration the overall resources in all the available RATs.

Under this framework, this chapter starts with an overview of the heterogeneous network concept in Section 6.1. Then, Section 6.2 provides a description of the main characteristics of the available RATs that will coexist with UMTS to form heterogeneous networks scenarios, with a particular focus on GERAN and WLAN. The envisaged architectures for interworking and coupling between UMTS, GERAN and WLAN will be discussed in Section 6.3, and a general framework for CRRM operation in advanced scenarios including flexible spectrum management is presented in Section 6.4. Finally, some specific implementations of CRRM algorithms will be discussed in Section 6.5.

6.1 HETEROGENEOUS NETWORKS

In parallel with the development of UMTS, other wireless access technologies have experienced significant growth and have now arrived in the mass-market. This is the case with the Wireless Local Area Networks (WLAN) and Personal Area Networks (PAN) technologies, with IEEE 802.11 and Bluetooth, respectively, being two of their most representative members. The use of non-licensed frequency bands together with the requirement of relatively simple network infrastructures, mainly relaying on the existing IP core network, has allowed a fast rollout of the WLAN technologies. In turn, these are able to provide much higher bandwidths than currently available in the existing cellular systems, though with reduced coverage areas. These aspects have contributed to an increase in the popularity of WLAN in recent years, which is expected to continue with the appearance of new multimedia services.

On the other hand, the extension of GSM to GPRS, including packet transmission capabilities in the radio interface, has been the first milestone in the evolutionary path of 2G cellular systems towards UMTS. Although GPRS penetration probably has not been as successful as was originally expected, its compatibility with the popular GSM technology together with the fact that initial UMTS releases will make use of the same GSM/GPRS core network infrastructure, provide the basis for believing that the coexistence and interaction between UMTS and GSM/GPRS technologies constitutes one of the key points for the success of 3G technologies.

Radio Resource Management Strategies in UMTS   J. Pérez-Romero, O. Sallent, R. Agustí and M. A. Díaz-Guerra
© 2005 John Wiley & Sons, Ltd

As a matter of fact, GSM/GPRS has also followed its independent path towards 3G systems, with the development of an improved radio access technology that allows higher bit rates thanks to the use of more efficient modulation schemes. The term EDGE (Enhanced Data rates for GSM Evolution) is used to refer to this improved system and GERAN (GSM/EDGE Radio Access Network) is the name of the evolved GSM radio access network which includes these new capabilities.

As a result, scenarios where UMTS is to be deployed will probably differ from initial expectations, and it will have to coexist not only with previous 2G and 2.5G systems but also with WLAN and WPAN systems. However, rather than being an inconvenience for UMTS deployment, these new scenarios should be regarded as a new challenge to offer users efficient and ubiquitous radio access by a coordinated use of the available Radio Access Technologies (RATs). In this way, not only the user can be served through the RAT that best fits the terminal capabilities and service requirements, but also a more efficient use of the available radio resources can be achieved from the operator's point of view. This challenge calls for the introduction of new RRM algorithms operating from a common perspective that take into account the overall amount of resources in the available RATs, and therefore are referred to as CRRM (Common Radio Resource Management) algorithms. Furthermore, for a proper support of such algorithms, the currently existing network architectures must be modified accordingly to ensure the desired interworking capabilities among the different technologies [1][2].

The heterogeneous network concept for beyond 3G systems is intended to propose a flexible and open architecture for a large variety of different wireless access technologies, applications and services with different QoS demands, as well as different protocol stacks. Figure 6.1 shows an example of one

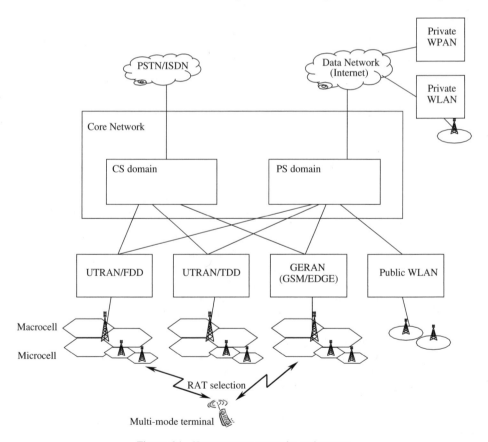

Figure 6.1   Heterogeneous networks environment

such heterogeneous networks scenario. It is constituted by several radio access networks (RAN) interfacing a common core network. Radio access networks include cellular networks (e.g. UTRAN with the two modes FDD and TDD, or GERAN), which may in turn be subdivided into different cellular layers (e.g. macro, micro or picocells) depending on the expected coverage area, and also other public non-cellular access networks (e.g. WLAN). Typically, the core network infrastructure is subdivided in the CS and PS domains providing access to external networks, for example, PSTN or Internet. These external networks can also include other private WLAN or WPAN networks, from which terminals may also have access to the core network. The scenario assumes the existence of multi-mode terminals, possibly with reconfigurability capabilities, which can be connected to multiple access networks either in different time instants or even simultaneously.

The common core network deals with all network functionalities and operates as a single network, while the RANs handle only those specific functions of the corresponding radio access technology, mainly related to physical and link layer issues. Ideally, services must be delivered through the most efficient radio access network according to the resource availability, the service QoS requirements, the terminal capabilities and preferences as well as the operator's policies, by means of a common manager of the radio resources in each RAN. In this way, the heterogeneous network becomes transparent to the final user and the so-called ABC (Always Best Connected) paradigm [3], which applies to the connection to the RAT that offers the most efficient radio access at each instant, can be achieved.

The achievement of these goals requires definition of the appropriate CRRM algorithms, but it also has implications for the network architectures in terms of interworking and coupling between RANs depending on how they are interconnected to the core network. These new network architectures must face requirements such as mobility management for seamless handover, including vertical handover between RATs, as well as authentication and billing mechanisms. Furthermore, convergence between radio access networks is necessary, which requires a standardisation effort and business commitment to support it.

6.2 RADIO ACCESS NETWORKS CHARACTERISATION

This section provides a brief description of the GERAN and WLAN radio access networks, which are the two RANs more usually combined with UTRAN in a heterogeneous network scenario. The purpose is not to give a detailed description of the radio interface, as was the case of UMTS in Chapter 3, but to provide an overview of the main aspects that have an impact on the design of the CRRM algorithms, as will be described in Section 6.5.

6.2.1 GERAN

6.2.1.1 GERAN Architecture

The architecture of the GSM/EDGE Radio Access Network (GERAN) is depicted in Figure 6.2. Essentially, the constituent elements are the same that exist in the GSM access network but they have been upgraded to include the new functionalities of GPRS for packet transmission and EDGE to increase the bit rates at the radio interface. The GERAN is composed of different entities that allow the connection of the mobile terminal to the core network, denoted as BSSs (Base Station Subsystems).

Each BSS is composed of a BSC (Base Station Controller), which is interconnected with the core network, and several BTSs (Base Transceiver Stations). The BTS is connected to the mobile terminals through the radio interface (denoted as Um) and handles the radio transmission procedures. In turn, the BSC controls the available resources at the BTSs and allocates and deallocates them depending on the service needs. Furthermore, it controls the handover procedures between the BTSs that are connected to it. The interconnection between BTS and BSC is done through the A-bis interface. With the introduction of GPRS functionalities, a new element was included in the BSC, the PCU (Packet Control Unit), responsible for the allocation of resources to GPRS users, according to packet scheduling algorithms. When comparing the GERAN architecture with the UTRAN architecture, it is worth noting that the BSC is equivalent to the RNC while the BTS is equivalent to the Node B.

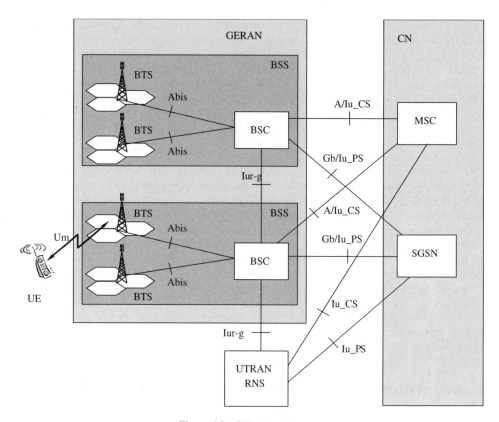

Figure 6.2 GERAN architecture

In the original GSM architecture, the BSC was interconnected through the A interface with the Core Network, particularly with the MSC (Mobile Switching Centre), responsible for handling the establishment and routing of CS calls to the external network. With the inclusion of GPRS, two new CN elements were added to deal with packet transmission. These are the SGSN (Serving GPRS Support Node), which handles the mobility functions for packet services, and the GGSN (Gateway GPRS Support Node), which allows the interconnection with the external IP networks. With this new infrastructure, a new interface – called the Gb interface – was developed to interconnect the BSC and the SGSN for GPRS users.

With the introduction of UMTS, the requirement to have access-independent services for multi-mode UMTS/GSM/GPRS terminals, leads to the modification of the interconnection between the BSC and the CN with two different alternatives [4]. The first one consists of further developing the capabilities of the A and Gb interfaces, originally defined in the GERAN architecture, to match UMTS requirements. This is the solution adopted for terminals not yet supporting UMTS release 5. The second alternative, for release 5 and release 6 terminals, consists of extending the functionalities of the GERAN BSS to connect it with the CN through the Iu interface (divided into Iu_CS and Iu_PS for CS and PS services, respectively) introduced in the UMTS architecture. This second alternative allows a full integration of UTRAN and GERAN under a single UMTS multi-access network. Furthermore, for a proper interworking between UTRAN and GERAN, the Iur-g interface is introduced to establish the interconnection between RNCs and BSCs as well as between BSCs.

The evolution of the different interfaces is mainly related to the support of new service types, in addition to the speech service supported by the A interface and the interactive and background services supported by the Gb interface. Specifically, the inclusion of streaming and conversational services is under consideration for the Gb interface, which would allow the provision of the same service classes through GERAN and through UTRAN. Additionally, some modifications are also introduced in GERAN to cope with the different services in a more efficient way. This includes the support at the BSC of header adaptation mechanisms for RTP/UDP/IP protocols (i.e. the PDCP functionality at the RNC in UTRAN) as well as the use of AMR (Adaptive Multi-Rate) speech encoding for enhanced speech quality.

From the QoS management point of view, full system integration between UTRAN and GERAN can be ensured by the adoption in GERAN of the same service classes and the same QoS attributes defined in UMTS, which is possible thanks to the use of the Iu interface in the BSC.

6.2.1.2 Description of the GERAN Radio Access Technology

This section presents the fundamental characteristics of the GSM/GPRS radio transmission, since they are required for a proper understanding of the CRRM operation between UTRAN and GERAN. The reader is referred to References 5–8 for more details.

GSM/GPRS makes use of a hybrid FDMA/TDMA multiple access system with FDD. There are 125 carrier frequencies in the band 890–915 MHz for uplink and 935–960 MHz for downlink and 375 carrier frequencies in the band 1710–1785 MHz for uplink and 1805–1880 MHz for downlink [9]. The separation between carriers is 200 kHz, and each carrier is organised into 4.615 ms frames, each of them having eight time slots, thus allowing the allocation of up to eight users per carrier.

Each BTS has one or more carriers allocated. One of them, denoted as beacon frequency, has one slot reserved for the transmission of broadcast and control channels, and thus only seven users can be allocated in it. When there are several carriers in one BTS, other slots can also be reserved for the use of common and dedicated control channels.

Different types of logical channels, including both control and dedicated information, are multiplexed in the GSM/GPRS frame structure. The initial releases of GSM only included support for the transmission of speech with a bit rate of 13 kb/s and for Circuit Switched Data (CSD) with a bit rate of up to 9.6 kb/s. In both cases, this corresponds to the permanent allocation of a time slot to the user for the whole duration of the connection. Additionally, a SMS (Short Message Service) was also available, making use of some remaining capacity in control channels. In all the cases, the modulation scheme used is GMSK (Gaussian Minimum Shift Keying).

With the introduction of GPRS, certain slots of the frame structure are devoted to data transmission in PS mode. Typically, a packet session (e.g. a web browsing session) requires the transmission of different data flows (e.g. the different objects of a web page), denoted as TBFs (Temporary Block Flows), each one composed of a given number of radio blocks. A GPRS channel, denoted as PDCH (Packet Data Channel), must be allocated to the user for the transmission of a TBF either in the uplink or in the downlink direction. The PDCH is characterised by one slot in the uplink and the downlink. It is possible that several users share the same time slot (i.e. the same PDCH) during the transmission of their TBFs. In this case, a centralised scheduling algorithm multiplexes the TBFs of the different users in the PDCH and decides the order in which these users occupy the channel. The transmission of one radio block requires the allocation of four time slots (i.e. one time slot in four consecutive frames).

The rate of transmitted radio blocks in a PDCH is one radio block per 20 ms, so the available bit rate depends on the amount of application data that is included in the radio block that in turn depends on the RLC/MAC headers and the channel coding scheme. Four channel coding schemes exist in GPRS depending on the degree of protection of the transmitted data, leading to four different bit rates. They are listed in Table 6.1.

The selection of the appropriate modulation scheme should be done by taking into account the channel conditions of each mobile terminal to determine the adequate degree of protection and therefore maximise the application bit rate. As an example, if the channel conditions are very good, so that very few errors occur, CS-4, which provides the highest bit rate, could be the optimum scheme. However, if the channel is bad, CS-4 may not be suitable since it does not provide any type of protection to the transmitted data. The selection of the modulation is done by means of link adaptation algorithms.

Table 6.1 Modulation schemes in GPRS

Scheme	Maximum RLC/MAC data rate (kb/s)	Nominal data rate (kb/s)	Application data rate (kb/s)
CS-1	8	9.05	7.7
CS-2	12	13.4	11.5
CS-3	14.4	15.6	13.8
CS-4	20	21.4	19.2

The bit rates in Table 6.1 are the maximum that can be achieved by a terminal that occupies a PDCH throughout the whole period. Obviously, if several terminals are multiplexed on the same PDCH, the bit rates will be decreased depending on the scheduling algorithm behaviour, which may prioritise one user over another depending on their QoS requirements.

The application bit rates can be increased by allocating several PDCHs (i.e. several time slots in a given frame) to the same user. This depends on the terminal multislot capabilities, which determine the maximum number of time slots that can be allocated in the uplink and downlink directions. There are 29 multislot classes characterised by the maximum number of slots in the uplink, the downlink, and the total sum including both uplink and downlink. They are listed in Table 6.2. It is not necessary for

Table 6.2 Multislot capability classes

Multislot class	Maximum number of slots			Type
	Rx (DL)	Tx (UL)	Sum (UL + DL)	
1	1	1	2	1
2	2	1	3	1
3	2	2	3	1
4	3	1	4	1
5	2	2	4	1
6	3	2	4	1
7	3	3	4	1
8	4	1	5	1
9	3	2	5	1
10	4	2	5	1
11	4	3	5	1
12	4	4	5	1
13	3	3	N/A	2
14	4	4	N/A	2
15	5	5	N/A	2
16	6	6	N/A	2
17	7	7	N/A	2
18	8	8	N/A	2
19	6	2	N/A	1
20	6	3	N/A	1
21	6	4	N/A	1
22	6	4	N/A	1
23	6	6	N/A	1
24	8	2	N/A	1
25	8	3	N/A	1
26	8	4	N/A	1
27	8	4	N/A	1
28	8	6	N/A	1
29	8	8	N/A	1

Table 6.3 Modulation and Coding schemes in EDGE

Scheme	Modulation	Maximum net RLC/MAC data rate (kb/s)	Mean application data rate (kb/s)
MCS-9	8-PSK	59.2	56.8
MCS-8		54.4	52.2
MCS-7		44.8	43.0
MCS-6		29.6	28.4
MCS-5		22.4	21.5
MCS-4	GMSK	17.6	16.9
MCS-3		14.8	14.2
MCS-2		11.2	10.7
MCS-1		8.8	8.4

type 1 terminals to transmit and receive at the same time, but type 2 terminals are required to transmit and receive simultaneously. As an example, a terminal belonging to class 10 could have up to four slots in the downlink direction and up to two slots in the uplink. However, since the total number of slots must not exceed 5 and, for example, 2 slots are already allocated in the uplink, only three can be allocated in the downlink. For this terminal, the maximum RLC/MAC bit rate that could be achieved in the uplink would be $2 * 20\,\text{kb/s} = 40\,\text{kb/s}$ using CS-4 (see Table 6.1), while in the downlink it would be $4 * 20\,\text{kb/s} = 80\,\text{kb/s}$.

EDGE (Enhanced Data rates for GSM Evolution) is an evolution of GSM/GPRS radio interface that enhances the bit rates both for circuit switched services (ECSD – Enhanced CSD) and packet switched services (EGPRS – Enhanced GPRS) using a different modulation technique (i.e. 8-PSK) in conjunction with new channel coding schemes.

A total of nine modulation and coding schemes (MCS) have been defined and are listed in Table 6.3. Note that with these new schemes, the maximum theoretical MAC/RLC bit rate that could be achieved would be $59.2 * 8 = 473.6\,\text{kb/s}$, corresponding to a terminal with eight time slots (e.g. of multislot class 29) operating with MCS-9. In any case, note that the increase in the bit rate is achieved at the expense of an increase in power consumption, since the higher modulation and coding schemes require higher signal to noise ratios.

6.2.2 WLAN

Wireless Local Area Networks (WLAN) were originally developed as an alternative to wired networks to allow data transmission in environments where the installation of wired infrastructures was hard to achieve. Then, they were thought of simply as a wireless extension of the traditional wired networks like Ethernet or Token Ring. One of their main features was the ease of the installation because they do not require their own core infrastructure but they are simply extensions of other existing networks.

Several standards have been developed for Wireless Local Area Networks (WLAN). The dominating one is the IEEE 802.11, whose specification was approved in 1997 as a member of the IEEE 802 family of standards for local and metropolitan area networks [10]. The European response to the WLAN challenge was the HIPERLAN family, developed by ETSI, and including HIPERLAN1 and HIPER-LAN2 standards [11]. The latter included support for QoS and was able to handle larger traffic volumes than the 802.11. Furthermore, the HIPERLAN2 specifications included interworking concepts with 2.5G and 3G systems, which make this standard a good candidate to be used in heterogeneous networks. Nevertheless, the main problem with HIPERLAN2 was not the technology itself but the lack of commercial support from WLAN vendors. As a result, when the WLAN market took off in 2002, the only products available belonged to the IEEE 802.11 family, which has therefore become the most popular standard.

In fact, IEEE 802.11 is a family of standards, denoted as 802.11x, which covers the specifications of the MAC and physical layers for WLAN transmission. Since WLAN are only wireless extensions of fixed networks, the standards do not cover layer 3 and above aspects.

6.2.2.1 WLAN Architecture

In terms of network architecture, WLANs based on IEEE 802.11x are composed of Basic Service Sets (BSS), constituted by a set of stations (either mobile or fixed) that communicate with each other. The simplest WLAN configuration is achieved when all the stations are mobile and there is no connection to a wired network, so that only communication between stations is possible. This special type of BSS is called independent BSS (IBSS) and constitutes an ad hoc network. The association is the dynamic procedure used by a station to become member of a BSS. When a mobile station moves out of the IBSS, communication with other stations is not possible.

Although ad hoc networks are a possible WLAN configuration, the most usual case is to have infrastructure mode WLANs, shown in Figure 6.3. In this case, there is a special type of station, called Access Point (AP), that interconnects the BSS with the Distribution System (DS), typically formed by a wired infrastructure (e.g. a wired LAN). The DS allows communication among stations belonging to different BSSs.

The whole set of interconnected BSSs through the DS is called Extended Service Set (ESS), and is seen by the upper layers (i.e. the LLC) as a single LAN that can be connected to an external data network (see Figure 6.3).

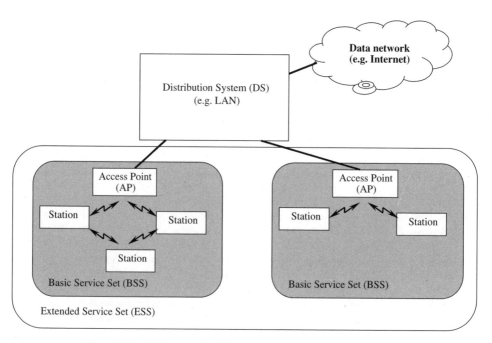

Figure 6.3   Architecture of a IEEE 802.11 WLAN in infrastructure mode

IEEE 802.11 specifies a set of services provided by the stations and by the Distribution System (DS). For the stations, the following set of services is identified:

- Authentication. This service is used to prove the identity of one station to another of the BSS. Without this proof of identity, the station is not allowed to use the WLAN.
- Deauthentication. This is used to eliminate a previously authorised user from any further use of the network.
- Privacy. This service provides a protection level for the data transmitted through the WLAN equivalent to that existing in a wired network.
- MSDU (MAC Service Data Unit) delivery. This service provides a reliable delivery of data frames from the MAC layer in one station to the MAC layer in other stations.

In turn, the services provided by the DS are as follows:

- Association. This is used to make a logical connection between a mobile station and an AP. This logical connection is necessary for the DS to know where and how to deliver data to the mobile station. The logical connection is also necessary for the access point to accept data frames from the mobile station and to allocate resources to support the mobile station.
- Disassociation. This service is used to remove a current association. It can be requested by the mobile station if it no longer requires the WLAN services or it can be forced by the access point.
- Re-association. This service is used by the mobile stations as they move throughout the ESS when they lose contact with the current AP and need to be associated with a new AP, similar to a handover procedure in cellular networks. During the re-association procedure, the mobile station includes information about the old AP so that the new and the old AP can communicate in order to obtain data frames that might be waiting for delivery at the old station.
- Distribution. The distribution service is invoked by the AP when it receives a data frame from a given station. The purpose of this service is to determine if the frame should be either sent back into its own BSS, for delivery to another station associated with the AP, or into the DS for delivery to another station associated with a different AP or to a network destination outside the WLAN.
- Integration. This service connects the IEEE 802.11 WLAN to other LANs, either wired or wireless.

6.2.2.2 Description of the WLAN Radio Access Technology

With respect to the physical layer, three different standards exist in the IEEE 802.11 family – the IEEE 802.11a, IEEE 802.11b and IEEE 802.11g – leading to three different radio technologies, all of them operating under a common MAC layer [12]. The main features of these three standards are shown in Table 6.4.

Table 6.4   Characteristics of the IEEE 802.11x WLAN standards

	802.11b	802.11a	802.11g
Frequency	2.4 GHz	5 GHz	2.4 GHz
Physical layer	Direct Sequence Spread Spectrum (DS-SS)	Orthogonal Frequency Division Multiplexing (OFDM)	Orthogonal Frequency Division Multiplexing/ Complementary Code Keying OFDM/CCK
Channel bandwidth	22 MHz	20 MHz	20 MHz
Typical range	30–40 m indoor 80–120 m outdoor	20–30 m indoor 60–100 m outdoor	25–35 m indoor 60–100 m outdoor
Transmission rates	1,2,6,11 Mb/s	6,9,12,18,36,54 Mb/s	6,9,12,18,36,54 Mb/s
Data throughput (on top of MAC)	0.8–6 Mb/s	4–20 Mb/s	4–20 Mb/s
MAC protocol	CSMA/CA in Distributed Coordinated Function Mode (DCF) (optional) Polling Based in Point Coordination Function (PCF)		

One of the most widely used standards by current WLANs is the IEEE 802.11b standard, usually known as Wi-Fi technology. It operates in the 2.4 GHz band supporting transmission rates up to 11 Mb/s with a corresponding data throughput of around 6 Mb/s. Transmission is based on Direct Sequence Spread Spectrum with CCK (Complementary Code Keying) modulation, leading to a total bandwidth of 22 MHz. Due to the use of the same frequency band as Bluetooth devices, it is susceptible to suffering interference from these devices, which is problematic mainly for high speed data applications.

IEEE 802.11a operates in the 5 GHz band under an OFDM physical layer. It supports transmission rates up to 54 Mb/s, and the maximum data throughput is 20 Mb/s, although the channel bandwidth is smaller than in IEEE 802.11b, thus achieving a higher spectral efficiency.

Finally, IEEE 802.11g was the last standard to be approved, and it operates in the 2.4 GHz band with OFDM/CCK. It provides backward compatibility with IEEE 802.11b.

With respect to the ETSI standards, HIPERLAN2 is also based on OFDM with a maximum transmission rate of 54 Mb/s, operating in the 5 GHz frequency band. Therefore, the physical technology is similar to that of IEEE 802.11a, although the upper layers of both systems are incompatible.

The MAC layer defined for the IEEE 802.11x standards is based on the CSMA/CA (Carrier Sense Multiple Access with Collision Avoidance) protocol to regulate the access of the different terminals to the shared radio transmission medium. This is a listen-before-talk protocol which means that the different mobile stations sense the channel and only start transmission if no other station is already transmitting. This protocol does not require centralised control and is therefore a suitable solution for ad hoc WLANs. Nevertheless, most WLANs operate under a centralised architecture with different terminal stations transmitting towards a single access point connected to the wired network.

The use of a completely distributed MAC protocol may not be an efficient solution because the access point that carries the downlink traffic of all the mobile stations should have a certain priority to transmit over the rest of mobile stations. Moreover, the CSMA/CA protocol is not able to make distinctions between different types of traffic so it cannot take into account delay sensitive or high priority traffic. As a result, the MAC protocol standardised for IEEE 802.11 provides an optional centralised control in addition to the distributed access principles based on CSMA/CA. This is achieved by subdividing the MAC layer into two differentiated functions, as depicted in Figure 6.4. At the lower layer is the DCF

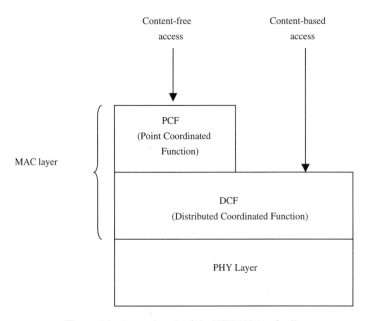

Figure 6.4 Protocol stack of the IEEE 802.11x family

(Distributed Coordinated Function), which provides a content-based access according to CSMA/CA, and is used by non delay sensitive and low priority traffic (e.g. asynchronous traffic). In contrast, the PCF (Point Coordinated Function) operates above the DCF, providing content-free access to high priority traffic by means of a centralised manager that has the ability to block the content-based traffic.

The functions of the MAC layer include also authentication, association and re-association services, as well as optional encryption/decryption procedures and power management to reduce power consumption in mobile stations.

Further enhancements of the MAC layer are included in the IEEE 802.11e standard, with the goal of providing QoS support based on the PCF/DCF functionalities. Similarly, the task group k within IEEE 802.11 is trying to standardise metrics to make easier the deployment of larger WLANs. This includes an interface to the core network for radio resource measurements that can be used as triggers for RRM procedures such as vertical handovers and RAT selection within a heterogeneous network.

6.3 INTERWORKING AND COUPLING AMONG RADIO ACCESS NETWORKS

The interworking among the different RANs in a heterogeneous wireless network can be realised with different degrees of coupling. These depend on how the access networks are interconnected and how aspects such as common operations or exchange of measurements can be carried out to achieve a common control of the radio access networks. This section describes the interworking approaches that are being standardised for UTRAN, GERAN and WLAN.

6.3.1 UTRAN/GERAN INTERWORKING

As was mentioned in Chapter 1, the UMTS architecture was devised initially from the perspective of having the same basic Core Network architectural elements that existed in GSM/GPRS systems in both the CS and the PS domains. In subsequent releases, the core network was modified with the inclusion of the IP protocol at the different elements of the network in both CS and PS domains, leading to the existence of specific 3G core networks. In all the releases, the interconnection of the RNS with the core network is done by means of the Iu interface, split in Iu_CS and Iu_PS for circuit and packet switched services. In turn, with respect to the GSM/GPRS access network, and as explained in Section 6.2.1, it was originally connected to the 2G core network by means of the A and Gb interfaces for circuit and packet switched services, respectively. In this case, starting with release 5, the evolution also introduced the support of the Iu interface in the GERAN BSS, thus allowing the support of conversational and streaming services in GPRS networks (except the interactive and background services already supported by the Gb interface) and opening up the possibility for multimode terminals to receive the same type of services independent of the RAN to which they are connected.

Under the above considerations, the interworking between UTRAN and GERAN is obtained through the connection of the RNC and the BSC to the same 3G core network, particularly to the SGSN node through the Iu interface (see Figure 6.2). This achieves a tight coupling between both access networks, and allows the use of common radio resource management strategies that try to optimise the network performance from an overall perspective. An additional degree of coupling is also possible if the Iur-g interface is available to interconnect the BSC of GERAN and the RNC of UTRAN (see Figure 6.2). This is usually referred to as a very tight coupling approach and it allows the execution of common resource management strategies directly between controllers of the access networks, without relying on the communication with entities in the core network.

6.3.1.1 Common Radio Resource Management Architectures

Common Radio Resource Management (CRRM) strategies are intended to achieve an efficient utilisation of the radio resources in heterogeneous scenarios, by means of a coordination of the available resources

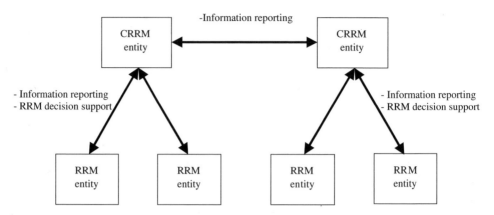

Figure 6.5   CRRM functional model

in the existing Radio Access Technologies. Therefore, CRRM is a general concept, applicable to any combination of RATs, although the specific implementation and the degree of coordination depends highly on the degree of coupling that exists between the specific radio access networks. In the case of GERAN and UTRAN, where a tight coupling exists, several possibilities are being standardised for the operation of CRRM strategies [1].

The functional model assumed in 3GPP for CRRM operation considers the total amount of resources available for an operator divided into radio resource pools. Each radio resource pool consists of the resources available in a set of cells, typically under the control of a RNC or a BSC in UTRAN and GERAN, respectively. Two types of entities are considered for the management of these radio resource pools [2], as shown in Figure 6.5:

- The RRM entity, which carries out the management of the resources in one radio resource pool of a certain radio access network. This functional entity involves different physical entities in the RNS or BSS depending on the specific considered functions, although for representation purposes it is usual to assume that the RRM entity resides in the RNC or the BSC. Note that different RRM entities do not necessarily belong to different radio access technologies, but it is possible to have different RRM entities in the same radio access technology.
- The CRRM entity, which executes the coordinated management of the resource pools controlled by different RRM entities, ensuring that the decisions of these RRM entities also take into account the resource availability in other RRM entities. Each CRRM entity controls several RRM entities and may communicate with other CRRM entities as well, thus collecting information about other RRM entities that are not under its direct control.

The interactions among RRM and CRRM entities mainly involve two types of functions:

(a) Information reporting function. This allows the RRM entity to indicate to its controlling CRRM entity either static or dynamic information such as the cell capacity or different types of measurements. The reporting is controlled by the CRRM entity, which can request it either at given instants or according to periodical or event-triggered reports. The exchange of information is also possible among different CRRM entities to learn the status of their corresponding RRM entities.

There are mainly two types of information to be reported to the CRRM entity:

- Dynamic common measurements on cells controlled by a distant RRM entity. These measurements include the current 2G cell load in both the CS and PS domains measured as a fraction of the occupied cell capacity, the 3G cell load, information such as the transmitted carrier power, the received total wideband power, interference measurements, etc.
- Static information on cells controlled by a distant RRM entity. This includes knowledge about the cell relations (e.g. if they are overlapped or if they belong to different HCS layers), the cell capabilities (e.g. whether a cell supports GPRS, EDGE, etc.), the cell capacities in the CS and PS domains (e.g. the number of available time slots) or the available QoS in the PS domain (e.g. maximum bit rate per service).

(b) RRM decision support function. This function describes how the CRRM entity affects the decisions taken by the RRM entities under its control. Depending on how the CRRM is implemented in the network, several possibilities exist for the RRM decision support function. For example, it is possible that the CRRM simply advises the RRM entity, so that the RRM remains the master of the decisions, or it is possible that the CRRM is the master so that its decisions are binding on the RRM entity. Similarly, several degrees of coupling or interaction exist between the CRRM and the RRM entities, ranging from where the CRRM is involved in any RRM decision (e.g. in every intersystem handover) to when the CRRM simply dictates policies for RRM operation and the RRM entity takes decisions according to these specific policies.

With respect to the network topologies supporting the previous CRRM functional model, there are two different approaches, namely the CRRM server and the integrated CRRM solutions, which impact the way the CRRM functions are realised in practice. The two approaches are now described starting with the CRRM server solution.

CRRM Server This approach introduces the CRRM functionality in a stand-alone node, denoted as CRRM server (CRMS), and common to the UTRAN and GERAN, thus constituting a centralised approach, as depicted in Figure 6.6. RRM and CRRM entities are then located in different physical nodes and interconnected through an open interface towards the RNC and the BSC. When considering heterogeneous networks with RATs other than GERAN or UTRAN, the CRMS can also be connected to the corresponding RRM entities in these networks.

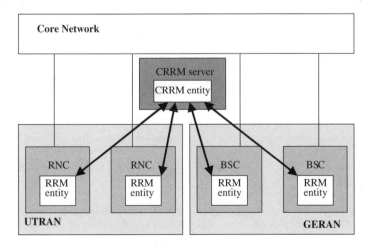

Figure 6.6   CRRM server approach

The CRRM server solution allows the taking of decisions, for example, of intersystem handover or call admission control, in a more optimal way because the CRMS can collect the information from all the available systems, layers and modes. In principle, the specific RRM algorithms are not completely moved to the CRMS but are supposed to be executed locally by the RRM entities of each RAN, while the CRRM server role is reserved for those procedures involving different systems, modes or layers.

Essentially, there are three types of services provided by the CRMS to other entities (i.e. a RNC, a BSC or another CRMS) in this centralised approach:

- Cell measurement gathering. This service belongs to the information reporting function defined between RRM and CRRM entities and is used to provide the CRMS measurements of the cells controlled by its RNCs or BSCs or the cells under the control of another CRMS. The reporting procedures are based on those of the UTRAN Iub and Iur interfaces, supporting on-demand, periodical or event-triggered reporting methods. There are four elementary procedures defined for this service: the measurement initiation, the measurement reporting, the measurement termination and the measurement failure.
- Prioritisation of the list of candidate cells of a UE for a specific operation. This service is requested by the BSC or the RNC whenever certain procedures such as handover or cell change order must be executed for a given terminal. The RNC or BSC sends the list of candidate cells of the UE to the CRMS, including the mobile measurements for these cells and information about the required quality of service. The CRMS applies specific CRRM algorithms and returns the prioritised list, which can be used by the RNC/BSC for example to select the cell to execute an intra-system handover. Note that this service is included in the RRM decision support function, allowing for the fact that local RRM algorithms are executed at the RNC or BSC taking into account the status of other RATs depending on the information provided by the CRMS in the prioritised list. It is also worth mentioning that the use that the local RRM makes of the prioritised list is not specified, so that this information can be considered simply as advice (i.e. the RRM is the master) or it can be binding on the RRM (i.e. the CRRM is the master).
- Error indication. This service is used to report errors detected in a received message exchanged either between CRMS and RNC/BSC or between two CRMSs.

Integrated CRRM The integrated CRRM approach is based on the fact that the current 3GPP standards already support most of the envisaged CRRM functionalities, such as intra-system and inter-frequency handovers or directed retry (see Section 5.3.3). Furthermore, the Iur and Iur-g interfaces already include almost all the necessary functions to support the CRRM procedures. Therefore, a natural approach consists of integrating the CRRM functionality in the existing UTRAN and GERAN nodes, leading to a distributed CRRM architecture as depicted in Figure 6.7.

Note in Figure 6.7, that the CRRM entity may be included either in all the RNC/BSCs or only in a subset of them. In the first case, only the reporting information function between different CRRM entities must be standardised, because the RRM decision support function between CRRM and RRM is done locally at the same physical entity. However, if only a sub-set of RNC/BSCs include the CRRM entity, the RRM decision support function must also be standardised [2].

Different procedures can be used in the integrated CRRM approach to exchange the common measurements and static information on cells controlled by the distant entities (RNC or BSC), depending on the considered interface.

The options provided by 3GPP recommendations for the exchange of common measurements are as follows (see Figure 6.8):

(a) Common Measurements over Iur-g (Figure 6.8a). This is the adopted solution when the BSC supports an Iur-g interface to interwork with the RNC. In this case, the measurement of common resources functionality of the RNSAP (Radio Network Subsystem Application Part) protocol is used (see [13] for details).

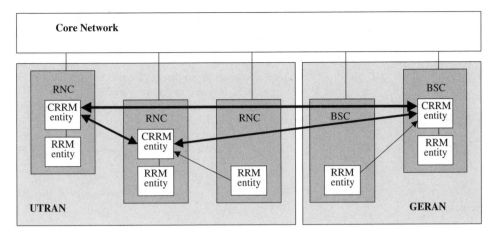

Figure 6.7   Integrated CRRM approach

(b) Common measurements over Iu/A interfaces (Figure 6.8b). This solution is used when the BSC does not support the Iur-g interface. In this case, communication between the BSC and the RNC is done through the MSC of the core network, using the A interface between BSC and MSC together with the Iu interface between the MSC and the RNC. Measurements are transmitted through the A interface in the messages of the BSSMAP (Base Station Subsystem Management Application Part) protocol handover procedures, inserted in some transparent information elements. Similarly, through the Iu interface, measurements are inserted in messages of the RANAP relocation procedure. Consequently, the availability of common measurements is limited by the periodicity of the above handover and relocation procedures in the network. For example, during high traffic situations, there

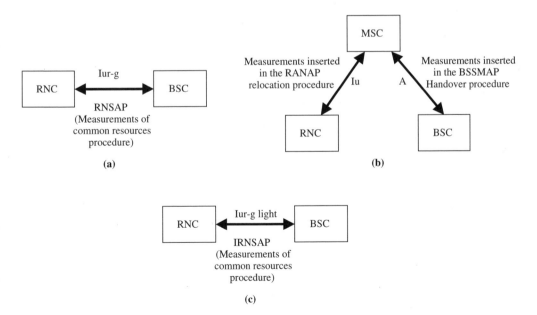

Figure 6.8   Procedures for common measurements exchange in the integrated CRRM approach

are more frequent inter-RAT handovers and thus more frequent measurements and more reliable information. In any case, this solution is not as efficient as the use of the Iur-g interface.

(c) Common measurements over a Iur-g light interface (Figure 6.8c). This solution is also used when the BSC does not support the Iur-g interface. It consists of the introduction of a new interface between the BSC and the RNC specifically designed to support the exchange of common measurements and information on cells controlled by distant entities. In order to ensure compatibility with the Iur-g interface, it is proposed that the Iur-g light interface is based on a RNSAP based protocol denoted as IRNSAP (Inter Radio Network Subsystem Application Part) using the measurement of common resources functionality (see [13]). Furthermore, the IP stack defined in release 5 for the Iur interface is proposed for use as the transport solution, in order to retain compatibility with 2G BSC equipments.

In turn, for the exchange of static information, two possibilities are envisaged in the 3GPP specifications:

(a) Use of existing Iur features. This solution is similar to the one for common measurements through the Iu/A interface in the sense that the static information is inserted in specific information elements of some messages of the RNSAP protocol, such as the Radio Link Setup Response or the Radio Link Addition Response.

(b) Use of the Iur-g or the Iur-g light interfaces. When using the Iur-g interface, the information exchange procedures of the RNSAP protocol are used. In the case of the Iur-g light interface it is envisaged these same procedures will introduced in the new IRNSAP protocol.

6.3.2 UTRAN/WLAN INTERWORKING

The interworking between UTRAN and WLAN is devised from a different perspective than between UTRAN and GERAN. The main reason is that WLANs are being generally deployed by parties not belonging to 3GPP, and consequently they do not follow the same networking architectures of 3GPP systems like UMTS or GSM/EDGE. Furthermore, the number of different WLAN standards existing today puts difficulties in developing general interworking mechanisms.

Despite the above inconveniences, the successful deployment of WLAN systems worldwide and the high data rates offered by such systems make them an interesting alternative for increasing the wireless coverage of cellular systems in hotspots locations. Consequently, efforts exist in the standardisation bodies to address the integration of 3GPP and WLAN technologies.

Such UTRAN/WLAN interworking should take into account both technical and non-technical aspects, because the environments where both systems coexist (i.e. public, corporate or residential environments) may involve different administrative domains and different WLAN owners, thus leading, for example, different security, billing or authentication requirements.

In the context of 3GPP standardisation, six different scenarios of UTRAN/WLAN interworking have been identified [14]. They are devised in a progressive way, so that each scenario represents a further step in the integration process with respect to previous ones. These scenarios are characterised by the following features:

- Scenario 1. Common billing and customer care. This scenario is the simplest one and in fact does not represent a real interworking between 3GPP and WLAN. In this case, the customer simply receives a single bill including 3GPP and WLAN services provided by the mobile operator, who offers a user name and password to the user to access the WLAN. The security level of 3GPP and WLAN systems may be independent. Therefore, this scenario does not require any particular standardisation activity.
- Scenario 2. 3GPP system based access control and charging. In this scenario, the authentication, authorisation and accounting (AAA) to the WLAN is provided by the 3GPP operator following the

same procedures as in UMTS/EDGE services. This means that the same USIM used in cellular systems can be used for WLAN access. Nevertheless, the WLAN access only includes the services that are normally accessible through Internet, and not the UMTS services. Therefore, if the access to the 3GPP and to the WLAN services is done from the same device (e.g. laptop), separate sessions must be used. From the operator perspective, it is an easy extension of the subscriber capabilities to include WLAN access since the same 3GPP customer base is used. In turn, from the subscriber point of view, a more secure access than simply a user name and password, as in scenario 1, is obtained.

- Scenario 3. Access to 3GPP system PS based services. Apart from including the features from scenario 1 and scenario 2, this scenario allows access to the 3GPP packet services (e.g. instant messaging, location-based services, etc.) through WLAN. Nevertheless, this scenario does not require service continuity, which means that if the user starts receiving a service through UMTS, it is not possible to move to WLAN and continue with the same service session.

- Scenario 4. Service continuity. This scenario extends scenario 3 by including a vertical handover procedure that allows transferring a given service session from WLAN to UMTS/EDGE or vice versa. Although the user does not need to re-establish the service when changing the access network, such change may be noticeable for the user (e.g. in the form of brief interruptions in the data flows), and even some change in the quality of service may occur due to the different capabilities of the access networks. Furthermore, depending on the services supported by the new access network, it is possible that some sessions are terminated during the handover (e.g. in the case of certain services with stringent delay constraints, which may not be supported by WLAN systems).

- Scenario 5. Seamless services. This scenario provides the same PS based services and vertical handover capabilities of scenario 4 but supports seamless service continuity, which means that the occurrence of events such as service interruptions or data loss during the network switch is minimised. In this way, the user does not notice significant differences in the service provision when changing the access technology.

- Scenario 6. Access to 3GPP CS systems. In this scenario, the operator provides the access to circuit switched services through the WLAN. Similar to the scenario 5 for PS services, seamless mobility should be also available for CS services. It is worth mentioning that the provision of CS services through WLAN does not necessarily require the packet access nature of WLAN being changed.

Note that the above scenarios are general enough from the point of view of ownership and do not assume any specific relationships between the 3GPP network operator and the WLAN owner. Therefore, the WLAN owner may be either a 3GPP or non-3GPP network operator or even an entity providing WLAN access in a local area (e.g. an hotel owner or an airport authority).

6.3.2.1 Coupling Alternatives

The above six scenarios identified in 3GPP for UMTS/EDGE and WLAN interworking reflect different degrees of coupling between the 3GPP and WLAN networks that can be mapped into different network architectures to support the interworking requirements of each scenario [15]. These degrees of coupling are now described.

(a) Open coupling. This corresponds to the simplest case in which two different access and transport networks are used for UMTS and for WLAN. As a result, there is no real interworking between both technologies and it would be the solution used for scenario 1. This architecture is shown in Figure 6.9. Note that the mobile terminal may access the external Internet either through WLAN or through the 3GPP network, and the data flows follow independent paths.

(b) Loose coupling. In the loose coupling architecture, the WLAN and the 3GPP networks are connected at the Gi interface, so the user data flow does not go through the GGSN or SGSN core network elements of the cellular operator network. This allows a relatively simple migration of the current open coupling architecture to include WLAN and 3GPP interworking facilities. Depending on the

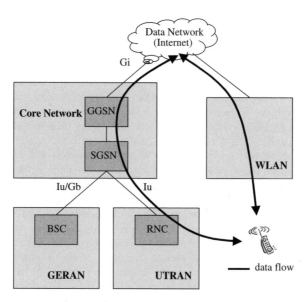

Figure 6.9   Open coupling WLAN/3GPP architecture

new core network elements and related control signalling and user planes, different possibilities can be considered under the loose coupling architecture, thus accommodating different interworking scenarios.

The first possibility, shown in Figure 6.10, agrees with 3GPP scenario 2, where 3GPP access control and charging mechanisms exist in the WLAN. In this case, the WLAN provides Internet access to the user while some control signalling is transferred to the operator CN. In particular, an additional 3GPP AAA server is located within the 3GPP network in order to retrieve authentication information and subscriber profile from the HSS of the subscriber's home 3GPP network. Based on this information and the authentication procedure, the AAA server authorises the WLAN access.

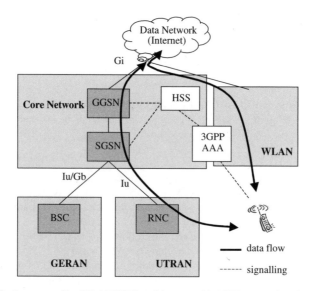

Figure 6.10   Loose coupling WLAN/3GPP architecture with 3GPP system based access control

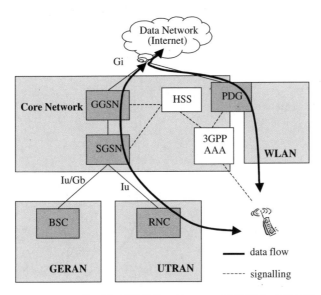

Figure 6.11 Loose coupling WLAN/3GPP architecture with PDG for 3GPP PS services

The second loose coupling approach agrees with 3GPP scenario 3, including the provision of 3GPP PS services through WLAN. To this end, as shown in Figure 6.11, a Packet Data Gateway (PDG) is used in the CN in addition to the AAA server, allowing access to the packet services. The PDG contains routing information for WLAN-3G connected users, routes the packet data from/to the user and performs address translation and mapping.

Finally, the third loose coupling alternative, illustrated in Figure 6.12, also agrees with 3GPP scenario 3, and the difference with the previous case is that Internet access is performed through

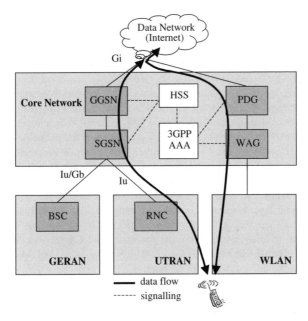

Figure 6.12 Loose coupling WLAN/3GPP architecture with access through the CN

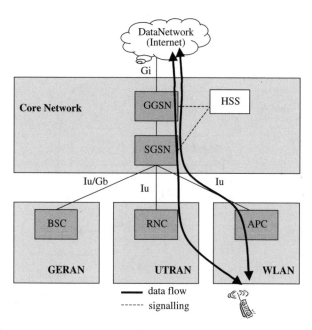

Figure 6.13   Tight coupling WLAN/3GPP architecture

the CN of the operator. The WLAN Access Gateway (WAG) is added to the CN and connected to the PDG. The WAG allows the routing of data from the WLAN via the PLMN by performing the proper IP tunnelling between the WAG and the PDG.

(c) Tight coupling. The rationale behind the tight coupling approach consists of devising an architecture that allows considering the WLAN as an additional UMTS access network like GERAN or UTRAN. This requires the WLAN to be connected to the UMTS core network through the Iu interface, as shown in Figure 6.13. As a result, the tight coupling approach should make it possible to extend to WLAN all the functionalities foreseen by the standard specifications in terms of interworking between UTRAN and GERAN (see Section 6.3.1). In particular, the role of the core network is to allow the exchange of information between the radio network controllers of these RANs, thus making possible the use of common radio resource management strategies.

Note in Figure 6.13 that in order to support RRM and CRRM functionalities, a new element is required in the WLAN with equivalent functions such as the RNC or the BSC for the UTRAN and GERAN [16]. This is the Access Point Controller (APC), which controls the radio resources of the access points to which the WLAN users are connected. The APC is linked to the SGSN of the mobile core network through an Iu or an Iu like interface.

With the tight coupling architecture, the mobility management of the different users can be performed through radio-layer handover, thus guaranteeing service continuity for the different UMTS service classes. Therefore this architecture can be suitable for the support of 3GPP scenarios 4, 5 and 6. In any case, whether seamless service continuity and CS services can be provided or not, depends on the specific CRRM and RRM functionalities as well as on the ability to map UMTS radio access bearers into the adequate WLAN parameters.

With respect to the different WLAN technologies, it is worth mentioning that the HIPERLAN2 technology specified by the ETSI BRAN group is fully compliant with the UMTS architecture, because the specification addresses the UMTS interworking. This allows the inclusion of HIPER-LAN2 as an additional RAN like GERAN or UTRAN with minor modifications. However, for other

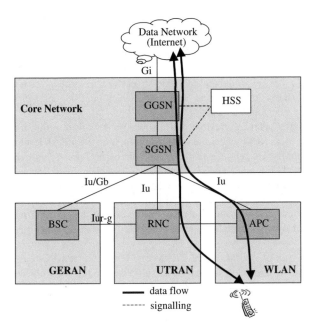

Figure 6.14   Very tight coupling WLAN/3GPP architecture

technologies such as the IEEE 802.11x, out of the scope of 3GPP standardisation, it is possible to
follow the tight coupling architecture by means of specific inter-working functions.

(d) Very tight coupling. This architecture offers the highest degree of coupling between 3GPP and
WLAN networks. As in the tight coupling approach, WLAN is seen as an additional RAN connected to the operator core network through the Iu interface. However, an additional interface between
the RNC and the APC is defined, similar to the Iur-g interface between the RNC and the BSC (see
Figure 6.14). This interface allows some CRRM decisions or procedures to be executed without
relying on SGSN functions. Information about the usage of both networks is available locally,
without having to obtain this information by explicit request to other entities. Note that with both the
tight and very tight coupling architectures, it is possible to extend to WLAN the same CRRM
architectures defined between GERAN and UTRAN and explained in Section 6.3.1.1.

6.4   FLEXIBLE RADIO RESOURCE AND SPECTRUM MANAGEMENT

The envisaged beyond 3G scenarios including heterogeneous networks, with a multiplicity of access
technologies as well as the diversity of terminals with reconfigurability capabilities, introduces a new
dimension into the radio resource management problem, in addition to the need for a proper interworking
of RATs through adequate architectures. Therefore, instead of performing the management of the radio
resources independently for each RAT, some form of overall and global management of the pool of
radio resources can be envisaged. Joint or Common Radio Resource Management (CRRM) is the process
envisaged to manage dynamically the allocation and de-allocation of radio resources (e.g. time slots,
codes, frequency carriers, etc.) within a single or between different radio access systems for the fixed
spectrum bands allocated to each of these systems. In this way, a more efficient usage of the available
radio resources can be achieved.

   In addition, the traditional concept of static allocation of licensed spectrum resources to networks
operators in wireless communications seems not to be the most suitable approach in beyond 3G

scenarios, characterised by changing traffic throughout time and space, changing availability of RATs, etc. In order to overcome these constraints and to achieve a better utilisation of the scarce spectrum, Advanced Spectrum Management (ASM) techniques are envisaged: ASM enables the dynamic management of the allocation, de-allocation and sharing of spectrum blocks within a single or between different radio access systems so that spectrum bands allocated to each of the systems are not fixed. In this context, there are approaches such as the Spectrum Brokerage, which consider spectrum to be a tradable economic good similar to stocks or real estate [17].

On the other hand, the adaptive, flexible and tunable framework resulting from the conjunction of ASM, CRRM and the local RRM techniques applied at the individual RAT level may suggest revisiting the static network planning concept. Certainly, reconfigurable technology will significantly change the operational mechanisms. Dynamic Network Planning and flexible network Management (DNPM) refers to the radio network planning, self-tuning network parameters and flexible management processes interworking with CRRM and ASM processes. It can be envisaged that an operator can expect that in its operational area some of the coverage will be offered using the classical method, (e.g., single air interface, fixed functionality and capability of base station), whereas in some special areas, DNPM will be applied.

The ultimate realisation of DNPM, ASM, CRRM and RRM in a consistent and coherent way would allow the achievement of unprecedented spectrum efficiencies and high efficiency in radio resource usages on top of the potential capabilities provided by the physical layer design of the involved RATs.

The high level relationships among the different elements described above are summarised in Figure 6.15. The main distinguishing factor in this context is the time scale or frequency at which the interactions between elements occur and/or actions from a given element are taken. In particular, it can be envisaged that:

(a) Network deployment (i.e. the number of cell sites and their locations) can be seen as static for the study of CRRM purposes, since it can change in the order of months/weeks depending on the network maturity status.
(b) DNPM and ASM act in a rather long-term scale (e.g. typically once or twice a day), in response to very significant demand profiles changes. An example of a situation triggering DNPM would be a temporal event such as, for example, a mass meeting, a football match, etc.

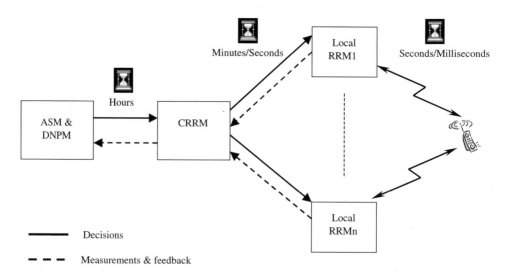

Figure 6.15   Framework for CRRM operation

(c) CRRM. For a given configuration in the scenario and for the period of time that all the RATs and amount of radio resources assigned to the cells in the scenario remain fixed, it will be the responsibility of CRRM to achieve good efficiency in the overall usage of the radio resource pool. Given that CRRM has the perspective of several RATs, it is expected that interactions occur in the order of minutes/seconds, thus responding to some higher-level objectives such as load balancing among RATs.

(d) Local RRM. This element will cope with the most dynamic elements of the scenario, such as traffic variability (i.e. short term variations on the offered load), user mobility (i.e. the different amount of radio resources needed as the user moves closer or farther from the cell site and at a certain mobile speed) and propagation/interference conditions within a given RAT. Actions of the corresponding functionalities may occur in a very short time scale, in the order of seconds (e.g. handover between contiguous cells belonging to the same RAT) or milliseconds (e.g. packet scheduling).

The ASM&DNPM, CRRM and Local RRM model can be seen as a hierarchical structure, where the underlying level is characterised by a reduced set of parameters that are made visible to the overlying level. Thus, provided that the set of parameters are suitably chosen to capture the essential of a given entity, the overlying level does not need to be aware of the detailed behaviour of the underlying level in order to make suitable decisions affecting the underlying level. These parameters, provided in the form of feedbacks/measurements, can be either configured periodically or be event-triggered.

For the variety of entities that may potentially be involved (i.e. ASM&DNPM, CRRM and several Local RRM such as UMTS, GSM, GPRS, WLAN, etc.) it is important to assure consistency among the decisions taken at the different hierarchical levels in order to achieve an overall coherent behaviour. The parameters deemed feasible to be exchanged depend on the heterogeneous network architecture and the coupling scheme.

Within the identified general framework, the focus of this chapter will be on the CRRM aspects, including the relationships between CRRM and local RRM as well as specific implementation issues, as will be detailed in the next section.

6.5 CRRM ALGORITHM IMPLEMENTATION

6.5.1 INTERACTIONS BETWEEN CRRM AND LOCAL RRM

Different possibilities exist for the implementation of CRRM procedures in heterogeneous networks. From the point of view of network topology, the possible alternatives depend on the physical node where the CRRM entity resides, as was discussed in Section 6.3.1.1. In turn, from the functional point of view, different degrees of interaction exist between CRRM and local RRM entities. The first aspect to take into account is the specification of the master of the different decisions. One possibility is that the CRRM simply advises the RRM entity, so that the local RRM is mainly responsible for the decisions taken (i.e. the RRM is the master). Another possibility is that the CRRM decisions are binding on the local RRM entity, so that the CRRM is the master and responsible for these decisions [2].

On the other hand, the split of functionalities between RRM and CRRM also depends on the degree of interaction between both entities, often referred as RRM-CRRM coupling. The lowest degree of interaction is shown schematically in Figure 6.16. In this case, CRRM only dictates policies for RRM operation.

A policy-based approach is usually assumed for CRRM operations. Policy-based management in IP-based multiservice networks [18] has been the subject of extensive research during the last few years, and may also be considered as a possibility for CRRM design. A policy can be defined as a high-level declarative directive that specifies some criterion to guide the behaviour of a network responding to some network operator preferences. As shown in Figure 6.16, CRRM operation responds to an external policy repository that contains the set of high-level policies to be applied when managing radio resources. In this approach, the CRRM is considered simply as a policy consumer that translates the specific policies into an adequate configuration of the RRM algorithms. Note that almost all functionalities reside in the

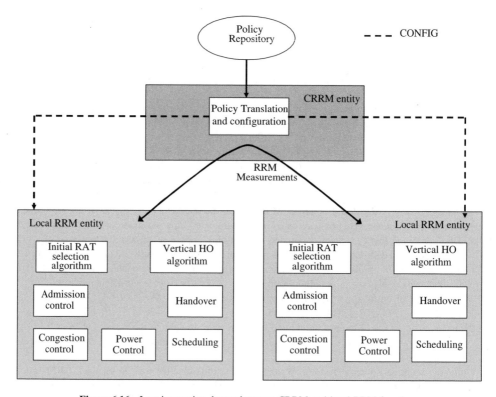

Figure 6.16 Low interaction degree between CRRM and local RRM functions

local RRM entity, which is responsible for the initial RAT selection at the beginning of a session and the decision to execute an intersystem or vertical handover, taking into account the intra and intersystem measurements provided by the mobile terminals as well as the cell measurements from other RRM entities. Communication between RRM entities can be done either through direct interfaces (i.e. in case of integrated CRRM approaches) or through a CRRM server. The interaction between CRRM and RRM is limited to the policy specification and update, so it is expected to occur at a long-term time scale.

Other approaches with a higher degree of interaction between RRM and CRRM entities are depicted in Figures 6.17, 6.18 and 6.19. In the first case, called intermediate interaction degree, CRRM not only provides the policies that configure the local RRM algorithms but is also involved in the RAT selection and vertical handover algorithms (see Figure 6.17) by deciding the appropriate RAT to be connected. The local RRM entities provide RRM measurements including the list of candidate cells for the different RATs and cell load measurements, so that the CRRM can take into account the availability of each RAT for the corresponding mobile terminal. The RAT selection either during vertical handover or in the initial RAT selection case will respond to the specific policies, for example, establishing correspondences between RATs and different services or user types. Some examples would be to allocate www services to UTRAN and voice services to GERAN, or to allocate business users to UTRAN and consumer users to GERAN. Similarly, other possible policies for selecting the adequate RAT could be related to having load balancing between the different access networks or could respond to congestion situations detected in any of the access technologies, for which status measurements are necessary.

Once the RAT has been selected, the local RRM algorithms deal with the specific admission control and intrasystem handovers. Similarly, fast resource allocation by means of scheduling algorithms is also handled by the local RRM to ensure the specific QoS requirements. Note in any case that, according to the scheme shown in Figure 6.17, the CRRM and local RRM operations can be quite independent

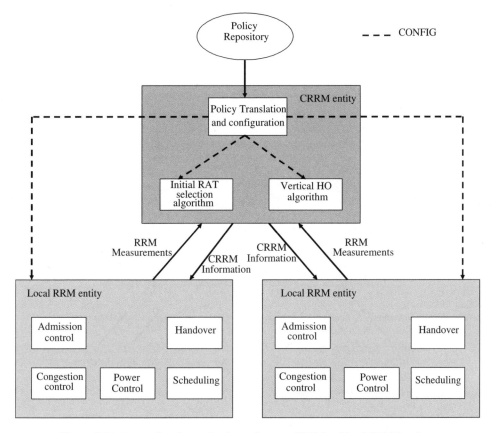

Figure 6.17   Intermediate interaction degree between CRRM and local RRM functions

and interactions between them are expected to occur at a relatively long-term scale in the order of minutes or seconds. The exchange of measurements can be periodic or event-triggered.

Figure 6.18 shows an even stronger interaction between CRRM and RRM. In this case, CRRM is involved in most of the local RRM decisions, so interaction between RRM and CRRM is expected to occur at a shorter-term scale than in the previous cases. CRRM responsibilities rely not only on selecting the appropriate RAT but also the specific cell for the selected RAT. Thus, CRRM is involved in each intrasystem handover procedure and requires a more frequent measurement exchange. Similarly, joint congestion control mechanisms could be envisaged to avoid overload situations in any of the underlying access networks. Only the fastest allocation mechanisms based on scheduling algorithms and operating at a millisecond time scale would remain at the local RRM entities.

Finally, the highest degree of interaction between CRRM and local RRM would be the introduction of joint scheduling algorithms that take into account the status of the different networks to allocate resources to the most appropriate RAT. This solution is shown schematically in Figure 6.19, where the local RRM functionality would remain at a minimum, limited to the transfer of adequate messages to CRRM and some specific technology dependent procedures that occur in very short periods of time (e.g. inner loop power control in the case of UTRAN, which occurs with periods below 1 ms). This solution would require that CRRM decisions be taken at a very short time scale in the order of milliseconds, with the possibility of executing frequent RAT changes for a given terminal. Consequently, this poses hard requirements to the reconfigurability capabilities of the mobile terminals and can be regarded as a long-term implementation of CRRM functionalities.

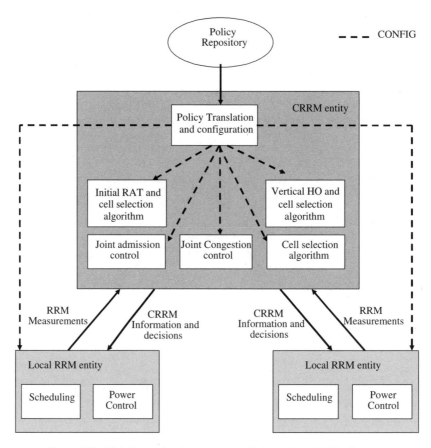

Figure 6.18   High interaction degree between CRRM and local RRM functions

It is worth mentioning that the interaction between local RRM and CRRM is not directly related to the coupling architectures between radio access networks. The latter refer to how the different networks are interconnected and to the level of interworking between them, while the former refer only to the operation of radio resource management procedures. Nevertheless, and due to the delays in communication between entities that result from the different coupling architectures, it is envisaged that the approaches for frequent CRRM-RRM interactions will only be feasible with tight and very tight coupling architectures.

Table 6.5 shows the reference time scales at which the CRRM-RRM interactions are expected to occur in the different approaches.

Table 6.5   Time scale for the CRRM-RRM operations

Typical time scale of the interactions CRRM-RRM	CRRM-RRM operation
Hours/days	Low interaction
Minutes	Intermediate interaction
Seconds	High interaction
Milliseconds	Very high interaction

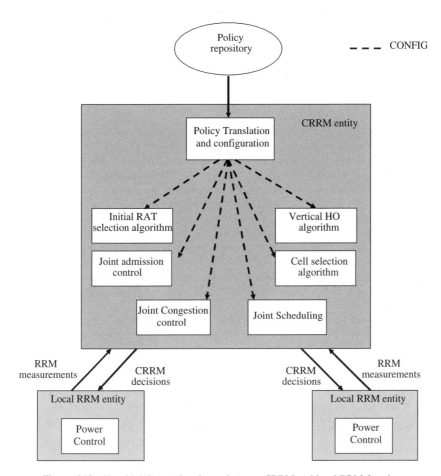

Figure 6.19 Very high interaction degree between CRRM and local RRM functions

6.5.2 RAT SELECTION SCHEMES

According to the framework for RRM and CRRM operation presented in the previous section, we now present a couple of examples corresponding to two different RAT selection schemes that involve two different interaction degrees between RRM and CRRM.

6.5.2.1 Example 1: Policy-Based CRRM Algorithms

An example of a CRRM algorithm is now presented to provide a better understanding of the CRRM framework described above and to move from a conceptual plane to a more specific one.

Policies can be defined at very different levels, from very generic high-level declarative directives to more detailed ones coupled in some extent to the radio access network architecture and/or deployment. For example, policies can be established at user type level (e.g. better QoS performance for business versus consumer users), at service type level (e.g. higher priority for conversational versus interactive traffic) or at radio network level (e.g. send the traffic towards the RAT requiring less radio resources). Combinations of policies belonging to the different levels are also possible. More specifically, let us consider the following sample cases:

1. Service-based policy.

  - VG (Voice GERAN) policy. Voice service is allocated to GERAN while the interactive service is allocated to UTRAN.
  - VU (Voice UTRAN) policy. Voice service is allocated to UTRAN while the interactive service is allocated to GERAN.

  In order to stress the vital importance on the achieved performance of the radio interface configuration and RAT characteristics, two situations are considered with this policy: the case where there is transport channel type switching available for interactive users in UTRAN and the case where only DCH channels are used without transport channel type switching. In the case of transport channel type switching, users occupy a DCH during activity periods (i.e. during page downloads), while during inactivity periods (e.g. during reading times between pages), no dedicated resources are allocated to the terminals.

2. Radio network-based policy. Given that the amount of radio resources necessary for an indoor user in UTRAN is considerably higher than for an outdoor user [19], the policy would define that indoor users are allocated to GERAN while outdoor users are allocated to UTRAN. For comparison purposes, a random RAT selection mechanism will be considered (i.e. both an outdoor and an indoor connection request are assigned to UTRAN or GERAN with equal probability).

The algorithm has been evaluated by means of a dynamic system level simulator including detailed UTRAN and GERAN features. Complete UTRAN/GERAN co-siting has been considered, with seven omnidirectional cells in the scenario. Two cell layouts have been evaluated, corresponding to a cell radius of 500 m and 1 km, respectively. Mobile speed is 3 km/h.

A mix of voice and interactive users is considered. For voice users, the call rate is 10 calls/hour per user and the average call duration is 180 s. In turn, interactive users follow the www model described in Appendix 5.3.3 with an average of 5 pages per www session and 30 s reading time between pages. In the downlink, the average offered bit rate during the activity periods (i.e. during a page download) is 128 kb/s, while in the downlink it is 24 kb/s. The average www session rate is 18 sessions/hour per user.

The RABs for GERAN and UTRAN are approximately equivalent in terms of bit rates. Specifically, the interactive RAB for UTRAN has maximum bit rates of 64 kb/s in the uplink and 128 kb/s in the downlink. In GERAN, the multislot capability assumes that up to 2 slots can be used in uplink and up to 3 in downlink, provided that the sum of uplink and downlink slots is not higher than 4 slots (i.e. this corresponds to class 6 mobiles, as indicated in Table 6.2). The Modulation and Coding Scheme (MCS) is changed dynamically between MCS1 and MCS7 (see Table 6.3), leading to a maximum bit rate of 44.8 kb/s per slot. Then, taking into account the multi-slot capability, the maximum bit rate is 89.6 kb/s for uplink and 134.4 kb/s for downlink. The BLER target is 1% for voice users and 10% for www browsing users.

On the other hand, three carriers per cell are assumed for GERAN (i.e. a total of 21 carriers in the scenario) and a single UTRAN FDD carrier is assumed for UTRAN. Note that, in this way, the amount of occupied bandwidth by UTRAN and GERAN is similar (i.e. the total GERAN bandwidth is 4.2 MHz, assuming 200 kHz per carrier, while the UTRAN bandwidth is about 4.69 MHz, assuming a chip rate of 3.84 Mchips/s and a roll-off factor of 0.22 for pulse shaping).

It is assumed that in GERAN, interactive traffic is scheduled within the available slots, provided that voice traffic has priority in slot assignment in front of interactive traffic.

Table 6.6 presents the aggregate throughput (in Mbit/s) achieved with the sum of both RATs (GERAN and UTRAN) and with the sum of both services (voice and www) for the different service-based policies (VU and VG). Simulations consider a total of 400 voice users in the scenario together with three different interactive load levels, corresponding to 200, 600 and 1000 www users in the scenario. Additionally, the case of VG without Transport Channel Type Switching mechanisms (i.e. an interactive user keeps a dedicated OVSF code even during page reading times) is also presented. In this respect, it is shown

Table 6.6 Aggregate throughput (Mbit/s) for the different policies and 400 conversational users in the scenario

	VU				VG				VG (no TrCH switch)			
	UL		DL		UL		DL		UL		DL	
www users	0.5 km	1 km	0.5 km	1 km	0.5 km	1 km	0.5 km	1 km	0.5 km	1 km	0.5 km	1 km
200	2.18	2.08	2.22	2.17	2.14	2.14	2.20	2.22	2.03	2.01	2.08	2.07
600	3.01	2.88	3.15	3.09	2.96	2.95	3.16	3.15	2.06	2.05	2.11	2.11
1000	3.80	3.64	4.05	3.96	3.77	3.76	4.08	4.08	2.08	2.05	2.14	2.13

that the throughput is greatly reduced in this later case, so that it is advisable for the operator to take full advantage of the transition to the RACH/FACH state if DCH are used for interactive users. With respect to VU and VG policies comparison, there are no substantial differences on the overall achieved throughput for the case of 500 m cell radius. Nevertheless, for the 1 km radius, the VG policy achieves somewhat better throughput as long as the shorter coverage range for UTRAN is causing some quality problems on voice users (i.e. increase in the block error rate and eventually call dropping) when the offered load is high.

Table 6.7 shows the average page delay when 400 voice users are in the scenario together with a variable number of web browsing users. The delay is presented for both uplink and downlink and for two different cell radii for each of the two service-based policies. It can be observed that VG (i.e. voice users through GERAN while interactive users through UTRAN) tends to provide lower delays. This is due to the higher efficiency for non real time traffic transmission in UTRAN achieved in the VG case, since web browsing traffic is supported by means of dedicated channels whereas in VU a packet scheduling algorithm must be implemented in GERAN. It is also worth noting that, for 1 km cell radii, delay increase in VG compared to 500 m cell radii is almost negligible. However, more noticeable page delay increase is found in VU. This is because in VU (i.e. web supported by GERAN), the link adaptation mechanisms force the use of modulation and coding schemes with lower associated transmission rates, thus increasing the delay. Note that for VG (i.e. web supported by UTRAN), the higher coverage radius causes some increase in the BLER beyond the target value, which causes some moderate delay increase due to increase in packet retransmissions.

The presented results reveal the advisability of using VG policy, i.e. to allocate voice users to GERAN and high bit rate interactive users to UTRAN, provided that transport channel type switching strategies are used.

Finally, Figure 6.20 plots the voice dropping probability in a scenario with only voice traffic and with 30% of users located indoors, in order to see the impact of the radio network-based policies. It can be observed that the RAT selection according to the policy that allocates indoor users to GERAN provides better performance as long as the radio resource consumption results one much lower. The comparison is presented with the case where RAT selection between UTRAN and GERAN is random.

Table 6.7 Average page delay (seconds) for www users with the different policies and 400 conversational users

	VU				VG			
	UL		DL		UL		DL	
www users	0.5 km	1 km	0.5 km	1 km	0.5 km	1 km	0.5 km	1 km
200	2.91	3.09	0.74	0.76	2.89	2.88	0.76	0.76
600	2.94	3.15	0.77	0.83	2.90	2.90	0.76	0.76
1000	3.03	3.74	0.99	1.26	2.91	2.93	0.76	0.77

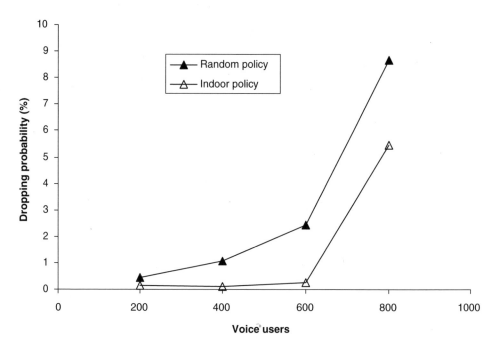

Figure 6.20 Voice dropping probability according to indoor policy and random policy

Note that there would be several mechanisms suitable for estimating whether a user requesting a service is indoor or outdoor (e.g. location-aided mechanisms, comparison between estimated path loss and reported path loss, etc.). Notice also that the indoor policy could be also considered for users with a high path loss, even if they are located outdoors.

6.5.2.2 Example 2: Fuzzy-Neural CRRM Algorithms

The fuzzy subset methodology has proven to be good at explaining how to reach the decisions from imprecise information by using the fuzzifier and defuzzifier rules and the inference engine concept [20][21]. The use of this methodology has been widely proposed in different fields of the literature [22–24]. In the framework of heterogeneous networks, one of the problems that CRRM algorithms must face is the existence of uncertainties when comparing different measurements belonging to different RATs that are necessarily of a different nature together with subjective criteria that have to do with techno-economic issues. As a result, the use of fuzzy logic as a robust decision making procedure becomes a possible solution for CRRM algorithm development. However, pattern aspects such as the selected membership functions and their particular shapes are still rather subjective in this solution.

On the other side, the use of neural networks, which are good at recognising patterns by means of learning procedures, could also be considered and, as a matter of fact, they have been proposed for use in hybrid fuzzy-neural based systems [25][26]. Taking these considerations into account, a fuzzy neural framework might be a good candidate for the solution of CRRM related issues.

The objective of the problem considered here is to select the most appropriate RAT taking into account different algorithm inputs that include system measurements in a scenario with three RATs, namely UTRAN, GERAN and WLAN. Furthermore, a certain amount of resources (i.e. bit rate or bandwidth) are also allocated in the selected RAT by the algorithm, which reflects the requirement for a high degree of interaction between CRRM and RRM. According to Figure 6.21, three main blocks are identified; fuzzy neural, reinforcement learning and multiple decision making. These blocks represent a general

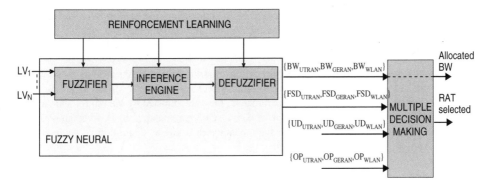

Figure 6.21 Block diagram of the proposed CRRM algorithm

framework including both technical and economical aspects [27]. Brief descriptions of the fuzzy neural blocks and reinforcement learning algorithm now follow. The reader is referred to References 27 and 28 for more details.

Fuzzy Neural Block   The purpose of the fuzzy neural algorithm is to obtain, for each RAT, a numerical indication (denoted as Fuzzy Selected Decision: FSD) between 0 and 1 of the suitability of selecting the RAT. The decision is obtained from a set of input linguistic variables (LV_i), reflecting technical measurements. This decision is taken in three steps, as depicted in Figure 6.21.

Step 1.   Fuzzification. The objective of this process is to assign, for each input linguistic variable, a value between 0 and 1 corresponding to the degree of membership of this input to a given fuzzy subset. A fuzzy subset is a linguistic subjective representation of the input variable. Some examples of input variables are the signal strength (SS) and the resource availability (RA), for each of the considered RATs, and the mobile speed.

   As an example, a fuzzy subset for the resource availability RA (e.g. for the number of available time slots in GERAN) could be formed by the possibilities H (high), M (medium) and L (low). One membership function exists for each one of the three terms H, M and L, reflecting the degree of membership of the RA value to each term. For example, if the value obtained for the term H is 0.9, it means that the resource availability is high, and if the value for the term M is 0.2, it means that the assertion 'the resource availability is medium' is likely to be false.

Step 2.   Inference Engine. For each combination of fuzzy subsets from step 1, the inference engine makes use of some predefined fuzzy rules to indicate, for each RAT, the suitability of selecting it. So, at the output of this step, there will be a combination of three output linguistic variables D (D_{UTRAN}, D_{GERAN}, D_{WLAN}) each one with four fuzzy subsets: Y (yes), PY (probably yes), PN (probably not) and N (not), with different degrees of membership for each of them.

   An example of inference rule could be: if ($SS_{UTRAN} = H$, $SS_{GERAN} = L$, $SS_{WLAN} = L$, $RA_{UTRAN} = H$, $RA_{GERAN} = H$, $RA_{WLAN} = M$, $MS = L$) then ($D_{UTRAN} = Y$, $D_{GERAN} = N$, $D_{WLAN} = N$). This means that if the signal strength of UTRAN is high and the signal strength of GERAN and WLAN is low, and since there is a high availability of resources in UTRAN, the decision to select UTRAN is yes (Y) while the decision to select the other RATs is no (N).

   As well as the decision about the selected RAT, with the fuzzy subsets H (high), M (medium) and L (low), the inference engine can also determine a level of allocated bandwidth in it.

Step 3.   Defuzzification. This procedure converts the outputs of the inference engine into a number ranging between 0 and 1, named Fuzzy Selected Decision: FSD_{UTRAN}, FSD_{GERAN} and FSD_{WLAN} for each RAT that reflects the suitability of select it. The algorithm can also provide the bandwidth allocated in each one: BW_{UTRAN}, BW_{GERAN} and eventually BW_{WLAN}.

At this point, the selected RAT could be the one having the highest FSD. Nevertheless, as will be explained later, it is also possible to take into account the FSD together with techno-economical aspects in a multiple decision making procedure.

Reinforcement Learning   The reinforcement learning algorithm procedure is used to suitably tune the parameters (means, deviations, shapes, etc.) of the different functions involved in the fuzzy logic controller. After the first selection of these parameters, they are adjusted by the reinforcement learning procedure [25] to ensure a certain target value of a given QoS parameter, such as, for example, the ratio of non-satisfied users (i.e. the users that receive a bandwidth below a certain desired value BW_D), the ratio of blocked users, dropping calls, etc.

Let us assume that the QoS parameter to ensure is the ratio of non-satisfied users. Therefore, the input signal for the reinforcement learning procedure would be:

$$r(t) = P^* - P_I(t) \tag{6.1}$$

where P^* is the target value of the ratio of non-satisfied users and $P_I(t)$ is the real value measured at time t. In order to ensure the corresponding target value, the reinforcement learning algorithm adjusts the different parameters to minimise the square error $E(t)$ defined as:

$$E(t) = \frac{1}{2}\left(P^* - P_I(t)\right)^2 \tag{6.2}$$

Therefore, the criterion to update a given parameter $w(t)$ of the fuzzy neural block is:

$$w(t+1) = w(t) + \varepsilon \cdot \left(-\frac{\partial E(t)}{\partial w(t)}\right) = w(t) + \varepsilon(P^* - P_I(t))\frac{\partial P_I(t)}{\partial w(t)} \tag{6.3}$$

where ε is the learning rate and the parameter $w(t)$ is varied in order to reduce the error $E(t)$.

An example of the performance that can be obtained by the reinforcement learning algorithm is shown in Figure 6.22. It presents the time evolution of the percentage of non-satisfied users in a simple scenario with three concentric cells: one UTRAN cell, one GERAN cell and one WLAN access point. In this example, the simulation time is measured in periods of 100 ms. Two values of the target ratio of

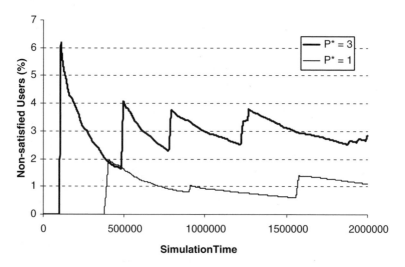

Figure 6.22   Evolution of the probability of non-satisfied users towards convergence

non-satisfied users are considered, namely $P^* = 1\%$ and $P^* = 3\%$, and it can be noticed that the algorithm is able to converge to the desired value under variable traffic and mobility conditions.

Multiple Decision Making Often, qualitative or techno-economic inputs are considered to make selections about the most suitable RAT or the allocated bandwidth in the above scenario. This is illustrated in Figure 6.21, where the Multiple Decision Making block decides on technical related inputs coming from the defuzzifier and from techno-economic related inputs such as User Demand (UD) and Operator Preferences (OP).

For each RAT, a membership value between 0 and 1 is used to define how well each criterion (i.e. technical criterion, user demand and operator preferences) is fulfilled: for the technical criterion, the FSD values obtained by the fuzzy-neural algorithm are considered, while for the user demand and operator preferences, the membership values may be set in a more subjective way according to cost-demand curves, operator agreements and/or policies. From these membership values, a multiple decision making strategy like the one defined in Reference 21 can be used to obtain the final decision of the selected RAT.

REFERENCES

[1] 3GPP TR 25.881 v5.0.0 'Improvement of RRM across RNS and RNS/BSS'
[2] 3GPP TR 25.891 v0.3.0 'Improvement of RRM across RNS and RNS/BSS (Post Rel-5) (Release 6)'
[3] G. Fodor, A. Eriksson, A. Tuoriniemi, 'Providing Quality of Service in Always Best Connected Networks', *IEEE Communications Magazine*, July 2003, pp. 154–163
[4] 3GPP TS 43.051 'TSG GSM/EDGE Radio Access Network; Overall Description – Stage 2'
[5] M. Mouly, M.B. Pautet, *The GSM System for Mobile Communications*, published by the authors' company, Ceel & Sys, 1992
[6] S.M. Redl, M.K. Weber, M.W. Oliphant, *An Introduction to GSM*, Artech House, 1995
[7] R.J. Bates, *GPRS General Packet Radio Service*, McGraw-Hill, 2002
[8] T. Halonen, J. Romero, J. Melero, *GSM,GPRS and EDGE Performance*, John Wiley & Sons Ltd, 2002
[9] GSM 05.05-DCS v3.3.0 'Radio Transmission and Reception', October 1993
[10] IEEE STD 802.1X 'Standards for Local and Metropolitan Area Networks: Port Based Access Control', 2001
[11] ETSI TR 101 683 'Broadband Radio Access Networks (BRAN); HIPERLAN Type 2; System Overview'
[12] ISO/IEC 8802-11 IEEE Std 802.11 'Information technology – Telecommunications and Information Exchange between Systems – Local and Metropolitan Area Networks – Specific Requirements. Part 11: wireless LAN Medium Access Control (MAC) and Physical Layer (PHY) specifications', 1999
[13] 3GPP TS 25.423 'UTRAN Iur interface RNSAP signalling'
[14] 3GPP TR 22.934 v6.2.0 'Feasibility study on 3GPP System to Wireless Local Area Network (WLAN) interworking'
[15] A.K. Salkintzis, C. Fors, R. Pazhyannur, 'WLAN-GPRS Integration for Next-Generation Mobile Data Networks', *IEEE Wireless Communications*, October, 2002, pp. 112–124
[16] P. Karlsson (editor) *et al.* 'Target Scenarios Specification: vision at project stage 1' Deliverable D05 of the EVEREST IST-2002-001858 project, April, 2004. Available at http://www.everest-ist.upc.es/
[17] IST End to End Reconfigurability (E2R) Project'. http://e2r.motlabs.com
[18] P. Flegkas, P. Trimintzios, G. Pavlou, 'A Policy-Based Quality of Service Management System for IP DiffServ Networks', *IEEE Network*, March–April, 2002, pp. 50–56
[19] J. Pérez-Romero, O. Sallent, R. Agustí, 'On The Capacity Degradation in W-CDMA Uplink/Downlink Due to Indoor Traffic', *IEEE 59th Semiannual Vehicular Technology Conference (VTC 2004 – Fall)*, Los Angeles, USA, 2004
[20] J.M. Mendel, 'Fuzzy Logic Systems for Engineering: A Tutorial', *Proceedings of the IEEE*, **83**(3), March, 1995, pp. 345–377
[21] R.R. Yager, 'Multiple Objective Decision Making using Fuzzy Sets', *Int'l Man–Machine Studies*, 9, 1977, pp. 375–382
[22] P.M.L. Chan, R.E. Sheriff, Y.F. Hu, P. Conforto, C. Tocci, 'Mobility Management Incorporating Fuzzy Logic for a Heterogeneous IP Environment', *IEEE Communications Magazine*, December, 2001, pp. 42–51

[23] P.M.L. Chan, Y.F. Hu, R.E. Sheriff, 'Implementation of Fuzzy Multiple Objective Decision Making Algorithm in a Heterogeneous Mobile Environment', *Wireless Communications and Networking Conference*, WCNC2002, pp. 332–336

[24] M. Singh, A. Prakash, D.K. Anvekar, M. Kapoor, R. Shorey, 'Fuzzy Logic Based Handoff in Wireless Networks', *51st IEEE VTC Spring Conference*, Tokyo, 2000, pp. 2375–2379

[25] C.T. Lin, C.S. George Lee, 'Neural-Network-Based Fuzzy Logic Control and Decision System', *IEEE Transactions on Computers*, **40**(12), December, 1991, pp. 1320–1336

[26] K.R. Lo, C.B. Shung, 'A Neural Fuzzy Resource Manager for Hierarchical Cellular Systems Supporting Multimedia Services', *IEEE Transactions on Vehicular Technology*, **52**(5), September, 2003, pp. 1196–1206

[27] R. Agusti, O. Sallent, J. Pérez-Romero, L. Giupponi, 'A Fuzzy-Neural Based Approach for Joint Radio Resource Management in a Beyond 3G Framework', *First International Conference on Quality of Service in Heterogeneous Wired/Wireless Networks, Qshine'04*, Dallas, USA, October, 2004

[28] L. Giupponi, R. Agusti, J. Pérez-Romero, O. Sallent, 'A Novel Joint Radio Resource Management Approach with Reinforcement Learning Mechanisms', *First IEEE International Workshop on Radio Resource Management for Wireless Cellular Networks (RRM-WCN)*, April, 2005, Phoenix, Arizona, USA

Index